CAMBRIDGE MONOGRAPHS ON
MATHEMATICAL PHYSICS

General editors: P. V. Landshoff, D. R. Nelson, D. W. Sciama, S. Weinberg

RELATIVISTIC FLUIDS AND MAGNETO-FLUIDS

With Applications in Astrophysics and Plasma Physics

RELATIVISTIC FLUIDS AND MAGNETO-FLUIDS

With Applications in Astrophysics and Plasma Physics

A. M. ANILE
University of Catania

The right of the University of Cambridge to print and sell all manner of books was granted by Henry VIII in 1534. The University has printed and published continuously since 1584.

CAMBRIDGE UNIVERSITY PRESS
Cambridge
New York Port Chester Melbourne Sydney

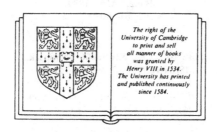

CAMBRIDGE UNIVERSITY PRESS
Cambridge, New York, Melbourne, Madrid, Cape Town, Singapore, São Paulo

Cambridge University Press
The Edinburgh Building, Cambridge CB2 2RU, UK

Published in the United States of America by Cambridge University Press, New York

www.cambridge.org
Information on this title: www.cambridge.org/9780521304061

First published 1989
This digitally printed first paperback version 2005

A catalogue record for this publication is available from the British Library

Library of Congress Cataloguing in Publication data

Anile, Angelo Marcello.
Relativistic fluids and magneto-fluids: with applications in
astrophysics and plasma physics / A. M. Anile.
p. cm. – (Cambridge monographs on mathematical physics)
Bibliography: p.
Includes index.
ISBN 0-521-30406-7
1. Plasma astrophysics. 2. Relativistic fluid dynamics.
3. Magnetohydrodynamics. I. Title. II. Series.
QB462.7.A55 1989
523.01–dc19 88-36742
CIP

ISBN-13 978-0-521-30406-1 hardback
ISBN-10 0-521-30406-7 hardback

ISBN-13 978-0-521-01812-8 paperback
ISBN-10 0-521-01812-9 paperback

TO MY PARENTS

La filosofia e' scritta in questo grandissimo libro che continuamente ci sta aperto innanzi agli occhi (io dico l'universo), ma non si puo' intendere se prima non s'impara a intender la lingua, e conoscer i caratteri, ne' quali e' scritto. Egli e' scritto in lingua matematica, e i caratteri sono triangoli, cerchi, ed altre figure geometriche, senza i quali mezzi e' impossibile a intenderne umanamente parola; senza questi e' un aggirarsi vanamente per un oscuro labirinto.

G. GALILEI, Il Saggiatore, *Opere*, Ed. Nazionale, vol. VI, p. 300.

Certo, volendo, uno puo' anche mettersi in testa di trovare un ordine nelle stelle, nelle galassie, un ordine nelle finestre illuminate dei grattacieli vuoti dove il personale della pulizia tra le nove e mezzanotte da' la cera agli uffici. Giustificare, il gran lavoro e' questo, giustificate se non volete che tutto si sfasci.

I. CALVINO, *Ti con zero*, 2nd edition, 1967, G. Einaudi, Torino.

Contents

Preface

Most of the matter in the universe can, in some form or other, be treated as a fluid and in several instances (supernova explosions, jets in extragalactic radio sources, accretion onto neutron stars and black holes, high-energy particle beams, high-energy nuclear collisions, etc.) undergoes relativistic motion. This consideration alone should be sufficient to motivate research in relativistic fluids. In the past most of the results on relativistic fluids have been obtained in a piecemeal way, in relation to a particular problem under consideration and using ad hoc techniques. Although this approach is perfectly legitimate in the process of research in the various areas of applications (astrophysics, plasma physics, nuclear physics), in the long run it is unsatisfactory because it tends to obscure the underlying unity of the subject and of the relevant techniques. In fact, a problem tailored approach (instead of a systematic and general one) necessarily precludes utilizing, in a particular area, results obtained in another area, and therefore hinders the cross fertilization of various techniques, a method which has been fruitful in several areas of science.

In 1967 the French mathematician André Lichnerowicz published a masterful monograph on relativistic fluid dynamics and magneto-fluid dynamics, which covered mainly existence and uniqueness results. Since then there has been no other attempt at a systematic development of the subject, although there have been several important developments in the field (particularly in shock wave theory). The aim of this book is to provide a unified and systematic treatment of the main results and techniques of relativistic fluid dynamics with an emphasis on nonlinear wave propagation in relativistic fluids and magneto-fluids. The book should be of interest to astrophysicists working in relativistic astrophysics, to plasma physicists working on relativistic plasmas, and to nuclear physicists interested in high-energy heavy ion collisions. The book should also be of some interest to the applied mathematician because it utilizes explicitly (whenever is convenient) covariant methods that render the treatment and the proofs more transparent and concise (this is, for instance, the case of the entropy principle, the introduction of the main field, and the symmetrization of the field equations).

This book draws heavily on the research of many scientists and I have tried to give credit to their work in the text and in the references. Some chapters of the book (in particular those treating the propagation and stability of relativistic shocks and relativistic simple waves in magneto-fluids) are almost entirely drawn from work by myself and my co-workers, in particular, A. Majorana, S. Motta, O. Muscato, and G. Russo. I am indebted to them for their constant advice and help in preparing the manuscript. I have discussed the contents and the style of the book with my former supervisor D. Sciama and J. Miller, and I gratefully acknowledge their constant encouragement and help.

Finally, I have to thank my wife, Marisa, and my children, Stefano and Giorgia, for accepting their share of the burden caused by the writing of this book.

1
Introduction

Relativistic fluid models are of considerable interest in several areas of astrophysics, plasma physics, and nuclear physics. Here we will mention briefly some of these areas and emphasize the problems which form the physical motivations for the theories expounded in this book.

Theories of gravitational collapse and models of supernova explosions (Van Riper, 1979; Chevalier, 1981; Shapiro and Teukolsky, 1983) are based on a relativistic fluid model for the star. In most models a key feature is the occurrence of an outward propagating relativistic shock. The precise conditions under which the shock forms at some point with exactly the necessary strength to expel the bulk of the star but still leave behind a remnant remain to be studied in detail and are the subject of current investigation. The effects of deviations from spherical symmetry due to an initial angular momentum and magnetic field must also be assessed. This requires the use of relativistic magneto-fluid dynamical models (Yodzsis, 1971; Maeda and Oohara 1982; Sloan and Smarr, 1986). The problem of the shock stability when traversing regions where the equation of state softens could be of interest for supernova models.

In the theories of galaxy formation, relativistic fluid models have been used in order to describe the evolution of perturbations of the baryon and radiation components of the cosmic medium (Peebles, 1980). Other components consisting of collisionless particles (such as massive neutrinos or photinos) are usually treated within a kinetic framework (Peebles, 1980). Such an approach is satisfactory for linear problems but encounters severe mathematical difficulties for nonlinear ones. In this case, at least for some problems, a generalized fluid model based upon an appropriate truncation of the kinetic moment equations might be adequate.

Theories of the structure and stability of neutron stars assume that the medium can be treated as a relativistic perfectly conducting magneto-fluid (Bekenstein and Oron, 1979). Theories of relativistic stars (which would be models for supermassive stars) are also based on relativistic fluid models (Zel'dovich and Novikov, 1978).

The problem of accretion onto a neutron star or a black hole is usually set in the framework of relativistic fluid models (Shapiro and Teukolsky, 1983; Abramowicz, Calvani, and Nobili, 1980; Lovelace, Mehanian, and Mobarry, 1986; Mobarry and Lovelace, 1986).

Several theories of jets and superluminal variations in extragalactic radio sources and quasars are based on relativistic magneto-fluid models (Begelmann, Blanford, and Rees, 1984). In particular, in some models, the observed variability is associated with the development of magneto-fluid dynamical waves or instabilities (Ferrari and Tsinganos, 1986).

In the field of plasma physics there are areas where relativistic fluid models are of interest.

Magnetohydrodynamic shock waves in the near relativistic regime have been obtained with the Columbia University Plasma Laboratory Electromagnetic High-Energy Shock Tube (Gross, 1971; Taussig, 1973). The theoretical analysis of these experiments is rather difficult and has been obtained by numerical simulation in a nonrelativistic framework (Liberman and Velikovich, 1985). A small increase in the attained speed would require a proper relativistic magneto-fluid dynamical calculation.

Intense relativistic electron beams (Davidson, 1974, Miller, 1985), which have very interesting applications such as the free electron laser, have also been modeled as relativistic fluids near thermal equilibrium (Toepfer, 1971). In situations far from thermal equilibrium, relativistic fluid models with generalized state equations have been proposed (Siambis, 1979; Newcomb, 1982) in a noncovariant framework. More satisfactory fully covariant models have also been introduced (Amendt and Weitzner, 1985; Anile and Pennisi, 1989).

In the field of nuclear physics, high-energy collisions among heavy nuclei have been modeled by using relativistic fluid dynamics (Amsden, Harlow, and Nix, 1977). When a nondissipative description applies and relativistic effects are not negligible, nuclear matter is described by the equations of relativistic fluid dynamics and all the details of nuclear interactions are incorporated in the state equation (Clare and Strottman, 1986). Many models predict the occurrence of a relativistic shock when two heavy nuclei collide (Sobel et al., 1975). Also, some models under current investigation predict that relativistic shocks (or detonation and deflagration waves) might be related to the phase transition from nuclear matter to quark-gluon plasma (Barz et al., 1985; Clare and Strottman, 1986; Cleymans, Gavai, and Suhonen, 1986). The questions related to the propagation and stability of relativistic shocks are obviously of relevance in this area.

The aim of this book is to develop and expound some theoretical methods and results which are relevant for the study of the relativistic fluid models used in applications to astrophysics, plasma physics, and nuclear physics. We will focus on the subject of nonlinear waves, which is fundamental for the understanding of dynamical theories and has diverse applications. In particular, we will emphasize the study of weak discontinuities, simple waves, and shock waves in relativistic fluid dynamics (RFD) and magneto-fluid dynamics (RMFD). Also, we will present some useful asymptotic methods (such as the method of asymptotic waves and the two-timing method) in a covariant framework and apply them to the study of high-frequency waves in RFD and RMFD, to Einstein's equations, and to the study of the interaction of locally plane electromagnetic waves with a relativistic plasma. Finally, we treat in detail the question of the stability of relativistic shocks.

In Chapter 2 we start with an introduction to the equations of relativistic fluid dynamics ad magneto-fluid dynamics in a given but arbitrary space-time. Then we perform a detailed analysis of the mathematical structure of the equations in the framework of the theory of symmetric hyperbolic systems. In Chapter 3 we give a covariant treatment of the theory of singular hypersurfaces in space-time, which will be used extensively in subsequent chapters. In Chapter 4 we discuss in detail the propagation of weak discontinuities in RFD and RMFD as well as electromagnetic and gravitational discontinuities. Chapter 5 is entirely devoted to the study of simple waves in RFD and RMFD. In Chapter 6 we introduce a covariant formulation of the two-timing method and apply it to the study of geometrical optics in a relativistic plasma. In Chapter 7 we introduce the method of asymptotic waves and apply it to RFD, RMFD, and Einstein's equations in vacuo. In Chapter 8 we discuss in detail the thermodynamic properties of shock waves in RFD and RMFD. In Chapter 9 we study the damping of a relativistic shock and present a method for weak shocks. In Chapter 10 we investigate the stability of plane relativistic shocks in a fluid with an arbitrary state equation and present some examples arising from astrophysics and nuclear physics.

The preliminary knowledge which is required by the reader is essentially a working knowledge of relativity theory and tensor calculus (at the level of Misner, Thorne, and Wheeler, 1973) and some previous acquaintance with hydrodynamical concepts.

2
Mathematical structure

2.0. Introduction

The simplest model for a relativistic medium is that of a relativistic fluid. When the medium interacts electromagnetically and is electrically highly conducting the simplest description is in terms of relativistic magneto-fluid dynamics.

From the mathematical viewpoint relativistic fluid dynamics (RFD) and magneto-fluid dynamics (RMFD) have mainly been treated in the framework of general relativity, that is, as describing possible sources of the gravitational field. This means that both the RFD and RMFD equations have been studied in conjunction with Einstein's equations.

In this framework Lichnerowicz (1967) has made a thorough and deep investigation of the initial value problem, and by using the theory of Leray systems, has obtained a local existence and uniqueness theorem in a suitable function class.

In many applications (particularly in plasma physics) one can neglect the gravitational field generated by the medium in comparison with the background gravitational field, or, in many cases, one can simply assume special relativity.

Mathematically this amounts to taking into account only the conservation equations for the matter, neglecting Einstein's equations. The resulting theory can be called test relativistic fluid dynamics or magneto-fluid dynamics. These theories are mathematically much simpler than the full general relativistic ones, and, consequently, stronger and more detailed results can be obtained.

In Section 2.1, following ideas originally introduced by Friedrichs (1974) and developed by Ruggeri and Strumia (1981a), we give a covariant definition of a quasi-linear hyperbolic system. The concept of systems of conservation laws is also introduced in this section.

In Section 2.2 we introduce the equations of test nondissipative (perfect) relativistic fluid dynamics. We give several examples of equations of state, for example, those describing a barotropic fluid, or a polytropic fluid, or a nondegenerate relativistic monatomic gas.

In Section 2.3 we study the mathematical structure of the equations of test perfect relativistic fluid dynamics. In particular, we prove that under a certain assumption on the equation of state they form a quasi-linear hyperbolic system. Finally, we give a covariant definition of the normal speed of propagation of a hypersurface with respect to a family of observers. By using this concept one can interpret the above mentioned restriction on the equation of state as the principle of relativistic causality.

In Section 2.4 we introduce the covariant equations of test relativistic magneto-fluid dynamics and study their mathematical structure. We perform a detailed study of the roots of the characteristic equation and of the associated left and right eigenvectors. We show that unless some (nonphysical) restrictions are imposed on the field variables, the system of equations is not hyperbolic. This is due to the fact that the covariant equations comprise the constraint part of Maxwell's equation.

A reduction to a (symmetric) hyperbolic system is possible by loosing manifest covariance and this is achieved in Section 2.7. Most of the results expounded in Sections 2.4 and 2.7 have been obtained jointly by the author and S. Pennisi (1987).

In Section 2.5 we present a concise account of a method, originally introduced by Friedrichs and Lax (1971) and Friedrichs (1974, 1978) and suitably modified and extended by Boillat (1974, 1976) and by Ruggeri and Strumia (1981a), aimed at obtaining a symmetric hyperbolic system from a given system of conservation laws when supplementary conservation laws (with suitable properties) exist.

Then one can apply to these systems the modern theory of quasi-linear hyperbolic symmetric equations in order to obtain (local in time) existence and uniqueness theorems for the initial value problem.

In Section 2.6 the method previously expounded is applied to test relativistic fluid dynamics. The results presented in this section have been obtained by Ruggeri and Strumia (1981a).

Finally, in Section 2.7 the above mentioned method is applied to test relativistic magneto-fluid dynamics.

2.1. Quasi-linear hyperbolic systems in conservation form

In order to introduce the concept of quasi-linear hyperbolic systems it is convenient to start with an example from Newtonian physics.

The equations for a perfect fluid are (Landau and Lifshitz, 1959a), in an inertial frame and Cartesian coordinates,

$$\frac{\partial \rho}{\partial t} + v^i \frac{\partial \rho}{\partial x^i} + \rho \frac{\partial v^i}{\partial x^i} = 0$$

$$\frac{\partial v^i}{\partial t}+v^j\frac{\partial v^i}{\partial x^j}+\frac{1}{\rho}\frac{\partial p}{\partial x^i}=0$$

$$\frac{\partial S}{\partial t}+v^i\frac{\partial S}{\partial x^i}=0$$

where ρ, p, S, v^i are, respectively, the mass density, pressure, specific entropy, and velocity. The pressure p is assumed to be given by a state equation of the form

$$p=p(\rho,S).$$

By introducing the field column vector

$$\mathbf{U}^T=(\rho,v^i,S)$$

these equations can be rewritten in matrix form

$$\frac{\partial \mathbf{U}}{\partial t}+A^i(\mathbf{U})\frac{\partial \mathbf{U}}{\partial x^i}=0,$$

where the matrixes A^i are

$$A^1=\begin{bmatrix} v^1 & \rho & 0 & 0 & 0\\ \dfrac{p'_\rho}{\rho} & v^1 & 0 & 0 & \dfrac{p'_s}{\rho}\\ 0 & 0 & v^1 & 0 & 0\\ 0 & 0 & 0 & v^1 & 0\\ 0 & 0 & 0 & 0 & v^1 \end{bmatrix},$$

$$A^2=\begin{bmatrix} v^2 & 0 & \rho & 0 & 0\\ 0 & v^2 & 0 & 0 & 0\\ \dfrac{p'_\rho}{\rho} & 0 & v^2 & 0 & \dfrac{p'_s}{\rho}\\ 0 & 0 & 0 & v^2 & 0\\ 0 & 0 & 0 & 0 & v^2 \end{bmatrix},$$

$$A^3=\begin{bmatrix} v^3 & 0 & 0 & \rho & 0\\ 0 & v^3 & 0 & 0 & 0\\ 0 & 0 & v^3 & 0 & 0\\ \dfrac{p'_\rho}{\rho} & 0 & 0 & v^3 & \dfrac{p'_s}{\rho}\\ 0 & 0 & 0 & 0 & v^3 \end{bmatrix}.$$

In this way the equations of Newtonian fluid mechanics have been written as a quasi-linear system.

A quasi-linear system will be said to be hyperbolic in the time direction (Courant and Hilbert, 1953) if the eigenvalue problem

$$(-\lambda I + A^i(\mathbf{U})n_i)\mathbf{d} = 0,$$

where $\mathbf{n}$ is an arbitrary unit vector and I is the identity matrix, has N real eigenvalues $\lambda^{(i)}$ (where N is the number of the field column vector components), and the corresponding set of eigenvectors $\mathbf{d}^{(i)}$ span the N-dimensional Euclidean space $\mathbb{R}^N$.

It is easy to check that, according to this definition, the equations of Newtonian fluid mechanics form a hyperbolic system (Jeffrey, 1976). The physical significance of the eigenvalues λ is such that they correspond to the propagation speeds of disturbances.

The definition of a quasi-linear hyperbolic system can be easily formulated in a covariant framework as follows.

Let $\mathscr{M}$ be a space-time, that is, a differentiable manifold of dimension 4, endowed with a Lorentz metric g of signature $+2$. Units will be such that the speed of light $c = 1$. We denote by ∇ the canonical Riemannian connection associated with g (De Felice and Clarke, 1989).

The field $\mathbf{U}$ representing physical quantities will consist of a set of (piecewise) differentiable tensor fields. In local coordinates (x^α) (lower-case Greek indices run from 0 to 3 and Latin ones from 1 to 3, except where stated otherwise), the field $\mathbf{U}$ will have components U^A (x^α), $A = 1, 2, \ldots, N$, and its covariant derivative $\nabla\mathbf{U}$ by definition will be a set of tensor fields with components $\nabla_\alpha U^A$.

In $\mathscr{M}$ we consider a quasi-linear system of N first order partial differential equations for the unknown field $\mathbf{U}$, which in local coordinates (x^μ) is written

$$A^{\alpha A}_{B}(U^C)\nabla_\alpha U^B = f^A(U^C), \tag{2.1}$$

where $A^{\alpha A}_{B}(U^C)$ and $f^A(U^C)$, which can be interpreted as components of $N \times N$-matrixes and N-vectors in $\mathbb{R}^N$, are differentiable functions of (U^C) in some open domain $\mathbf{D} \subseteq \mathbb{R}^N$.

Following Friedrichs (1954, 1974), one can introduce the following definitions.

DEFINITION 2.1. *Let ξ^α be a differentiable timelike unit vector field in $\mathscr{W}$, $\mathscr{W}$ an open connected subset of $\mathscr{M}$. The system* (2.1) *is called hyperbolic in the time direction defined by ξ^α if the following two conditons hold in $\mathscr{W}$:*

(i) $\det(A^\alpha \xi_\alpha) \neq 0$, where A is the matrix whose components in a local chart are $A^{\alpha A}_{B}$;

(*ii*) *For any spacelike vector field* ζ_α *on* $\mathscr{W}$, *the eigenvalue problem*

$$A^\alpha(\zeta_\alpha - \mu\xi_\alpha)\mathbf{d} = 0$$

has only real eigenvalues μ *and* N *linearly independent eigenvectors* $\mathbf{d}$.

Note that it is not restrictive to take ζ_α such that $\zeta_\alpha\xi^\alpha = 0$ and $\zeta_\alpha\zeta^\alpha = 1$. In fact, any spacelike ζ_α can be decomposed as

$$\zeta_\alpha = b(v_\alpha - a\xi_\alpha),$$

where $v^\alpha v_\alpha = 1$, $v^\alpha\xi_\alpha = 0$, a and b are real numbers, $b > 0$.

When the roots μ are all distinct the system will be said to be *strictly hyperbolic* (Friedrichs, 1954, 1974).

The vectors $\zeta_\alpha - \mu\xi_\alpha$ are called *characteristic* and ξ_α is called *subcharacteristic*.

Most of the systems of mathematical physics derive from conservation laws. For instance, the equations of Newtonian fluid mechanics derive, with some additional assumptions, from the laws of conservation of mass, energy, and momentum (Whitham, 1974; Jeffrey, 1976). This provides the motivation for the following definition.

DEFINITION 2.2. *The system* (2.1) *is said to be in conservation form if there exist* $\mathbf{F}^\alpha(\mathbf{U})$ *such that, in any local chart,*

$$A^{\alpha A}_B(U^c) = \frac{\partial F^{\alpha A}}{\partial U^B} \tag{2.2}$$

$F^{\alpha A}(U^c)$ *being differentiable functions of* $\mathbf{U} \in \mathbf{D} \subseteq \mathbb{R}^N$.

In the following we shall consider hyperbolic systems in conservation form. Also, as will be seen in what follows, in many important situations in mathematical physics, from the field equations (2.2), one can derive a *supplementary conservation law*

$$\nabla_\alpha h^\alpha(U^c) = g(U^c) \tag{2.3}$$

with h^α, g differentiable functions of $\mathbf{U} \in \mathbf{D} \subseteq \mathbb{R}^N$. This is the case of the equations of Newtonian fluid dynamics, from which we can obtain the energy conservation law

$$\frac{\partial}{\partial t}\rho(\varepsilon + v^2/2) + \frac{\partial}{\partial x^i}[\rho v^i(v^2/2 + \varepsilon + p)] = 0,$$

where ε is the specific internal energy. Alternatively, one might take this latter equation as part of the system describing perfect fluids [using ε instead of S in the field variables and expressing $p = p(\rho, \varepsilon)$] and interpret the entropy equation as a supplementary conservation law.

The existence of a supplementary conservation law with suitable properties will be a key ingredient for the theories of discontinuous solutions of the system (2.2). This question is deeply related to the concept of entropy for continuous media and the theory of shock waves, as will be seen in later chapters.

We remark that when $\mathcal{M}$ is Minkowski space-time, then we can take $\mathcal{W} = \mathcal{M}$ and (x^0, x^1, x^2, x^3) as inertial coordinates. For the choice $\xi^\alpha = (1, 0, 0, 0)$ the previous definitions reduce to the usual ones in Cartesian coordinates which we gave above, with $-\mu$ as eigenvalues.

It is trivial to give a coordinate-free and global version of the above definitions. What is needed is a global definition of the vector field ξ^α and a coordinate-free definition of the covariant derivative $\nabla \mathbf{U}$ (Anile, 1982).

A coordinate-free formulation could be useful for some important problems encountered in relativistic astrophysics. For instance, for the problem of the stability of black holes one generally needs several charts in order to cover the whole space-time (Hawking and Ellis, 1973). Therefore, in this case, a coordinate-free definition of a hyperbolic system is essential in order to interpret the mathematical nature of the perturbation equations.

In the next section we will introduce the equations of relativistic fluid dynamics and in Section 2.3 we will write them in the form of a quasi-linear system and prove that under reasonable conditions they form a hyperbolic system.

2.2 The equations of relativistic nondissipative fluid dynamics

The equations of relativistic fluid dynamics, known since the early stages of relativity theory, have usually been derived by analogy with Newtonian fluid dynamics with an appropriate identification for the relativistic quantities representing energy and momentum densities and fluxes (Synge, 1956, 1960; Landau and Lifshitz, 1959: Misner et al., 1973; Taub, 1978). A better justification, starting from first principles, can be obtained in the framework of relativistic continuum mechanics (Cattaneo, 1970; Ferrarese, 1982). A deep discussion of the derivation of the relativistic fluid dynamical equations when dissipation is also taken into account can be found in Dixon (1978).

Alternative approaches based on relativistic kinetic theory can also justify the equations of relativistic fluid dynamics (De Groot, Van der

Leeuw, and Van Weert, 1980). Admittedly, these approaches are more limited in scope, because kinetic theory is applicable only to a rarefied gas (whereas continuum thermodynamics can deal with a broad class of materials). However, this limitation is more than compensated by the fact that, in a kinetic theory framework, it is possible to calculate the equation of state and the transport coefficients from the particle collision model.

Because several derivations of the relativistic fluid dynamical equations already exist in the literature, we shall not attempt in this section to present yet another derivation, but we shall simply sketch the basic argument in the case of a perfect fluid.

We assume that the fluid is characterized by a four-velocity u^μ (representing the average of the microscopic velocities) and an energy-momentum tensor $T^{\mu\nu}$ [whose existence can be inferred from the relativistic counterpart of Cauchy's theorem of continuum mechanics (Dixon, 1878)].

Let ρ be the rest mass density measured in the local rest frame (or baryon number density, according to the circumstances). Then one of the equations of relativistic fluid dynamics represents the local law of mass conservation and reads

$$\nabla_\alpha(\rho u^\alpha) = 0. \tag{2.4}$$

The other equations represent the local laws of conservation of energy and momentum and read

$$\nabla_\alpha T^{\alpha\beta} = 0. \tag{2.5}$$

A perfect fluid is defined by the property that, in the local rest frame, it allows no energy fluxes and no anisotropic stresses. Therefore, at a given space-time point, in the local rest frame [in which the components of the four-velocity are $\hat{u}^\mu = (1, 0, 0, 0)$], the energy-momentum tensor components are

$$\hat{T}^{\mu\nu} = \begin{bmatrix} e & 0 & 0 & 0 \\ 0 & p & 0 & 0 \\ 0 & 0 & p & 0 \\ 0 & 0 & 0 & p \end{bmatrix},$$

where e is the total energy density, p is the pressure (both measured in the local rest frame), and $\hat{}$ denotes components with respect to the local rest frame coordinates.

Therefore, in general coordinates, the form of the energy-momentum tensor is

$$T^{\alpha\beta} = (e + p)u^\alpha u^\beta + p g^{\alpha\beta}. \tag{2.6}$$

One can write

$$e = \rho(1 + \varepsilon), \tag{2.7}$$

where ε is defined as the specific (per unit mass) internal energy, measured in the local rest frame.

Let $h_{\mu\nu}$ be the projection tensor onto the 3-space orthogonal to u^μ

$$h_{\mu\nu} = g_{\mu\nu} + u_\mu u_\nu. \tag{2.8}$$

Then, by contracting equation (2.5) with u_β one obtains the conservation of energy equation

$$u^\alpha \nabla_\alpha e = -(e + p)\nabla_\alpha u^\alpha \tag{2.9}$$

while by contracting it with $h_{\beta\mu}$ one obtains the conservation of momentum equation

$$(e + p)u^\alpha \nabla_\alpha u^\mu = -h^{\alpha\mu}\nabla_\alpha p. \tag{2.10}$$

For a perfect fluid the quantities ρ, ε, p must obey the first law of thermodynamics

$$T\,\mathrm{d}S = \mathrm{d}\varepsilon + pd(1/\rho), \tag{2.11}$$

where S is the specific entropy and T the absolute temperature.

By using the conservation of mass equation (2.4) and the first law of thermodynamics (2.11) one can deduce from the conservation of energy equation (2.9)

$$u^\alpha \nabla_\alpha S = 0, \tag{2.12}$$

which expresses the fact that the fluid flow is adiabatic.

The fluid quantities ρ, ε, p are related by an equation of state, which arises from considerations of kinetic theory or statistical mechanics.

An interesting class of fluids consists of the so-called barotropic fluids obeying the equation of state

$$e = e(p), \tag{2.13}$$

which is assumed to be invertible giving $p = p(e)$.

Since the rest mass density ρ does not intervene in the state equation, these fluids can be described solely by the conservation of energy and momentum equations (2.9)–(2.10). The mass conservation equation decouples from the others and can be solved after the fluid motion has been determined.

In this case the only role of the continuity equation is to provide a volume tracer. In this sense it can also be introduced even when there are no conserved charges (baryon number, lepton number, etc.). The introduction of the continuity equation in these situations (which might occur when discussing the early universe) greatly helps the physical interpretation of hydrodynamical calculations (such as the evolution of bubbles in the quark-hadron phase transition; Miller and Pantano, 1987).

An interesting class of barotropic fluids is that of fluids consisting of ultrarelativistic particles in thermal equilibrium. For these fluids it is possible to define an entropy function $H(T, V)$ (Weinberg, 1972), where V is the volume, such that

$$T\,\mathrm{d}H = \mathrm{d}(eV) + p\,\mathrm{d}V \tag{2.14}$$

with e the energy density. The integrability condition is

$$\frac{\partial}{\partial e}\left(\frac{e+p}{T}\right) = \frac{1}{T}. \tag{2.15}$$

In thermal equilibrium, for such a fluid, one has $e = e(T)$ and $p = p(T)$, and therefore equation (2.15) yields

$$p'(T) = (e+p)/T$$

and then it is obvious that

$$H(T, V) = (e+p)V/T.$$

The entropy density $\eta = H/V$ is then

$$\eta = (e+p)/T + \text{const.} \tag{2.16}$$

An equivalent expression for η which is independent of temperature is

$$\eta = \exp\left(\int \frac{\mathrm{d}e}{e+p}\right). \tag{2.17}$$

An example, of astrophysical interest, of these fluids is provided by a gas in local thermodynamical equilibrium with radiation when the radiation energy density greatly exceeds the total gas energy density. Since for radiation one has $e = a_R T^4$ (a_R being the Stefan–Boltzmann constant, $a_R = 7.56 \times 10^{-15}$ in cgs. units) and $p = \frac{1}{3}a_R T^4$, when the radiation energy density dominates, the fluid obeys the equation of state

$$e = 3p. \tag{2.18}$$

Also, from equation (2.16) one has

$$\eta = \tfrac{4}{3}a_R T^3.$$

Other examples are provided by a fluid of massless neutrinos, for which (Weinberg, 1972)

$$e = \tfrac{7}{16} a_R T^4$$

or a fluid of ultrarelativistic electron-positron pairs (Weinberg, 1972), for which

$$e = \tfrac{7}{8} a_R T^4.$$

In both cases the pressure is given by $p = \frac{1}{3}e$.

It is interesting to notice that in the early universe, at sufficiently high temperatures, all the particles become relativistic. Therefore, the equation of state

$$p = \tfrac{1}{3}e$$

would be applicable under these circumstances.

Another interesting case of barotropic fluids is provided by matter at zero temperature. In fact, at $T = 0$ the state equation depends on only one parameter. Usually the state equation is given as

$$p = p(\rho)$$

and

$$e = e(\rho).$$

When it is possible to eliminate ρ one obtains a barotropic fluid $p = p(e)$. For instance, this is the case for a completely degenerate cold neutron gas (Shapiro and Teukolsky, 1983), which is the crudest model for nuclear matter in the interior of neutron stars. More realistic examples comprise the cold e–n–p gas, the Harrison–Wheeler, and the Bethe–Baym–Pethick state equations. These examples will be discussed in detail in Section 10.4 in connection with the stability of shocks in nuclear matter.

Another class of fluids of astrophysical interest consist of those obeying a polytropic state equation of index γ,

$$p = K(S)\rho^{\gamma}. \tag{2.19}$$

An example is provided by a gas in local thermodynamical equilibrium with radiation when only the internal energy and pressure are dominated by radiation.

Then one has

$$\varepsilon = \frac{a_R T^4}{\rho}$$

and

$$p = \tfrac{1}{3} a_R T^4,$$

hence

$$p = \tfrac{1}{3} \rho \varepsilon. \tag{2.20}$$

From the first law of thermodynamics one obtains

$$S = \frac{4a_R T^3}{3\rho},$$

whence the polytropic state equation (2.19) with $\gamma = 4/3$ and

$$K(S) = \tfrac{1}{3} a_R \left(\frac{3S}{4a_R} \right)^{4/3}.$$

It is convenient to notice that, in this case, the polytropic state equation can be written in the form

$$p = p(e, S), \tag{2.21}$$

where, in equation (2.19), ρ has been expressed as a function of e and S by inverting the relationship

$$e = \rho + 3K(S)\rho^{4/3}.$$

The polytropic gas state equation is usually considered in the analytical properties of stellar properties. In fact, this state equation is a reasonable approximation to the average thermodynamic properties of stellar material and its simple expression makes it useful for analytical calculations (such as in stability analysis; Zel'dovich and Novikov, 1971).

For the applications in astrophysics and plasma physics, an important equation of state is that appropriate for a relativistic nondegenerate monatomic gas, studied by Chandrasekhar (1939) and Synge (1957), called the Synge gas for brevity. This state equation is the relativistic extension of the nonrelativistic perfect gas law. Its underlying kinetic description is in terms of a distribution function (the Juttner distribution function) which is the relativistic counterpart of the Maxwell–Boltzmann distribution. From relativistic kinetic theory it is possible to prove that, for an equilibrium distribution function,

$$p = \frac{k_B}{m} \rho T, \tag{2.22}$$

which is the usual perfect gas law with k_B the Boltzmann constant and m the gas particle mass.

Furthermore,

$$e + p = \rho G(z), \tag{2.23}$$

with

$$z = \frac{m}{k_B T} \tag{2.24}$$

and

$$G(z) = K_3(z)/K_2(z), \tag{2.25}$$

where $K_n(z)$ are the modified Bessel functions of the second kind (Synge, 1957)

$$K_n(z) = \int_0^\infty e^{-z \cosh y} \cosh ny \, dy.$$

Moreover the specific entropy S is given by

$$\exp\left(-\frac{m(S - S_0)}{k_B}\right) = \rho L(z), \tag{2.26}$$

with

$$\ln L(z) = -zG(z) - \ln\left(\frac{K_2(z)}{z}\right), \tag{2.27}$$

S_0 being a constant reference entropy.

It is useful to have asymptotic expansions of $G(x)$ and $L(x)$ in the cases of low and high temperatures (Synge, 1957).

One can show that for low temperatures, $z \to \infty$, one has, asymptotically,

$$G(z) \cong 1 + \frac{5}{2z}, \tag{2.28}$$

$$L(z) \cong \sqrt{2/\pi}\, z^{3/2}, \tag{2.29}$$

whereas for high temperatures, $z \to 0$, one obtains

$$G(z) \cong \frac{4}{z}, \tag{2.30}$$

$$L(z) \cong \tfrac{1}{2} e^{-4} z^3 \tag{2.31}$$

(where, in this formula, e denotes the Euler number).

For the Synge gas the ratio of specific heats is not constant. However, it is useful to introduce a parameter $\hat{\gamma}$,

$$\hat{\gamma} = 1 + \frac{p}{\rho\varepsilon}, \tag{2.32}$$

which is a function of z and has values in the range

$$4/3 \leq \hat{\gamma} \leq 5/3$$

(Fig. 2.1).

In the nonrelativistic limit, $z \to \infty$ and $\hat{\gamma} \to 5/3$, whereas in the ultra-relativistic limit, $z \to 0$ and $\hat{\gamma} \to 4/3$.

Also, in this case, by using some properties of the functions $G(z)$ and $L(z)$, it is possible to prove that one can obtain a state equation of the form (2.21), that is,

$$p = p(e, S).$$

In fact,

$$e = \rho(1 + \varepsilon) = pz(G(z) - 1).$$

Now

$$\exp\left(-\frac{m(S - S_0)}{k_B}\right) = zL(z)p$$

and it suffices to show that it is possible to invert the above relationship yielding $z = z(p, S)$.

Fig. 2.1. The quantity $\hat{\gamma} = 1 + p/(\rho\varepsilon)$ for a Synge gas as a function of $z = m/k_B T$.

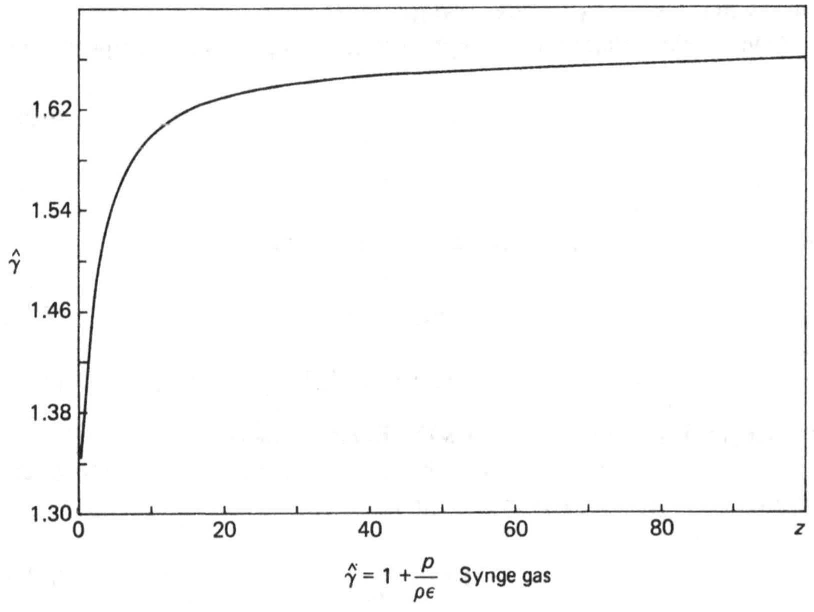

This is indeed possible because (Synge, 1957)

$$\frac{\mathrm{d}}{\mathrm{d}z}(zL(z)) = -z^2L(z)\frac{\mathrm{d}G}{\mathrm{d}z} \neq 0.$$

A little caution is necessary when applying the Synge state equation. In fact, in the relativistic regime ($kT > mc^2$) one cannot ignore photon production (when ions are also present) and, subsequently, pair production. Therefore, in general, for a mixture of (usually nonrelativistic) ions and relativistic electrons one must add to the Synge state equation for electrons the contribution of ions and that of photons and pairs (Lanza, Miller, and Motta, 1987).

In the following we shall always assume (except when stated otherwise) that the fluids under consideration obey the equation of state of the form (2.21); that is, we can take e and S as independent thermodynamical variables.

This assumption is verified by the state equations considered in the previous examples. It can be seen that more complex state equations, such as those appropriate for nuclear matter, can also be put in this form.

In the next section we shall write the equations of relativistic fluid dynamics in the form of a quasi-linear system and study its hyperbolicity.

2.3. Test relativistic fluid dynamics as a quasi-linear hyperbolic system

In many situations arising in astrophysical contexts or in laboratory physics (plasmas and nuclear matter) one can neglect the gravitational field produced by the fluid in comparison with the background gravitational field. This amounts to the test fluid approximation, where the motion of the fluid is assumed to occur in a given and preassigned space-time $\mathscr{M}$ endowed with a given Lorentz metric g. Then the equations governing the motion of the fluid can be taken to be the conservation of momentum equation

$$(e + p)u^\alpha \nabla_\alpha u^\mu + h^{\alpha\mu}\nabla_\alpha p = 0, \tag{2.33}$$

the conservation of energy equation

$$u^\alpha \nabla_\alpha e + (e + p)\nabla_\alpha u^\alpha = 0, \tag{2.34}$$

and the adiabaticity condition

$$u^\alpha \nabla_\alpha S = 0. \tag{2.35}$$

These equations must be supplemented by a general state equation

$$p = p(e, S), \tag{2.36}$$

assumed to be sufficiently differentiable. All the influence of the background gravitational field is included in the covariant derivatives. By introducing the field column vector U^A, with $A = 0, 1, 2, 3, 4, 5$,

$$\mathbf{U} = (u^\nu, e, S)^\mathrm{T}, \tag{2.37}$$

where $^\mathrm{T}$ denotes transposition, the equations of test relativistic fluid dynamics can be written as a quasi-linear system

$$A^{\alpha A}_{\mathbf{B}} \nabla_\alpha U^B = 0, \tag{2.38}$$

where the matrixes A^α are given by

$$A^\alpha = \begin{bmatrix} (e+p)u^\alpha \delta^\mu_\nu, & h^{\alpha\mu} p'_e, & h^{\alpha\mu} p'_s \\ (e+p)\delta^\alpha_\nu, & u^\alpha, & 0^\alpha \\ 0^\alpha_\nu, & 0^\alpha, & u^\alpha \end{bmatrix}, \tag{2.39}$$

where 0^α and 0^α_ν indicate the null vector and matrix, respectively, and p'_e, p'_s are the partial derivatives of $p(e, S)$. Now we shall assume the following restriction on the state equation

$$0 < p'_e \leq 1. \tag{2.40}$$

As we shall see this restriction has a deep physical meaning connected with causality and is verified by all the physically realistic state equations. Now we can prove that the equations of test relativistic fluid dynamics form a hyperbolic system. We have the following proposition.

PROPOSITION 2.1. *The quasi-linear system (2.38)–(2.39) with the restriction (2.40) is hyperbolic in the sense of Definition* 2.1.

Proof. We must show that both conditions (i) and (ii) of definition (2.1) hold. Let ξ^α be the timelike vector field of $\mathcal{M}$ which defines time-orientation. A simple calculation gives

$$\det(A^\alpha \xi_\alpha) = (e+p)^4 (u^\alpha \xi_\alpha)^4 ((u^\alpha \xi_\alpha)^2 - p'_e h^{\mu\nu} \xi_\mu \xi_\nu) \neq 0$$

because ξ^α is timelike and $p'_e \leq 1$.

Therefore condition (i) holds for any timelike ξ^α. In order to check condition (ii) let us consider the equation

$$\det(A^\alpha q_\alpha) = (e+p)^4 (u^\alpha q_\alpha)^4 ((u^\alpha q_\alpha)^2 - p'_e h^{\mu\nu} q_\mu q_\nu) = 0.$$

First of all, let q_α be a solution of

$$(u^\alpha q_\alpha)^2 - p'_e h^{\alpha\beta} q_\alpha q_\beta = 0.$$

By putting $q_\alpha = \zeta_\alpha - \mu\xi_\alpha$ in the local rest frame with $u^\alpha = \delta^\alpha_0$ and with the Minkowski metric $\eta_{\mu\nu}$, we obtain

$$(\zeta_0 - \mu\xi_0)^2 - p'_e(\underline{\zeta} - \mu\underline{\xi})^2 = 0,$$

where $\underline{\xi} = (\xi_i)$, $\underline{\zeta} = (\zeta_i)$ are Euclidean three-vectors and $(\underline{\xi})^2 = \xi_i\xi_i$.

This is a second degree equation in μ,

$$[\xi_0^2 - p'_e(\underline{\xi})^2]\mu^2 - 2[\zeta_0\xi_0 - p'_e\zeta_i\xi_i]\mu + \zeta_0^2 - p'_e(\underline{\zeta})^2 = 0$$

whose discriminant is

$$\begin{aligned} &p'_e(\zeta_0^2(\underline{\xi})^2 + \xi_0^2(\underline{\zeta}^2) - 2\xi_0\zeta_0\xi_i\zeta_i - [(\underline{\xi})^2(\underline{\zeta})^2 - (\zeta_i\xi_i)^2]p'_e) \\ &\quad > p'_e(\zeta_0^2(\underline{\xi})^2 + \xi_0^2(\underline{\zeta})^2 - 2\zeta_0\xi_0\zeta_i\xi_i - (\underline{\xi})^2(\underline{\zeta})^2 + (\zeta_i\xi_i)^2) \\ &\quad = p'_e([\xi_0^2 - (\underline{\xi})^2][(\underline{\zeta})^2 - \zeta_0^2] + (\zeta_0\xi_0 + \zeta_i\xi_i)^2) > 0, \end{aligned}$$

where use has been made of the restriction $0 < p'_e \leq 1$ and the fact that ξ^α is timelike, $\xi_0^2 > (\underline{\xi})^2, \zeta^\alpha$ is spacelike, and $\zeta_0^2 < (\underline{\zeta})^2$. Therefore we have two distinct real solutions μ_+ and μ_-. The corresponding linearly independent right eigenvectors $\mathbf{d}_\pm$ are easily seen to be

$$\mathbf{d}_\pm = \begin{pmatrix} -h^{\nu\alpha}(\zeta_\alpha - \mu_\pm\xi_\alpha)p'_e \\ (e+p)a_\pm \\ 0 \end{pmatrix}, \quad \text{with } a_\pm = u^\alpha(\zeta_\alpha - \mu_\pm\xi_\alpha).$$

Now let us consider the solution

$$(u^\alpha q_\alpha)^4 = 0,$$

which gives $\mu_{(I)} = \dfrac{u^\alpha\zeta_\alpha}{u^\alpha\xi_\alpha}$, $I = 1, 2, 3, 4$ (where $u^\alpha\xi_\alpha \neq 0$ because ξ^α is timelike), and has multiplicity four. The corresponding linearly independent right eigenvectors $\mathbf{d}_{(I)}$, $I = 1, 2, 3, 4$, are

$$\mathbf{d}_{(1)} = \begin{pmatrix} u^\alpha \\ 0 \\ 0 \end{pmatrix}, \quad \mathbf{d}_{(2)} = \begin{pmatrix} v_2^\alpha \\ 0 \\ 0 \end{pmatrix}, \quad \mathbf{d}_{(3)} = \begin{pmatrix} v_3^\alpha \\ 0 \\ 0 \end{pmatrix}$$

(where v_2^α, v_3^α are two linearly independent vectors orthogonal to u^α and q_α), and

$$\mathbf{d}_{(4)} = \begin{pmatrix} 0^\nu \\ -p'_s \\ p'_e \end{pmatrix}.$$

It is easy to check that $u^\alpha q_\alpha = 0$ is not a solution of

$$(u^\alpha q_\alpha)^2 - h^{\alpha\beta} q_\alpha q_\beta p'_e = 0.$$

Therefore, the eigenvectors $\mathbf{d}_-$, $\mathbf{d}_{(I)}$, $I = 1, 2, 3, 4$ form a basis of $\mathbf{R}^6$, for any spacelike ζ^α, and condition (ii) holds. Q.E.D.

Remark. The corresponding left eigenvectors can be taken to be

$$L_\pm = (a_\pm(\zeta_\nu - \mu_\pm \xi_\nu), \; -a_\pm^2, \; -h^{\alpha\beta}(\zeta_\alpha - \mu_\pm \xi_\alpha)(\zeta_\beta - \mu_\pm \xi_\beta)p'_s)$$

and

$$L_{(1)} = (u_\mu, 0, 0), \quad L_{(2)} = (v_{2\mu}, 0, 0),$$
$$L_{(3)} = (v_{3\mu}, 0, 0), \quad L_{(4)} = (0_\mu, 0, 1).$$

The eigenvalues $\mu_\pm$ and the corresponding eigenvectors are said to represent "acoustic waves," whereas $\mu_{(I)}$ and $\mathbf{d}_{(I)}$ represent "material waves."

As shall be seen in Chapter 4, the theory developed so far is strictly related to the theory of characteristic hypersurfaces for quasi-linear systems (Courant and Hilbert, 1953; Jeffrey, 1976).

Let Σ be a hypersurface in space-time $\mathscr{M}$, with the local equation in local coordinates

$$\phi(x^\mu) = 0. \tag{2.41}$$

Then Σ is said to be a characteristic hypersurface for the quasi-linear system (2.1) if

$$\det(A^\alpha \phi_\alpha) = 0, \tag{2.42}$$

where $\phi_\alpha = \nabla_\alpha \phi$, for short.

Now in our case ϕ satisfies either

$$u^\alpha \phi_\alpha = 0, \tag{2.43a}$$

or

$$(u^\alpha \phi_\alpha)^2 - p'_e h^{\alpha\beta} \phi_\alpha \phi_\beta = 0. \tag{2.43b}$$

It is easily seen that $\phi_\alpha \phi^\alpha \geq 0$ and therefore Σ is a timelike or null hypersurface. This is a fundamental causality requirement, as we shall see in later chapters. Now we can define the normal speed of propagation V_Σ of the hypersurface Σ with respect to an observer, described by a timelike world line of tangent vector field v^μ, $v^\mu v_\mu = -1$, as follows (Synge, 1960). Let us consider a particle $\mathscr{P}$ riding on the hypersurface Σ with four-velocity w^μ, $w^\mu \phi_\mu = 0$, passing through the event O^* of intersection of Σ with the world line of the observer v^μ. Then the square of the three-velocity w of

$\mathscr{P}$ in the rest frame of v^μ at the event O^* is

$$w^2 = \frac{(g^{\mu\nu} + v^\mu v^\nu) w_\mu w_\nu}{(v^\alpha w_\alpha)^2}$$

(this is easily checked in Minkowski coordinates with $v^\mu = \delta^\mu_0$). The normal speed of propagation (squared) of the hypersurface Σ with respect to v^μ at the event O^* is defined as the minimum of the above expression for w^2 under the constraint $w^\mu \phi_\mu = 0$. An easy calculation then gives

$$V_\Sigma^2 = \frac{(v^\mu \phi_\mu)^2}{(g^{\alpha\beta} + v^\alpha v^\beta)\phi_\alpha \phi_\beta}. \tag{2.44}$$

The choice of the sign for V_Σ can be done as follows. We define, conventionally, the *outward unit normal to* Σ in the rest frame of the observer v^μ as

$$\nu_\alpha = \frac{\phi_\alpha + v_\alpha v^\beta \phi_\beta}{(g^{\alpha\beta}\phi_\alpha\phi_\beta + (v^\alpha \phi_\alpha)^2)^{1/2}}.$$

Let w^α be the four-velocity of a particle $\mathscr{P}$ riding on the wavefront Σ, $w^\alpha \phi_\alpha = 0$, future directed with respect to v^μ (i.e., $w_\alpha v^\alpha < 0$). Then V_Σ will be positive if the wave is forward propagating with respect to ν_α in the rest frame of the observer v^μ, this is, if

$$(g^{\alpha\beta} + v^\alpha v^\beta)\nu_\alpha w_\beta > 0$$

which amounts to

$$v^\alpha \phi_\alpha < 0.$$

Then from equation (2.44) it follows that

$$V_\Sigma = -\frac{v^\mu \phi_\mu}{((g^{\alpha\beta} + v^\alpha v^\beta)\phi_\alpha\phi_\beta)^{1/2}}.$$

If one takes as an observer the one defined by ξ^μ, then

$$V_\Sigma = -\mu$$

and, therefore, the eigenvalues μ are the opposite of the normal speeds of propagation of the characteristic hypersurfaces relative to the observer defined by ξ^μ. If one takes as observer u^μ, then for the material waves one finds

$$V_\Sigma = 0,$$

which expresses the fact that the corresponding characteristic hypersurfaces are at rest with respect to the fluid. Similarly, for the acoustic waves, one finds

$$V_\Sigma = \sqrt{p'_e}.$$

Therefore the requirement $0 < p'_e \leq 1$ is equivalent to the statement that the acoustic surfaces have a nonvanishing propagation speed (the speed of sound) not exceeding the velocity of light relative to the fluid. Some authors (see Hartle, 1978 for a review of the different opinions) have claimed that the condition $p'_e \leq 1$ is too restrictive. Their argument is that in a realistic situation, with dissipative effects included, acoustic waves will be dispersive. Then the correct restriction to impose on the equation of state should be that the group speed of the waves be less than the speed of light, a condition which, in general, is less restrictive than $p'_e \leq 1$.

Although this argument has some merit, in fact it overlooks the important point that, once the mathematical model of nondissipative relativistic fluid dynamics has been assumed [as is done in several astrophysical problems, for example, in the analysis of equilibrium and stability of neutron stars (Hartle, 1978)], it is inconsistent not to impose the restriction $p'_e \leq 1$. In fact, it can be shown that violation of the latter condition implies the existence of complex eigenvalues μ and in its turn this would entail the nonlinear instability of the constant solution (Boillat, 1981).

The correct way of investigating the possibility of $p'_e > 1$ is within the framework of relativistic dissipative fluids (taking into account viscosity and heat conduction). Within this framework one could check whether the unstable mode (which exists in the dissipationless case under the condition $p'_e > 1$) still persists or is damped by dissipation. An assessment of this question would be of great significance for relativistic astrophysics and particularly for the problem of the maximum mass for a neutron star (Hartle, 1978). Violation of the condition $p'_e \leq 1$ could lead to a significant increase in the maximum neutron star mass and this is a key parameter for identifying black holes.

In many situations (neutron star interiors, accretion onto a magnetized black hole, etc.) the effects of magnetic fields on a highly conducting fluid cannot be neglected and this leads to the consideration of magnetohydrodynamical phenomena. In the next section we will introduce the basic equations describing magneto-fluid dynamics in a relativistic framework and investigate their mathematical structure.

2.4. Test relativistic magneto-fluid dynamics as a quasi-linear hyperbolic system

We start with a justification of the equations of relativistic magneto-fluid dynamics. This will be based on the phenomenological theory of an electromagnetically polarizable continuum.

Another approach which might be used could be based on a microscopic description in terms of the relativistic Vlasov equation, which would be appropriate for a rarefield plasma. Here we shall follow the first approach, because of its simplicity and greater generality.

Let us consider a fluid interacting with the electromagnetic field and let $T^{\alpha\beta}$ be the total energy-momentum tensor of the system (fluid and electromagnetic system). The standard energy-momentum conservation laws then are written

$$\nabla_\alpha T^{\alpha\beta} = 0. \tag{2.45}$$

The electromagnetic field will be described by two antisymmetric tensor fields $F^{\alpha\beta}$ (the electromagnetic field tensor) and $I^{\alpha\beta}$ (the electromagnetic induction tensor), obeying Maxwell's equations (Dixon, 1978)

$$\partial_{[\alpha} F_{\beta\gamma]} = 0, \tag{2.46}$$

$$\nabla_\beta I^{\alpha\beta} = 4\pi J^\alpha, \tag{2.47}$$

where J^α is the charge current four-vector and $[\alpha\beta\gamma]$ denotes antisymmetrization with respect to the indices $\alpha\beta\gamma$.

In addition to the energy-momentum conservation laws one has also mass conservation, which can be written in a local form as

$$\nabla_\alpha \rho^\alpha = 0,$$

where ρ^α is the mass flux four-vector. One can always introduce a scalar ρ and a velocity field u^α such that $u^\alpha u_\alpha = -1$ and $\rho^\alpha = \rho u^\alpha$, ρ being interpreted as the proper mass density. Then the mass conservation equation is

$$\nabla_\alpha(\rho u^\alpha) = 0. \tag{2.48}$$

The four-vector u^α can be called the fluid four-velocity, although it should be remarked that it does not represent, in general, the electron velocity.

With respect to the vector field u^α one can decompose $F^{\alpha\beta}$ and $I^{\alpha\beta}$ in the following way

$$\left.\begin{aligned} E^\alpha &\equiv F^{\alpha\beta}u_\beta, \quad B^{\alpha\beta} \equiv F^{\alpha\beta} - 2u^{[\alpha}E^{\beta]} \\ D^\alpha &\equiv I^{\alpha\beta}u_\beta, \quad H^{\alpha\beta} \equiv I^{\alpha\beta} - 2u^{[\alpha}D^{\beta]} \end{aligned}\right\} \tag{2.49}$$

where $[\alpha\beta]$ denotes antisymmetrization with respect to the indices α, β.

E^α and D^α are, respectively, the rest frame electric field and electric displacement, and $B^{\alpha\beta}$, $H^{\alpha\beta}$ represent the rest frame magnetic induction and magnetic field.

Clearly all these vectors and tensors are orthogonal to the four-velocity u^α.

In a medium the relationship between the electric displacement and electric field, and the magnetic field and magnetic induction, are characteristic of the medium (and are ultimately justified in a statistical mechanics approach), and are called phenomenological laws or constitutive equations.

For most applications we can assume linear isotropic constitutive relations for D^α, $H^{\alpha\beta}$ (Dixon, 1978)

$$D^\alpha = (1 + 4\pi k)E^\alpha \tag{2.50}$$

$$H^{\alpha\beta} = (1 - 4\pi\chi)B^{\alpha\beta} \tag{2.51}$$

with k and χ the electric and magnetic susceptibility.

These quantities, in general, are functions of the density and temperature of the fluid.

The charge current J^α may also be decomposed with respect to u^α, as

$$J^\alpha = qu^\alpha + j^\alpha, \tag{2.52}$$

where $j^\alpha u_\alpha = 0$ and $q = -J^\alpha u_\alpha$. Therefore, q is the proper charge density and j^α is the conduction current.

We shall assume a linear constitutive relation between j^α and E^α (Ohm's law),

$$j^\alpha = \sigma^{\alpha\beta}E_\beta, \tag{2.53}$$

where $\sigma^{\alpha\beta}$ is the conductivity tensor.

In general, in the presence of a magnetic field, the conductivity tensor will be anisotropic, and will be a function of density, temperature, and the magnetic field.

Here we shall assume that the magnetic field is sufficiently weak that the conductivity tensor reduces to

$$\sigma^{\alpha\beta} = \sigma g^{\alpha\beta} \tag{2.54}$$

with σ a function only of density and temperature.

Then, the limit of infinite conductivity $\sigma \to \infty$ and finite conduction current implies the usual approximation

$$E_\alpha = 0.$$

Notice that in the relativistic framework this is the only approximation which is made, as opposed to the nonrelativistic case where also the displacement currents are neglected in the Maxwell equations (2.46)–(2.47).

Under the assumption of local thermodynamic equilibrium a general expression can be found for the total energy-momentum tensor of the fluid and electromagnetic field. We shall adopt here the results of Dixon (1978), which are based on an elegant and deep theory. Dixon defines the local equilibrium states as those for which, locally, the entropy production rate $\nabla_\alpha s^\alpha$ (where s^α is the entropy flux) vanishes and is a minimum. By assuming that s^α depends only on the variables $T^{\alpha\beta}$, ρu^α, $F_{\alpha\beta}$, $I^{\alpha\beta}$, Dixon finds the following expression for $T^{\alpha\beta}$,

$$T^{\alpha\beta} = T_1^{\alpha\beta} + T_2^{\alpha\beta} + T_3^{\alpha\beta} + T_4^{\alpha\beta}, \tag{2.55}$$

with

$$T_1^{\alpha\beta} = \rho(1+\varepsilon)u^\alpha u^\beta + p h^{\alpha\beta} \tag{2.56}$$

the usual fluid's energy-momentum tensor,

$$T_2^{\alpha\beta} = \tfrac{1}{2}T\left(\frac{\partial k}{\partial T}E^2 + \frac{\partial \chi}{\partial T}B^2\right)u^\alpha u^\beta - \tfrac{1}{2}\rho\left(\frac{\partial k}{\partial \rho}E^2 + \frac{\partial \chi}{\partial \rho}B^2\right)h^{\alpha\beta}, \tag{2.57}$$

and

$$E^2 \equiv E_\alpha E^\alpha, \quad B^2 \equiv \tfrac{1}{2}B^{\alpha\beta}B_{\alpha\beta},$$

$$T_3^{\alpha\beta} = \frac{1}{2\pi}u^\beta F_\gamma^{[\alpha}I^{\mu]\gamma}u_\mu, \tag{2.58}$$

$$T_4^{\alpha\beta} = \frac{1}{4\pi}\left[F_\gamma^\alpha I^{\beta\gamma} - \tfrac{1}{4}F_{\gamma\delta}I^{\gamma\delta}g^{\alpha\beta}\right]. \tag{2.59}$$

Now, whereas the tensors $T_1^{\alpha\beta}$ and $T_4^{\alpha\beta}$ must be obviously attributed to the fluid and electromagnetic field, respectively, there is no unique logical way of assigning $T_3^{\alpha\beta}$ and $T_2^{\alpha\beta}$ to either components.

Notice that $T_4^{\alpha\beta}$ corresponds to the Minkowski tensor, while $T_3^{\alpha\beta} + T_4^{\alpha\beta}$ corresponds to the Abraham tensor (Dixon, 1978).

The equations of relativistic magneto-fluid dynamics are usually derived by assuming the Minkowski form for the energy-momentum tensor of the electromagnetic field (Lichnerowicz, 1967). Here we shall see that the same equations are obtained if we start from the full energy-momentum tensor

(2.55). In fact, we shall assume (as is usually done and it is a good approximation for plasmas) that the susceptibilities k and χ are constant, hence we can drop $T_2^{\alpha\beta}$. Furthermore, from the requirement $E^\alpha = 0$, it follows by equation (2.50), $D^\alpha = 0$, hence $T_3^{\alpha\beta} = 0$.

Finally, from equations (2.49)–(2.50) we have also

$$I^{\alpha\beta} = (1 - 4\pi\chi)F^{\alpha\beta},$$

which can be rewritten as

$$I^{\alpha\beta} = \frac{1}{\mu}F^{\alpha\beta}, \tag{2.60}$$

where μ is the (constant) magnetic permeability.

It is convenient to introduce the magnetic induction field B^α, as measured in the local rest frame,

$$B_\alpha = \tfrac{1}{2}\eta_{\alpha\beta\gamma\delta}u^\beta F^{\gamma\delta}, \tag{2.61}$$

where $\eta_{\alpha\beta\gamma\delta}$ is the Levi–Civita alternating tensor (Synge, 1960; Misner et al., 1973).

Then, because in our case $F^{\alpha\beta}u_\beta = 0$, from equation (2.61) one obtains

$$F^{\nu\sigma} = \eta^{\alpha\mu\nu\sigma}B_\alpha u_\mu. \tag{2.62}$$

By using equations (2.60)–(2.62) in equation (2.59) we obtain

$$4\pi\mu T_4^{\alpha\beta} = -B^\alpha B^\beta + \tfrac{1}{2}B_\sigma B^\sigma g^{\alpha\beta} + B_\sigma B^\sigma u^\alpha u^\beta, \tag{2.63}$$

and, therefore, by introducing the vector

$$b^\alpha = \frac{1}{\sqrt{4\pi\mu}}B^\alpha \tag{2.64}$$

the total energy-momentum tensor $T^{\alpha\beta}$ can be written as

$$T^{\alpha\beta} = (e + p + |b|^2)u^\alpha u^\beta + (p + \tfrac{1}{2}|b|^2)g^{\alpha\beta} - b^\alpha b^\beta, \tag{2.65}$$

where $e = \rho(1 + \varepsilon)$ and $|b|^2 = b_\alpha b^\alpha$ (notice that $b^\alpha u_\alpha = 0$, hence b^α is a spacelike vector and $|b|^2 > 0$).

Substituting equation (2.62) into the Maxwell equation (2.46), and taking the dual (multiplying by $\eta^{\alpha\mu\beta\gamma}$) yields

$$\nabla_\alpha(u^\alpha b^\beta - b^\alpha u^\beta) = 0. \tag{2.66}$$

The other Maxwell equation (2.47) can be used in order to calculate the charge current in terms of the given fields b^α.

The equations of relativistic magneto-fluid dynamics are then the conservation of energy-momentum

$$\nabla_\alpha T^{\alpha\beta} = 0, \tag{2.67}$$

the conservation of mass

$$\nabla_\alpha(\rho u^\alpha) = 0, \tag{2.68}$$

and the relevant Maxwell equations

$$\nabla_\alpha(u^\alpha b^\beta - b^\alpha u^\beta) = 0, \tag{2.69}$$

together with the state equation in the form $e = e(p, S)$.

The equations we have obtained are in conservation form. Now we will develop some consequences of these equations and obtain an equivalent set of equations (not in conservation form) which, in many cases, are more convenient to utilize.

By contracting equation (2.69) with u_β and b_β, respectively, we obtain

$$u^\alpha u^\beta \nabla_\alpha b_\beta + \nabla_\alpha b^\alpha = 0 \tag{2.70}$$

$$\tfrac{1}{2} u^\alpha \nabla_\alpha |b|^2 + |b|^2 \nabla_\alpha u^\alpha - b^\alpha b^\beta \nabla_\alpha u_\beta = 0. \tag{2.71}$$

From equation (2.67), by contracting with u_β and using equation (2.71) we obtain the fluid conservation of energy equation

$$u^\alpha \nabla_\alpha e + (e + p)\nabla_\alpha u^\alpha = 0. \tag{2.72}$$

From this equation, the conservation of mass equation (2.68), and the first law of thermodynamics (2.11) we obtain in the usual way the adiabaticity condition

$$u^\alpha \nabla_\alpha S = 0. \tag{2.73}$$

By using the adiabaticity condition, equation (2.72) is equivalent to

$$e'_p u^\alpha \nabla_\alpha p + (e + p)\nabla_\alpha u^\alpha = 0. \tag{2.72'}$$

From equation (2.67), by contracting with b_β and using (2.70) we obtain

$$u^\alpha u^\beta \nabla_\alpha b_\beta = (e + p)^{-1} b^\alpha \nabla_\alpha p. \tag{2.74}$$

From equation (2.67), by contracting with $h_{\beta\mu}$ and using equations (2.70)–(2.72) and (2.74), we obtain the conservation of momentum equation in the form

$$\begin{aligned} &(e + p + |b|^2) u^\alpha \nabla_\alpha u^\mu - b^\alpha \nabla_\alpha b^\mu + (h^{\mu\alpha} + u^\mu u^\alpha) b_\nu \nabla_\alpha b^\nu \\ &\quad + \frac{1}{(e + p)}((e + p)h^{\mu\alpha} - e'_p |b|^2 u^\mu u^\alpha + b^\mu b^\alpha)\nabla_\alpha p = 0. \end{aligned} \tag{2.75}$$

Finally, the Maxwell equations (2.69) can be rewritten in the form

$$u^\alpha \nabla_\alpha b^\beta - b^\alpha \nabla_\alpha u^\beta + \frac{1}{(e+p)}(-e'_p b^\beta u^\alpha + u^\beta b^\alpha)\nabla_\alpha p = 0. \tag{2.76}$$

We remark that equations (2.67)–(2.69) are equivalent to equations (2.75), (2.76), (2.72′), (2.73), and (2.70). In fact, let G^μ, H^β, $\tilde{G}$, K, H denote the left-hand side of equations (2.75), (2.76), (2.72′), (2.73), and (2.70), respectively. Then it is easy to check the following identities:

$$G^\mu = \nabla_\alpha T^{\alpha\mu} + (u_\beta \nabla_\alpha T^{\alpha\beta} + b_\beta \nabla_\alpha(u^\alpha b^\beta - u^\beta b^\alpha))u^\mu\left(1 + \frac{|b|^2}{(e+p)}\left(1 - \frac{e'_s}{T\rho}\right)\right)$$

$$+ \frac{b^\mu}{(e+p)}(b_\beta \nabla_\alpha T^{\alpha\beta} + (e+p+|b|^2)u_\beta \nabla_\alpha(u^\alpha b^\beta - u^\beta b^\alpha))$$

$$- u^\mu \frac{|b|^2}{T\rho^2} e'_s \nabla_\alpha(\rho u^\alpha),$$

$$H^\beta = \nabla_\alpha(u^\alpha b^\beta - b^\alpha u^\beta) + \frac{u^\beta}{(e+p)}(b_\gamma \nabla_\alpha T^{\alpha\gamma} + (e+p+|b|^2)u_\gamma \nabla_\alpha(u^\alpha b^\gamma - b^\alpha u^\gamma))$$

$$+ (u_\gamma \nabla_a T^{\alpha\gamma} + b_\gamma \nabla_\alpha(u^\alpha b^\gamma - b^\alpha u^\gamma))\frac{b^\beta}{(e+p)}(1 - e'_s/T\rho)$$

$$- \frac{b^\beta e'_s}{T\rho^2} \nabla_\alpha(\rho u^\alpha),$$

$$\tilde{G} = (u_\beta \nabla_\alpha T^{\alpha\beta} + b_\beta \nabla_\alpha(u^\alpha b^\beta - b^\alpha u^\beta))\left(-1 + \frac{e'_s}{T\rho}\right) + \frac{(e+p)}{T\rho^2} e'_s \nabla_\alpha(\rho u^\alpha),$$

$$K = -\frac{1}{T\rho}\left(u_\beta \nabla_\alpha T^{\alpha\beta} + b_\beta \nabla_\alpha(u^\alpha b^\beta - b^\alpha u^\beta) + \frac{(e+p)}{\rho}\nabla_\alpha(\rho u^\alpha)\right),$$

$$H = u_\beta \nabla_\alpha(u^\alpha b^\beta - u^\beta b^\alpha).$$

From these identities it is apparent that equations (2.67)–(2.69) hold if and only if equations (2.75), (2.76), (2.72′), (2.73), and (2.70) hold. Notice also that one has

$$\nabla_\beta\left[H^\beta - \frac{u^\beta b_\mu G^\mu}{e+p+|b|^2} + \frac{b^\beta \tilde{G}}{e+p}\right] = u^\mu \nabla_\mu H + H \nabla_\mu u^\mu = 0.$$

Hence, if equation (2.70) holds on a hypersurface $\mathscr{F}$ transverse to the vector field u^μ, it will hold also in a neighborhood of $\mathscr{F}$ as a consequence of the remaining equations.

Now we will investigate the mathematical structure of the above equations. This is a problem which is important not only conceptually but also for practical reasons as in the case of numerical computations (the numerical techniques can vary widely for different classes of equations). The question we shall investigate is that of the hyperbolicity of the equations of relativistic magneto-fluid dynamics. In order to answer this question it is necessary to analyze in detail the various modes of wave propagation and this analysis gives considerable insight into the physical content of the theory. First of all, we shall write the equations of relativistic magneto-fluid dynamics as a quasi-linear system.

We take equations (2.75), (2.76), (2.72′), (2.73) as the field equations for the field unknown

$$\mathbf{U} = (u^\nu, b^\nu, p, S)^{\mathrm{T}}. \tag{2.77}$$

These field equations can be written in matrix formulation

$$A_B^{\alpha A}\nabla_\alpha U^B = 0, \quad A, B = 0, \ldots, 9, \tag{2.78}$$

with

$$A^\alpha = \begin{bmatrix} Eu^\alpha\delta^\mu_\nu, & -b^\alpha\delta^\mu_\nu + p^{\mu\alpha}b_\nu, & 1^{\mu\alpha}, & 0^{\mu\alpha} \\ b^\alpha\delta^\mu_\nu, & -u^\alpha\delta^\mu_\nu, & f^{\alpha\mu}, & 0^{\mu\alpha} \\ \eta\delta^\alpha_\nu, & 0^\alpha_\nu, & e'_p u^\alpha, & 0^\alpha \\ 0^\alpha_\nu, & 0^\alpha_\nu, & 0^\alpha, & u^\alpha \end{bmatrix}, \tag{2.79}$$

where $\eta = e + p$, $E = \eta + |b|^2$, $p^{\mu\alpha} = h^{\mu\alpha} + u^\mu u^\alpha$,

$$1^{\mu\alpha} = \frac{1}{\eta}(\eta h^{\mu\alpha} - e'_p|b|^2 u^\mu u^\alpha + b^\mu b^\alpha), \quad \text{and} \quad f^{\alpha\mu} = -\frac{1}{\eta}(-u^\alpha b^\mu e'_p + u^\mu b^\alpha).$$

The characteristic matrix $A^\alpha\phi_\alpha$ then is written

$$A^\alpha\phi_\alpha = \begin{bmatrix} Ea\delta^\mu_\nu, & m^\mu_\nu, & 1^\mu, & 0^\mu \\ B\delta^\mu_\nu, & -a\delta^\mu_\nu, & f^\mu, & 0^\mu \\ \eta\phi_\nu, & 0_\nu, & e'_p a, & 0 \\ 0_\nu, & 0_\nu, & 0, & a \end{bmatrix},$$

where $a = u^\alpha\phi_\alpha$, $B = b^\alpha\phi_\alpha$,

$$1^\mu = \phi^\mu + \left(1 - \frac{e'_p|b|^2}{\eta}\right)au^\mu + \frac{B}{\eta}b^\mu,$$

$$f^\mu = -\frac{1}{\eta}(-ae'_p b^\mu + Bu^\mu),$$

$$m^\mu_\nu = (\phi^\mu + 2au^\mu)b_\nu - B\delta^\mu_\nu.$$

Now it is easy to show that

$$\det(A^\alpha \phi_\alpha) = Ea^2 A^2 N_4, \tag{2.80}$$

where

$$A = Ea^2 - B^2 \tag{2.81}$$

and

$$N_4 = \eta(e'_p - 1)a^4 - (\eta + e'_p|b|^2)a^2 G + B^2 G \tag{2.82}$$

with

$$G = g^{\mu\alpha}\phi_\mu\phi_\alpha.$$

Now we discuss some properties of the various modes.

LEMMA 2.1. *For any timelike vector ξ^α the condition (i) for hyperbolicity [i.e., $\det(A^\alpha\xi_\alpha) \neq 0$] holds, provided that $e'_p - 1 \geq 0$.*

Proof. From equation (2.80), by putting $\phi_\alpha = \xi_\alpha$, we have that det $(A^\alpha\phi_\alpha)$ can vanish only if

$$a = 0,$$

or

$$A = 0$$

or

$$N_4 = 0,$$

all these quantities being evaluated for $\phi_\alpha = \xi_\alpha$.

Without loss of generality, at a given point, it is possible to choose an orthonormal frame such that

$$u^\alpha = (1, 0, 0, 0), \quad b^\alpha = (0, |b|, 0, 0), \quad \xi_\alpha = (\xi_0, \xi_1, \xi_2, 0).$$

Then

$$a = \xi_0 \neq 0$$

and

$$A = \eta\xi_0^2 + |b|^2(\xi_0^2 - \xi_1^2) > 0$$

because ξ_α is timelike.

Furthermore,

$$N_4 = e'_p E\xi_0^2 - |b|^2\xi_1^2 > 0$$

because $e'_p - 1 \geq 0$. Q.E.D.

Note. If $e'_p < 1$ then $N_4 = 0$ can have roots in which hyperbolicity is violated.

PROPOSITION 2.2. *For the material and Alfvén modes one has $G > 0$, that is, the corresponding hypersurfaces $\phi = \text{const.}$ are timelike.*

Proof. At a given space-time point (x^μ) construct the orthonormal frame $(u^\mu,\ \hat{b}^\mu,\ e_2^\mu,\ e_3^\mu)$, where $\hat{b}^\mu = b^\mu/|b|$ and $e_2^\mu u_\mu = e_2^\mu b_\mu = e_3^\mu b_\mu = e_3^\mu u_\mu = 0$, $e_2^\mu e_{3\mu} = 0$, $e_2^\mu e_{2\mu} = e_3^\mu e_{3\mu} = 1$. Then

$$\phi^\mu = -au^\mu + \frac{B}{|b|}\hat{b}^\mu + c_2 e_2^\mu + c_3 e_3^\mu.$$

The *material waves* are defined as the solutions $\phi = \text{const.}$ of $a = 0$, hence

$$G = \frac{B^2}{|b|^2} + c_2^2 + c_3^2 > 0.$$

The *Alfvén waves* are defined as the solutions $\phi = \text{const.}$ of

$$a^2 = \frac{B^2}{E},$$

hence

$$G = -\frac{B^2}{E} + \frac{B^2}{|b|^2} + c_2^2 + c_3^2 > 0. \qquad \text{Q.E.D.}$$

Note. For the *magnetoacoustic waves*, defined as the solutions $\phi = \text{const.}$ of $N_4 = 0$, it is easy to check that, under the assumption $e_p' \neq 1$, one has $G \neq 0$.

In fact $G = 0$ in $N_4 = 0$ implies $a = 0$ or $e_p' - 1 = 0$, and by the previous proposition, $a = 0$ is incompatible with $G = 0$.

Now let us introduce the quantity

$$b_n^2 = \frac{B^2}{a^2 + G}. \tag{2.83}$$

It is easy to check that, under the assumption $e_p' > 1$, one has $a^2 + G > 0$, for all ϕ satisfying the characteristic equation.

In fact, by using an orthonormal frame as in the previous proposition it is seen that

$$a^2 + G = 0$$

iff $u^\alpha = -a\phi^\alpha$.

Therefore, ϕ^α would be timelike, and by Lemma 2.1, it cannot be a solution of the characteristic equation.

PROPOSITION 2.3. *One has $b_n^2 \leq |b|^2$ and $b_n^2 = |b|^2$ if and only if $\phi^\mu = -au^\mu + \frac{B}{|b|^2} b^\mu$.*

Proof. In local Minkowski coordinates (x^μ) such that $u^\mu = \delta_0^\mu$, $g_{\mu\nu} = \eta_{\mu\nu}$, one has $b^0 = 0$. Also,

$$(b^i \phi_i)^2 \leq \sum_i (b^i)^2 \sum_i (\phi^i)^2, \quad \text{hence } B^2 \leq |b|^2 \sum_i (\phi^i)^2.$$

But $a^2 + G = \sum (\phi^i)^2$, and the proposition follows.

Also, from the proof of Proposition 2.2, $\phi^\mu = -au^\mu + \frac{B}{|b|}\hat{b}^\mu + c_2 e_2^\mu + c_3 e_3^\mu$, hence

$$G + a^2 = \frac{B^2}{|b|^2} + c_2^2 + c_3^2, \quad \text{therefore,} \quad b_n^2 = \frac{B^2 |b|^2}{B^2 + |b|^2(c_2^2 + c_3^2)} = |b|^2$$

iff

$$c_2 = c_3 = 0. \qquad \text{Q.E.D.}$$

Now we investigate the normal speeds of propagation of the various waves with respect to the observers moving with the fluid. We recall that

$$V_\Sigma^2 = \frac{(u^\mu \phi_\mu)^2}{(g^{\alpha\beta} + u^\alpha u^\beta)\phi_\alpha \phi_\beta} = \frac{a^2}{a^2 + G}.$$

PROPOSITION 2.4. *The magnetoacoustic wave fronts, solutions of $N_4 = 0$, under the assumption $e'_p \geq 1$, have four normal speeds of propagation $\pm V_{\Sigma s}$ (slow waves), $\pm V_{\Sigma f}$ (fast waves), such that $0 \leq V_{\Sigma s} \leq c_s \leq V_{\Sigma f} \leq 1$, where $c_s = \sqrt{p'_e}$ is the hydrodynamical speed of sound. When $e'_p = 1$, then $V_{\Sigma s} = \frac{b_n}{\sqrt{E}}$ and $V_{\Sigma f} = 1$.*

Proof. It is easily seen that

$$N_4 = \frac{G^2}{(1 - V_\Sigma^2)^2} P(V_\Sigma^2),$$

where

$$P(V_\Sigma^2) = \eta(e'_p - 1)V_\Sigma^4 - (\eta + e'_p |b|^2 - b_n^2) V_\Sigma^2 (1 - V_\Sigma^2) + b_n^2 (1 - V_\Sigma^2)^2.$$

Now

$$P(0) = b_n^2 \geq 0, \quad P(1) = \eta(e'_p - 1) \geq 0,$$

and $P(c_s^2) = (1 - c_s^2)(b_n^2 - |b|^2) \leq 0$.

If $e'_p = 1$, then

$$P(V_\Sigma^2) = (1 - V_\Sigma^2)(-EV_\Sigma^2 + b_n^2)$$

which has the following zeros

$$V_\Sigma^2 = 1, \quad V_\Sigma^2 = \frac{b_n^2}{E} < 1.$$

If $e'_p > 1$, $b_n^2 = |b|^2$, then by Proposition 2.3, $\phi^\mu = -au^\mu + \dfrac{B}{|b|^2} b^\mu$ and $P(V_\Sigma^2)$ has two zeros, c_s^2 and $|b|^2/E$, both less than 1.

If $e'_p > 1$, $b_n^2 < |b|^2$, then $P(1) > 0$, $P(c_s^2) < 0$, $P(0) \geq 0$, and therefore $P(V_\Sigma^2)$ has two real roots, between 0 and 1, which are denoted by $V_{\Sigma s}^2$, $V_{\Sigma f}^2$, such that

$$0 \leq V_{\Sigma s} < c_s < V_{\Sigma f} < 1. \qquad \text{Q.E.D.}$$

COROLLARY. *The roots of* $N_4 = 0$ *can coincide if and only if* $e'_p > 1$, $b_n^2 = |b|^2$, *and*

$$\eta = (e'_p - 1)|b|^2.$$

Note. The normal speeds of propagation of the Alfvén wave fronts are $\pm V_{\Sigma A}$, where

$$V_{\Sigma A}^2 = \frac{b_n^2}{E}.$$

Also,

$$P(V_{\Sigma A}^2) = V_{\Sigma A}^2(e'_p - 1)(b_n^2 - |b|^2).$$

and therefore $P(V_{\Sigma A}^2) \leq 0$, hence $V_{\Sigma s} \leq V_{\Sigma A} \leq V_{\Sigma f}$.

It is interesting to have a graphical representation of the various wave front speeds (Lichnerowicz, 1971).

At a given point, consider the orthonormal frame $(u^\mu, \hat{b}^\mu, e_2^\mu, e_3^\mu)$ introduced in the proof of Proposition 2.2. We have seen that

$$\phi^\mu = -au^\mu + \frac{B}{|b|}\hat{b}^\mu + c_2 e_2^\mu + c_3 e_3^\mu.$$

For the sake of notation write $t = -a$, $x = \dfrac{B}{|b|}$, $y = c_2$, and $z = c_3$. Then the light cone $G = g^{\alpha\beta}\phi_\alpha\phi_\beta = 0$ is written as

$$\Gamma : -t^2 + x^2 + y^2 + z^2 = 0.$$

The material waves cone $u^\alpha \phi_\alpha = 0$ is

$$\Gamma_M : t = 0.$$

The Alfvén waves cone $A^2 = 0$ is

$$\Gamma_A : (\eta + |b|^2)t^2 - |b|^2 x^2 = 0,$$

whereas the magnetoacoustic waves cone $N_4 = 0$ is

$$\Gamma_H : \eta(e'_p - 1)t^4 - (\eta + e'_p|b|^2)t^2(-t^2 + x^2 + y^2 + z^2) + |b|^2 x^2(-t^2 + x^2 + y^2 + z^2) = 0.$$

Since these cones admit the axis Ox as an axis of rotation it is convenient to study their intersections with the planes $t = 1$ and $y = 0$. We obtain the following curves

$$S : x^2 + z^2 = 1 \quad \text{from the light cone } \Gamma,$$

$$S_A : x = \pm\sqrt{\hat{E}} \text{ from the Alfvén waves cone, and}$$

$$S_H : \hat{E}e'_p - (\hat{E} + e'_p - 1)(x^2 + z^2) + x^2(-1 + x^2 + z^2) = 0$$

Fig. 2.2. The curves drawn are as follows:

$S \,: x^2 + y^2 - 1 = 0$ Light cone,
$S_A : x = \pm(\hat{E})^{1/2}$ Alfvén cone,
$S_H : \hat{E}e'_p - (\hat{E} + e'_p - 1)(x^2 + z^2) + x^2(-1 + x^2 + z^2) = 0$ magnetoacoustic cone, $\hat{E} \equiv (\eta + |b|^2)/|b|^2$

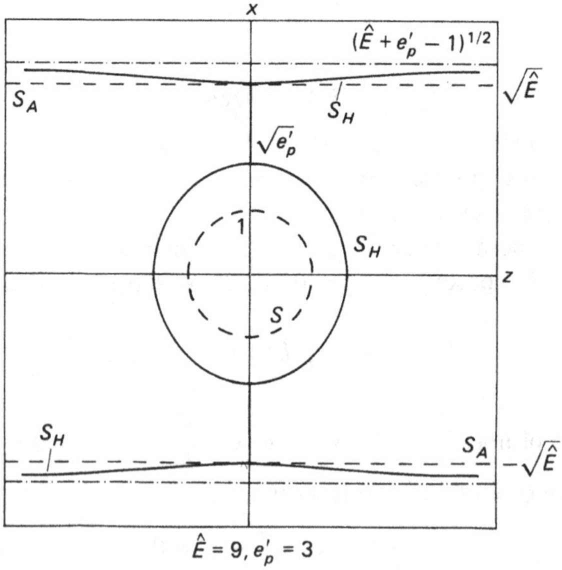

from the magnetoacoustic waves cone, where

$$\hat{E} = \frac{\eta + |b|^2}{|b|^2}.$$

Now we determine the right and left eigenvectors corresponding to the various modes of propagation.

At a given point, without loss of generality, we choose an orthonormal basis such that

$$u^\alpha = (1,0,0,0), \quad b_\alpha = (0,|b|,0,0),$$
$$\xi_\alpha = (\xi_0, \xi_1, \xi_2, 0), \tag{2.84}$$

and $\xi_\alpha \xi^\alpha = -1$.

Let $\phi_a = \zeta_\alpha - \mu\xi_\alpha$. First of all, we consider the case where

$$\det\begin{pmatrix} \zeta_0 & \zeta_1 \\ \xi_0 & \xi_1 \end{pmatrix} = 0.$$

The physical interpretation of such a ζ_α is the following. Let H^ν be the magnetic field in the rest frame of the observer with four-velocity ξ^μ, given by

$$H^\nu = -F^{*\nu\sigma}\xi_\sigma,$$

where $F^{*\nu\sigma}$ is the dual of $F^{\alpha\beta}$. Then it is easy to check that $\zeta_\nu H^\nu = 0$.

Fig. 2.3. As in Fig. 2.2, the singular case with $\hat{E} = e'_p$

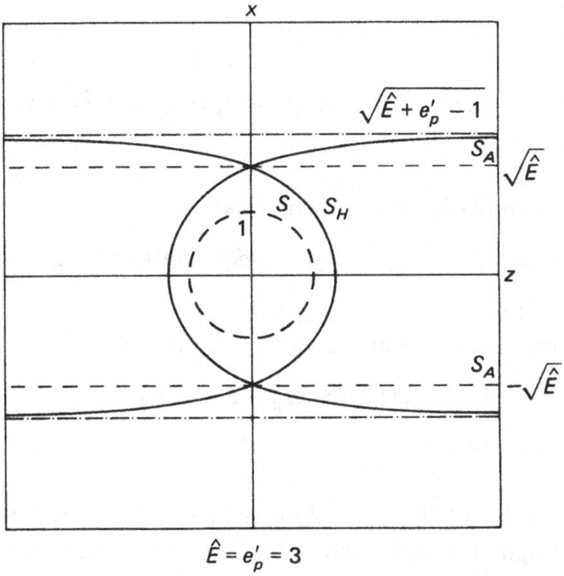

In the case under consideration one has

$$\zeta_0 = \beta\xi_0,$$
$$\zeta_1 = \beta\xi_1$$

for some factor β, $\phi_\alpha = ((\beta-\mu)\xi_0, (\beta-\mu)\xi_1, \zeta_2 - \mu\xi_2, \zeta_3)$. Then we have

$$a = (\beta-\mu)\xi_0,$$
$$B = |b|(\beta-\mu)\xi_1,$$
$$A = (\beta-\mu)^2(E\xi_0^2 - |b|^2\xi_1^2),$$
$$N_4 = (\beta-\mu)^2([-(\eta + e'_p|b|^2)\xi_0^2 + |b|^2\xi_1^2]G + \eta(e'_p - 1)(\beta-\mu)^2\xi_0^4).$$

Obviously, $a = 0$ iff $\mu = \beta$, and it is easily checked that this eigenvalue has multiplicity 8 and the corresponding right eigenspace has dimension 8. A basis is given by the vectors

$$\{(0^\alpha, 0^\alpha, 0, 1)^T; (u^\alpha, 0^\alpha, 0, 0)^T; (b^\alpha, 0^\alpha, 0, 0)^T; (\eta^\alpha_{\beta\gamma\delta}u^\beta b^\gamma \phi^\delta, 0^\alpha, 0, 0)^T;$$
$$(0^\alpha, u^\alpha, 0, 0)^T; (0^\alpha, \phi^\alpha, 0, 0)^T; (0^\alpha, \eta^\alpha_{\lambda\sigma\rho}u^\lambda b^\sigma \phi^\rho, 0, 0)^T; (0^\alpha, b^\alpha, -|b|^2, 0, 0)^T\}.$$

The corresponding basis of left eigenvectors is

$$\{(u_\mu, 0_\mu, 0, 0); (b_\mu, 0_\mu, 0, 0); (\eta_{\mu\beta\gamma\delta}u^\beta b^\gamma \phi^\delta, 0_\mu, 0, 0); (0_\mu, u_\mu, 0, 0); (0_\mu, b_\mu, 0, 0)$$
$$(0_\mu, \eta_{\mu\beta\gamma\delta}u^\beta b^\gamma \phi^\delta, 0, 0); (0_\mu, \phi_\mu, 0, 0); (0_\mu, 0_\mu, 0, 1)\}.$$

The remaining two eigenvalues are the solutions of

$$\frac{N_4}{(\beta-\mu)^2} = 0. \tag{2.85}$$

PROPOSITION 2.5. *Under the assumption $e'_p \geq 1$, the solutions of (2.85) are real and distinct.*

Proof. Equation (2.85) can be rewritten as

$$(\beta-\mu)^2[\eta(e'_p - 1)\xi_0^4 - D] + 2D\beta(\beta-\mu) + D(1-\beta^2) = 0,$$

where $D = -(\eta + e'_p|b|^2)\xi_0^2 + |b|^2\xi_1^2 < 0$.

The discriminant of the above second degree equation is

$$D[(1-\beta^2)\eta(e'_p - 1)\xi_0^4 - D]$$

and is therefore positive. Q.E.D.

To these real and distinct solutions there correspond two linearly independent right (left) eigenvectors which, together with those previously found, form a basis of $\mathbb{R}^{10}$. Their expression is the following

$$\mathbf{d} = (d^\alpha, d^{\alpha+4}, d^8, d^9)^{\mathrm{T}},$$

with

$$\begin{aligned} d^\alpha &= Ea^2(Bf^\alpha - a1^\alpha) + \frac{Ea}{\eta}(B^2 - e_p'|b|^2a^2)(2au^\alpha + \phi^\alpha), \\ d^{\alpha+4} &= \frac{B}{a}d^\alpha + EaAf^\alpha, \\ d^8 &= Ea^2A, \\ d^9 &= 0, \end{aligned} \tag{2.86}$$

where in equation (2.86) one must substitute $\phi_\alpha = \zeta_\alpha - \mu\xi_\alpha$, with μ the two roots of equation (2.85).

The corresponding left eigenvectors are

$$\mathbf{s} = (s_\nu, s_{\nu+4}, s_8, s_9),$$

with

$$\begin{aligned} s_\nu &= B\eta(G + 2a^2)b_\nu - Ea^2\eta\phi_\nu, \\ s_{4+\nu} &= -Ea\eta(G + 2a^2)b_\nu + EaB\eta\phi_\nu, \\ s_8 &= AEa, \\ s_9 &= 0. \end{aligned} \tag{2.87}$$

Now we investigate the case where

$$\det\begin{pmatrix} \zeta_0 & \zeta_1 \\ \xi_0 & \xi_1 \end{pmatrix} \neq 0.$$

In the basis (2.84) previously introduced we have

$$a = (\zeta_0 - \mu\xi_0).$$

Therefore, the solution of $a = 0$ (material wave) is then

$$\mu = \frac{\zeta_0}{\xi_0}. \tag{2.88}$$

It is easy to see that this solution is not a root of $A = 0$ or of $N_4 = 0$. Therefore, $\mu = \dfrac{\zeta_0}{\xi_0}$ is a double eigenvalue of the characteristic matrix. The corresponding two linearly independent right eigenvectors are

$$\begin{aligned} \mathbf{d}_1 &= (0^\mu, \phi^\mu, 0, 0)^{\mathrm{T}}, \\ \mathbf{d}_2 &= (0^\mu, 0^\mu, 0, 1)^{\mathrm{T}}. \end{aligned} \tag{2.89}$$

The corresponding two linearly independent left eigenvectors are

$$\mathbf{s}_1 = \left(0_\mu, -\frac{\eta}{B}\phi_\mu, 1, 0\right),$$
$$\mathbf{s}_2 = (0_\mu, 0_\mu, 0, 1). \tag{2.90}$$

For the Alfvén waves we have

$$A = (E\xi_0^2 - |b|^2\xi_1^2)\mu^2 - 2(E\xi_0\zeta_0 - |b|^2\xi_1\zeta_1)\mu + E\zeta_0^2 - |b|^2\zeta_1^2.$$

The discriminant of the above polynomial is

$$E|b|^2(\xi_1\zeta_0 - \xi_0\zeta_1)^2 > 0$$

and, therefore, $A = 0$ admits two real and distinct roots.

When these solutions are not roots of $N_4 = 0$, the corresponding two linearly independent right eigenvectors are

$$\mathbf{d} = \left(d^\mu, \frac{B}{a}d^\mu, 0, 0\right)^{\mathrm{T}}, \tag{2.91}$$

where d^μ is subject only to the constraints

$$d^\mu\phi_\mu = 0, \quad d^\mu b_\mu = 0.$$

When these solutions are also roots of $N_4 = 0$ we have

$$\mathbf{d} = \left(d^\mu, \frac{B}{a}d^\mu - \frac{\eta}{e'_p a^2}\phi_\nu d^\nu f^\mu, -\frac{\eta}{e'_p a}\phi_\nu d^\nu, 0\right), \tag{2.92}$$

with d^ν subject only to the constraint

$$d^\nu\left(b_\nu - \frac{B}{e'_p a^2}\phi_\nu\right) = 0.$$

The two linearly independent left eigenvectors, in the case $N_4 \neq 0$, are

$$\mathbf{s} = \left(s_\mu, -\frac{B}{a}s_\mu, 0, 0\right), \tag{2.93}$$

where s_μ is subject to the constraints

$$s_\mu(\phi^\mu + 2au^\mu) = 0 = s_\mu(Bb^\mu + |b|^2 au^\mu).$$

In the case when also $N_4 = 0$, the left eigenvectors remain the same as before. The only difference is that in this case the two vectors $\phi^\mu + 2au^\mu$, $Bb^\mu + |b|^2 au^\mu$ are no longer linearly independent and therefore the two constraints reduce to only one.

Now we investigate the solutions of $N_4 = 0$.

LEMMA 2.2. *In the basis (2.84) the following inequality holds*

$$\zeta_0^2 \le (\xi_1^2 + \xi_2^2)(1 - \zeta_3^2).$$

Proof. The orthonormality relations are

$$-\zeta_0\xi_0 + \zeta_1\xi_1 + \zeta_2\xi_2 = 0, \tag{2.94}$$

$$-\zeta_0^2 + \zeta_1^2 + \zeta_2^2 + \zeta_3^2 = 1. \tag{2.95}$$

If $\xi_1 = \xi_2 = 0$, then $\zeta_0 = 0$ and the inequality is satisfied. If $\xi_1 \ne 0$, $\xi_2 = 0$, from (2.94) we obtain ζ_1, which after substitution into (2.95) yields the inequality. If $\xi_2 \ne 0$, from (2.94) we get ζ_2, which after substitution into (2.95) gives a second degree equation for ζ_1, whose discriminant must be nonnegative (because ζ_1 is real) and this inequality is equivalent to the statement of the lemma. Q.E.D.

LEMMA 2.3. *Let* $c \ge 1$, $\mu_1(c)$, *and* $\mu_2(c)$ *be*

$$\mu_1(c) = p - \sqrt{q}, \quad \mu_2(c) = p + \sqrt{q},$$

with

$$p = \frac{\zeta_0\xi_0(c-1)}{1 + \xi_0^2(c-1)}, \quad q = \frac{1 + (c-1)(\xi_0^2 - \zeta_0^2)}{[1 + \xi_0^2(c-1)]^2}.$$

Then $\mu_1(c)$ *and* $\mu_2(c)$ *are real, distinct, and satisfy*

$$-1 < \mu_1(c) < \frac{\zeta_0}{\xi_0} < \mu_2(c) < 1.$$

Proof. From Lemma 2.2 we obtain $\xi_0^2 - \zeta_0^2 > 0$, hence $q > 0$ and $\mu_1(c)$, $\mu_2(c)$ are real and distinct. Furthermore, the function

$$f(\mu) = \mu^2[1 + (c-1)\xi_0^2] + 2\mu\xi_0\zeta_0(1-c) - 1 + \zeta_0^2(c-1)$$

has the coefficient of μ^2 positive and its roots are $\mu_1(c)$, $\mu_2(c)$. Also, it is easy to check that

$$f(\zeta_0/\xi_0) < 0, \quad f(1) > 0, \quad f(-1) > 0$$

whence the inequality of the lemma. Q.E.D.

LEMMA 2.4. *Let* V_Σ *be the normal speed of propagation of the characteristic hypersurfaces* $\phi = $ *const. with respect to the medium, which in the basis (2.84) is*

$$V_\Sigma^2 = \frac{(\zeta_0 - \mu\xi_0)^2}{(\zeta_1 - \mu\xi_1)^2 + (\zeta_2 - \mu\xi_2)^2 + \zeta_3^2}. \tag{2.96}$$

Then

$$V_\Sigma^2(\mu_1(c)) = V_\Sigma^2(\mu_2(c)) = \frac{1}{c}.$$

Proof. By direct calculation. Q.E.D.

PROPOSITION 2.6. *If* $e'_p > 1$, $\eta \neq (e'_p - 1)|b|^2$, *then* $N_4 = 0$ *has four real and distinct roots.*

Proof. If $e'_p = 1$, then $N_4 = -AG$ and its roots are those of A (which are real and distinct) and of G, which are also real and distinct.

Furthermore, from Proposition (2.2), we have $G > 0$ when $A = 0$. Now we discuss the case $e'_p > 1$.

N_4 is a polynomial of 4th degree in μ. One checks that

$$N_4(1) > 0, \quad N_4(-1) > 0.$$

It is easy to check that the four real numbers

$$\mu_1(e'_p), \quad \mu_2(e'_p), \quad \mu_1\left(\frac{E}{|b|^2}\right), \quad \mu_2\left(\frac{E}{|b|^2}\right)$$

are all distinct, as a consequence of Lemmas (2.3)–(2.4).

Furthermore,

$$\begin{aligned} N_4[\mu_1(e'_p)] = {} & \frac{|b|^2}{e'_p}(1 - e'_p)[(\zeta_1 - \mu_1(e'_p)\xi_1)^2 + (\zeta_2 - \mu_1(e'_p)\xi_2)^2 + \zeta_3^2] \\ & \cdot[(\zeta_2 - \mu_1(e'_p)\xi_2)^2 + \zeta_3^2] \le 0, \\ N_4[\mu_2(e'_p)] = {} & \frac{|b|^2}{e'_p}(1 - e'_p)[(\zeta_1 - \mu_2(e'_p)\xi_1)^2 + (\zeta_2 - \mu_2(e'_p)\xi_2)^2 + \zeta_3^2] \\ & \cdot[(\zeta_2 - \mu_2(e'_p)\xi_2)^2 + \zeta_3^2] \le 0, \\ N_4\left[\mu_1\left(\frac{E}{|b|^2}\right)\right] = {} & -\frac{\eta|b|^2}{E}\left[\left(\zeta_1 - \mu_1\left(\frac{E}{|b|^2}\right)\xi_1\right)^2 \right. \\ & \left. + \left(\zeta_2 - \mu_1\left(\frac{E}{|b|^2}\right)\xi_2\right)^2 + \zeta_3^2\right] \\ & \cdot\left[\left(\zeta_2 - \mu_2\left(\frac{E}{|b|^2}\right)\xi_2\right)^2 + \zeta_3^2\right] \le 0, \end{aligned}$$

$$N_4\left[\mu_2\left(\frac{E}{|b|^2}\right)\right] = -\frac{\eta|b|^2}{E}\left[\left(\zeta_1 - \mu_2\left(\frac{E}{|b|^2}\right)\xi_1\right)^2 + \left(\zeta_2 - \mu_2\left(\frac{E}{|b|^2}\right)\xi_2\right)^2 + \zeta_3^2\right] \cdot\left[\left(\zeta_2 - \mu\left(\frac{E}{|b|^2}\right)\xi_2\right)^2 + \zeta_3^2\right] \leq 0,$$

$$N_4\left(\frac{\zeta_0}{\xi_0}\right) > 0.$$

When

$$P_1 = [(\zeta_2 - \mu_1(e_p')\xi_2)^2 + \zeta_3^2][(\zeta_2 - \mu_2(e_p')\xi_2)^2 + \zeta_3^2] > 0$$

then

$$N_4[\mu_1(e_p')] < 0,$$
$$N_4[\mu_2(e_p')] < 0,$$

and, therefore, $N_4 = 0$ admits four real and distinct roots.

When

$$P_2 = \left[\left(\zeta_2 - \mu_1\left(\frac{E}{|b|^2}\right)\xi_2\right)^2 + \zeta_3^2\right]\left[\left(\zeta_2 - \mu_2\left(\frac{E}{|b|^2}\right)\xi_2\right)^2 + \zeta_3^2\right] > 0$$

then we also have four real and distinct roots.

When $P_1 = 0$ and $P_2 = 0$, then

$$\xi_2 = \zeta_2 = \zeta_3 = 0$$

and the roots of $N_4 = 0$ are

$$\mu_1(e_p'), \quad \mu_2(e_p'), \quad \mu_1\left(\frac{E}{|b|^2}\right), \quad \mu_2\left(\frac{E}{|b|^2}\right). \qquad \text{Q.E.D.}$$

The right and left eigenvectors, in the case when all the roots of N_4 are distinct and when none of the roots of $N_4 = 0$ coincide with any of $A = 0$, are

$$\mathbf{d} = (d^\alpha, d^{\alpha+4}, d^8, d^9)^T,$$

with

$$d^\alpha = Ea^2(bf^\alpha - al^\alpha) + \frac{Ea}{\eta}(B^2 - e_p'|b|^2a^2)(2au^\alpha + \phi^\alpha),$$

$$d^{\alpha+4} = \frac{B}{a}d^\alpha + EaAf^\alpha,$$

$$d^8 = Ea^2A, \qquad (2.97)$$

$$d^9 = 0,$$
$$\mathbf{s} = (s_\alpha, s_{\alpha+4}, s_8, s_9),$$

with

$$\begin{aligned} s_\alpha &= -B\eta(G+2a^2)b_\alpha + a^2\eta E\phi_\alpha, \\ s_{\alpha+4} &= Ea\eta[(G+2a^2)b_\alpha - B\phi_\alpha], \\ s_8 &= -EaA, \\ s_9 &= 0. \end{aligned} \tag{2.98}$$

When one or two of the roots of $N_4 = 0$ coincide with one or two roots of $A = 0$, then the corresponding eigenvectors are those given by equations (2.91)–(2.93).

Now we treat the case $\eta = (e'_p - 1)|b|^2$.

PROPOSITION 2.7. *When $\eta = (e'_p - 1)\ |b|^2$, $e'_p > 1$, there exists ξ^α such that both roots of A are roots of N_4 with multiplicity 2, and a basis of eigenvectors does not exist. For such ξ^α the hyperbolicity condition is violated.*

Proof. In fact, by choosing, in the frame (2.84),

$\zeta_3 = \zeta_2 = \xi_2 = 0$, we have

$$N_4 = \frac{A^2}{|b|^2},$$

and therefore the roots μ_1, μ_2 of A have multiplicity 4 for the characteristic equation. But to each of them there correspond only three linearly independent eigenvectors, equation (2.92). Q.E.D.

A detailed and thorough examination of all cases where the hyperbolicity condition is violated can be found in the article by Anile and Pennisi (1987). The lack of hyperbolicity of the covariant equations of relativistic magneto-fluid dynamics in the form (2.78)–(2.79) is an indication that great care must be exercised when performing numerical calculations with these equations.

It is interesting to remark that the field variables (2.77) are not all independent because they satisfy the algebraic constraints $u^\alpha u_\alpha = -1$, $b^\alpha u_\alpha = 0$. Therefore, the lack of hyperbolicity of the system (2.78)–(2.79) is plausibly related to a choice of the field variables that are not all independent. Hence, after having solved the system of field equations (2.78)–(2.79) only those solutions satisfying the above constraints are acceptable. This could be achieved by imposing the above constraints on a given noncharacteristic initial hypersurface $\mathscr{F}$ because it is possible then to show that the constraints are satisfied in a neighborhood of $\mathscr{F}$ (Anile and Pennisi, 1987).

For both conceptual and practical reasons, such as in numerical calculations, it is important to deal with hyperbolic systems. In the next section we will introduce a method, based on the entropy principle, by which a given set of evolution equations, if suitable conditions are met, can be transformed into a hyperbolic system. In Section 2.6 this method will be applied to relativistic fluid dynamics and in Section 2.7 to relativistic magneto-fluid dynamics.

2.5. Supplementary conservation laws and symmetrization

In this section we shall discuss a rather deep and elegant theory of the hyperbolic systems of mathematical physics, which originated with Friedrichs and Lax (1971) and Friedrichs (1974) and which has been extended by Boillat (1974, 1976) and Ruggeri and Strumia (1981a).

Let us consider a quasi-linear system of conservation laws

$$\nabla_\alpha F^{\alpha A} = f^A. \tag{2.99}$$

For the moment we do not assume that (2.99) forms a hyperbolic system when written in the form

$$A_B^{\alpha A} \nabla_\alpha U^B = f^A, \tag{2.100}$$

with $A_B^{\alpha A} = \dfrac{\partial F^{\alpha A}}{\partial U^B}$, and where the field $\mathbf{U}$ varies in an open domain $\mathbf{D}$ of $\mathbb{R}^N$.

We shall assume only the first condition of hyperbolicity, that is,

$$\det(A_B^{\alpha A} \xi_\alpha) \neq 0. \tag{2.101}$$

We shall also assume the existence of a supplementary conservation law of the form

$$\nabla_\alpha h^\alpha = g, \tag{2.102}$$

which holds as a consequence of the system (2.100), that is, for any solution of (2.100) which is sufficiently differentiable.

More precisely, we shall assume that the supplementary conservation law is obtained from the system (2.100) by linear combinations of the form

$$U'^A A_B^{\alpha A} = \frac{\partial h^\alpha}{\partial U^B}, \tag{2.103}$$

$$U'^A f^A = g, \tag{2.103'}$$

where $U'^A = U'^A(U^B)$ represents a set of multipliers. Examples of these systems are the equations of Newtonian fluid dynamics and of relativistic fluid and magneto-fluid dynamics.

Since $A_B^{\alpha A} = \dfrac{\partial F^{\alpha A}}{\partial U^B}$, equation (2.103) can be written, in terms of differentials, as

$$U'^A \, \mathrm{d}F^{\alpha A} = \mathrm{d}h^\alpha \tag{2.104}$$

from which we see that $\mathbf{U}'$ is determined only by the structure of the system of conservation laws (2.99) and the supplementary conservation law (2.102) and does not depend on the particular field $\mathbf{U}$ chosen in order to write (2.99) as a quasi-linear system (2.100).

For this reason, in order to obtain explicit expressions for $\mathbf{U}'$ from the relations (2.104) (also called the compatibility relations), it is possible to start with a field $\mathbf{U}$ which is more suitable. Because of the condition $\det(A_B^{\alpha A}\xi_\alpha) \neq 0$, we can take as a field

$$U^A = F^{\alpha A}\xi_\alpha, \tag{2.105}$$

which implies

$$A_B^{\alpha A}\xi_\alpha = \delta_B^A.$$

By multiplying equation (2.103) with ξ_α, we obtain

$$U'^A = \frac{\partial h}{\partial U^A}, \tag{2.106}$$

where

$$h = h^\alpha \xi_\alpha. \tag{2.107}$$

Equation (2.106) gives simple expressions for $\mathbf{U}'$ in terms of the field $U^A = F^{\alpha A}\xi_\alpha$.

Now we shall assume that $\mathbf{U}$ is defined in an open convex subset $\mathbf{D}$ of $\mathbb{R}^N$ and that h is a convex function of $U \in \mathbb{R}^N$. Systems of conservation laws possessing this property are called *convex covariant density systems* (Ruggeri and Strumia, 1981a).

Then for such systems the Jacobian matrix

$$\frac{\partial U'^A}{\partial U^B} = \frac{\partial^2 h}{\partial U^A \partial U^B}$$

is symmetric and positive definite in the convex domain $\mathbf{D}$ of $\mathbb{R}^N$. Therefore, a theorem of analysis (Berger and Berger, 1968) shows that the mapping $U \Leftrightarrow U'$ is globally invertible in $\mathbf{D}$. Hence, U' can be taken as a bona fide new field in $\mathbf{D} \subseteq \mathbb{R}^N$.

This remarkable property is the underlying motivation for calling $\mathbf{U}'$ the *main field associated (uniquely) with the convex covariant density system* (2.99) and (2.102).

It is of great interest to investigate the form taken by the system (2.100) when expressed in terms of the main field $\mathbf{U}'$. One has the following proposition (Ruggeri and Strumia, 1981a).

PROPOSITION 2.8. *A convex covariant density system is a conservative hyperbolic symmetric system in the field* $\mathbf{U}'$.

Proof. Let us consider the following Legendre transformation

$$h'^{\alpha} = U'^A F^{\alpha A} - h^{\alpha}, \quad \text{with } h'^{\alpha} = h'^{\alpha}(U'^B).$$

Then

$$\frac{\partial h'^{\alpha}}{\partial U'^B} = F^{\alpha B}$$

because

$$U'^A \frac{\partial F^{\alpha A}}{\partial U'^B} = \frac{\partial h^{\alpha}}{\partial U'^B}.$$

It follows that

$$A'^{\alpha A}_B = \frac{\partial F^{\alpha A}}{\partial U'^B} = \frac{\partial^2 h'^{\alpha}}{\partial U'^A \partial U'^B},$$

which is a symmetric matrix.

The first condition of hyperbolicity

$$\det(A'^{\alpha A}_B \xi_{\alpha}) \neq 0$$

is easily verified. In fact, let $h' = h'^{\alpha}\xi_{\alpha}$; then

$$h' = U'^A U^A - h$$

and h' is the Legendre transform of h. Therefore, from well-known properties of the Legendre transform of a convex function, h' is also a convex function.

The second condition of hyperbolicity, [Definition 2.1., (ii)] holds necessarily as a consequence of the symmetry of $A'^{\alpha A}_B$. Q.E.D.

The reduction of a quasi-linear symmetric hyperbolic system is very important from the mathematical viewpoint because a powerful theory is available to deal with the Cauchy problem. In particular, the following theorem due to Fisher and Marsden (1972) is one of the simpler and more useful results.

THEOREM. *Let $H^s(\mathbb{R}^n, \mathbb{R}^m)$ denote the H^s maps from $\mathbb{R}^n$ to $\mathbb{R}^m$ [for each integer $s \geq 0$, $H^s(\mathbb{R}^n, \mathbb{R}^m)$ is the space of all $u \in L^2(\mathbb{R}^n, \mathbb{R}^m)$ such that all distribution derivatives $D^\alpha u$ with $|\alpha| \leq s$ belong to $L^2(\mathbb{R}^n, \mathbb{R}^m)$], endowed with the usual Sobolev norm*

$$\| u \|_s^2 = \sum_{0 \leq |\alpha| \leq s} \int |D^\alpha u(x)|^2 \, dx,$$

and let $U^s \subseteq H(\mathbb{R}^n, \mathbb{R}^m)$ be an open set. Let $\delta > 0$, and for $(t, x, u) \in (-\delta, \delta) \times \mathbb{R}^n \times U^s$, let $A^i(t, x, u)$ be symmetric $m \times m$ matrixes and $\mathbf{B}(t, x, u)$ an m-component vector, assumed to be H^s-functions of (t, x) and rational functions of u with nonzero denominators. Given $u_0 \in U^s$, $s > n/2 + 2$, there exist $0 < \varepsilon < \delta$ and a unique $u(t, x)$, $|t| < \varepsilon$, $x \in \mathbb{R}^n$, which is H^s in (t, x) and which satisfies the following initial value problem

$$u(0, x) = u_0(x),$$

$$\frac{\partial u}{\partial t} = A^i(t, x, u)\frac{\partial u}{\partial x^i} + \mathbf{B}(t, x, u).$$

Furthermore, $u(t, x)$ depends continuously on u_0 in the H^s-topology.

More general results have also been obtained.

This theorem (and generalizations thereof) ensures that the differentiability properties of $u(t, x)$ are the same as those of $u_0(x)$. This is to be contrasted with the much weaker results which can be obtained for quasi-linear hyperbolic systems without the symmetry assumptions. In particular, for quasi-linear hyperbolic systems which have multiple characteristics (are not strictly hyperbolic) the Leray–Ohyia theory (Lichnerowicz, 1967) requires infinite differentiability for the initial datum, an assumption that is too restrictive for many physical applications.

In the forthcoming section we will apply the theory expounded in this section to relativistic fluid dynamics and determine the main field.

2.6. Main field for relativistic fluid dynamics

The fundamental equations of relativistic fluid dynamics can be written in the form (2.99), that is,

$$\nabla_\alpha F^{\alpha A} = 0, \tag{2.108}$$

with

$$F^\alpha = \begin{pmatrix} T^{\alpha\beta} \\ \rho u^\alpha \end{pmatrix}, \tag{2.109}$$

and the supplementary conservation law can be taken to be

$$\nabla_\alpha h^\alpha = 0, \tag{2.110}$$

with

$$h^\alpha = -\rho S u^\alpha. \tag{2.111}$$

The main field $\mathbf{U}'$ is determined by equation (2.104).
Let

$$\mathbf{U}' = \begin{pmatrix} w_\beta \\ \psi \end{pmatrix}. \tag{2.112}$$

Then $\mathbf{U}'$ obeys

$$w_\beta \, \mathrm{d}T^{\alpha\beta} + \psi \, \mathrm{d}(\rho u^\alpha) = -d(\rho S u^\alpha). \tag{2.113}$$

Let

$$f = 1 + \varepsilon + \frac{p}{\rho} \tag{2.114}$$

be the relativistic enthalpy of the fluid. Then,

$$T^{\alpha\beta} = \rho f u^\alpha u^\beta + p g^{\alpha\beta}$$

and one obtains from (2.113), after contracting with u_α, and using the thermodynamic relation

$$\mathrm{d}f = T\,\mathrm{d}S + \frac{\mathrm{d}p}{\rho}, \tag{2.115}$$

that

$$-(w_\beta v^\beta T + \rho)\,\mathrm{d}S + \left(\frac{\psi + S}{\rho} v_\beta - f w_\beta\right)\mathrm{d}v^\beta = 0, \tag{2.116}$$

where

$$v^\mu = \rho u^\mu.$$

Equation (2.16) yields

$$w_\beta = \frac{1}{T} u_\beta,$$

$$\psi = \frac{f - TS}{T},$$

whence

$$\mathbf{U}' = \frac{1}{T}\begin{pmatrix} u_\beta \\ G+1 \end{pmatrix}, \tag{2.117}$$

where

$$G = f - TS - 1 \tag{2.118}$$

is the relativistic free enthalpy.

Having determined the main field $\mathbf{U}'$, the next step is to show that

$$h = h^\alpha \xi_\alpha$$

is a convex function of $\mathbf{U}'$, that is, that the quadratic form

$$Q = \frac{\partial^2 h}{\partial U'^A \partial U'^B} \delta U'^A \delta U'^B \tag{2.119}$$

is positive definite for all variations $\delta\mathbf{U}'$ of $\mathbf{U}'$.

Now

$$Q = \delta U'^A \frac{\partial}{\partial U'^A}\left(\frac{\partial h}{\partial U'^B}\delta U'^B\right) = \delta U'^A \frac{\partial}{\partial U'^A}\left(\frac{\partial h}{\partial U^B}\delta U^B\right).$$

However, as we have seen in the previous section, it is possible to take as the field

$$U^A = F^{\alpha A}\xi_\alpha,$$

hence, by (2.105)–(2.106),

$$\frac{\partial h}{\partial U^B}\delta U^B = U'^B \delta U^B.$$

It follows that

$$Q = \delta U'^A \delta U^A. \tag{2.120}$$

We remark that, because of the symmetry of expression (2.120) with respect to the transformation $\mathbf{U} \Leftrightarrow \mathbf{U}'$, h will be a convex function of $\mathbf{U}$ also.

Then,

$$T^2 Q = (-u_\beta \delta T + T\delta u_\beta)\delta(T^{\alpha\beta}\xi_\alpha) + [-(G+1)\delta T + T\delta G]\delta(\rho u^\alpha \xi_\alpha).$$

A tedious but straightforward calculation shows that

$$TQ = \rho f v \delta u_\beta \delta u^\beta + 2\delta p \delta v + \rho v \delta S \delta T + \frac{v}{\rho}\delta p \delta \rho, \tag{2.121}$$

where

$$v = u^\alpha \xi_\alpha.$$

Now the thermodynamic stability condition for a fluid element in thermal and mechanical equilibrium with its surroundings is that the internal energy ε, considered as a function of specific entropy S and density ρ, be a minimum (McLellan, 1980). This stability condition, expressed in terms of the free enthalpy G, is equivalent to the convexity of the function $-G(T,p)$.

It follows that the quadratic form

$$K^2 = -\rho^2\left(\delta\left(\frac{\partial G}{\partial p}\right)\delta p + \delta\left(\frac{\partial G}{\partial T}\right)\delta T\right) \tag{2.122}$$

is positive definite.

A quick calculation shows that

$$K^2 = \rho^2\left(\frac{1}{\rho^2}\delta\rho\delta p + \delta S\delta T\right),$$

hence

$$TQ = \rho f v \delta u_\beta \delta u^\beta + 2\delta p\delta v + \frac{v}{\rho}K^2. \tag{2.123}$$

Now the quantity on the right-hand side is a invariant scalar and, therefore, can be evaluated in the rest frame of the fluid, where $u^\alpha = (1,0,0,0)$, using the Minkowski metric $\eta_{\mu\nu}$.

A simple calculation gives

$$TQ = \rho f\left(v\delta u_i + \xi_i\frac{\delta p}{\rho f}\right)^2 + \frac{K^2}{\rho} + \frac{(v^2-1)}{\rho}\left(K^2 - \frac{(\delta p)^2}{f}\right). \tag{2.124}$$

Now

$$v = u^\alpha \xi_\alpha = \xi_0$$

and from $\xi_\alpha \xi^\alpha = -1$ it follows that $v^2 > 1$.

Therefore, TQv is positive definite provided that

$$K^2 - \frac{(\delta p)^2}{f} \geq 0. \tag{2.125}$$

Let

$$\Delta = \frac{(\delta p)^2 - K^2 f}{f\rho^2}. \tag{2.126}$$

Then

$$\Delta = \left(\frac{\partial^2 G}{\partial p^2} + \frac{1}{f\rho^2}\right)(\delta p)^2 + 2\frac{\partial^2 G}{\partial p\partial T}\delta p\delta T + \frac{\partial^2 G}{\partial T^2}(\delta T)^2. \tag{2.127}$$

From the convexity of $-G$ one obtains

$$\frac{\partial^2 G}{\partial T^2} < 0. \tag{2.128}$$

Therefore, $\Delta \leq 0$ iff

$$\Delta' = \frac{\partial^2 G}{\partial T^2}\left(\frac{\partial^2 G}{\partial p^2} + \frac{1}{f\rho^2}\right) - \left(\frac{\partial^2 G}{\partial p \partial T}\right)^2 \geq 0. \tag{2.129}$$

Now

$$\Delta' = \frac{\partial^2 G}{\partial T^2}\frac{\partial^2 G}{\partial p^2} - \left(\frac{\partial^2 G}{\partial p \partial T}\right)^2 + \frac{1}{f\rho^2}\frac{\partial^2 G}{\partial T^2}$$

$$= \frac{D\left(\dfrac{\partial G}{\partial T}, \dfrac{\partial G}{\partial p}\right)}{D(T,p)} + \frac{1}{f\rho^2}\frac{D\left(\dfrac{\partial G}{\partial T}, p\right)}{D(T,p)}$$

in terms of Jacobian determinants.

It follows that

$$\Delta'' = \Delta' \frac{D(T,p)}{D\left(\dfrac{\partial G}{\partial T}, \dfrac{\partial G}{\partial p}\right)} = 1 + \frac{1}{f\rho^2}\frac{D\left(\dfrac{\partial G}{\partial T}, p\right)}{D\left(\dfrac{\partial G}{\partial T}, \dfrac{\partial G}{\partial p}\right)} \tag{2.129'}$$

because

$$J = \frac{D\left(\dfrac{\partial G}{\partial T}, \dfrac{\partial G}{\partial p}\right)}{D(T,p)} > 0$$

from the convexity of $-G$.

A simple calculation then shows that

$$\Delta'' = 1 - \frac{1}{f}\frac{1}{\left(\dfrac{\partial \rho}{\partial p}\right)_s}. \tag{2.130}$$

However, from the first law of thermodynamics

$$\mathrm{d}e - f\,\mathrm{d}\rho = \rho T\,\mathrm{d}S$$

one has

$$\left(\frac{\partial e}{\partial p}\right)_s = f\left(\frac{\partial \rho}{\partial p}\right)_s.$$

Finally,

$$\Delta'' = 1 - \left(\frac{\partial p}{\partial e}\right)_s \tag{2.131}$$

and, therefore, $\Delta'' \geq 0$ iff

$$\left(\frac{\partial p}{\partial e}\right)_s \leq 1. \tag{2.132}$$

Therefore, the following conditions:

(i) convexity of $-G$, that is, $\frac{\partial^2 G}{\partial T^2} < 0$ and $J > 0$;

(ii) causality, that is, $\left(\frac{\partial p}{\partial e}\right)_s \leq 1$

are sufficient to ensure that Qv is positive definite. Without loss of generality we can take $v = \xi_0 > 0$ and Q will then be positive definite. The case when $\xi_0 < 0$ can be obtained by simply replacing h^α with $-h^\alpha$.

It is interesting to remark that it is sufficient to assume only that

(a) $$\frac{\partial^2 G}{\partial T^2} < 0,$$

(b) $$0 < \left(\frac{\partial p}{\partial e}\right)_s \leq 1.$$

In fact, from (a) and (b) it follows that $J > 0$, as can be derived immediately from the thermodynamical identity

$$\left(\frac{\partial \rho}{\partial p}\right)_s = -\rho^2 \left(\frac{\partial V}{\partial p}\right)_s = -\rho^2 \frac{D(V,S)}{D(p,S)} = \rho^2 \frac{D\left(\frac{\partial G}{\partial p}, \frac{\partial G}{\partial T}\right)}{D(p,S)}$$

$$= \rho^2 \frac{D\left(\frac{\partial G}{\partial p}, \frac{\partial G}{\partial T}\right)}{D(p,T)} \frac{D(p,T)}{D(p,S)} = -\frac{J\rho^2}{\frac{\partial^2 G}{\partial T^2}},$$

where $V = 1/\rho$ and the relationship $\left(\frac{\partial e}{\partial p}\right)_s = f\left(\frac{\partial \rho}{\partial p}\right)_s$ has been used.

Therefore, the theory of the previous section can be applied and the system of equations of relativistic fluid dynamics can be written in a symmetric conservative form

$$A'^{\alpha A}_B \nabla_\alpha U'^B = 0, \tag{2.133}$$

with

$$A'^{\alpha A}_B = \frac{\partial^2 h'^\alpha}{\partial U'^A \partial U'^B}$$

and

$$h'^{\alpha} = U'^{A} F^{\alpha A} - h^{\alpha},$$

where a simple calculation shows that

$$h'^{\alpha} = \frac{p}{T} u^{\alpha}. \tag{2.134}$$

The symmetric form for the equations of relativistic fluid dynamics can be useful both for theoretical analysis (existence theory, shock waves) and for numerical computations.

In the next section we will obtain a similar formulation for relativistic magneto-fluid dynamics. In this case, excluding the aforementioned advantage, the most important result is that we obtain automatically a hyperbolic system.

2.7. Main field for relativistic magneto-fluid dynamics

The fundamental equations of relativistic magneto-fluid dynamics have been introduced in Section 2.4 and are

$$\nabla_{\alpha} T^{\alpha\beta} = 0, \tag{2.135}$$

$$\nabla_{\alpha}(\rho u^{\alpha}) = 0, \tag{2.136}$$

$$\nabla_{\alpha} \psi^{\alpha\beta} = 0, \tag{2.137}$$

where

$$\psi^{\alpha\beta} = u^{\alpha} b^{\beta} - u^{\beta} b^{\alpha}. \tag{2.138}$$

The energy-momentum tensor $T^{\alpha\beta}$ is decomposed into a "fluid" part $T_f^{\alpha\beta}$ and a "magnetic" part $T_m^{\alpha\beta}$ given by

$$T_f^{\alpha\beta} = (e + p) u^{\alpha} u^{\beta} + p g^{\alpha\beta}$$

and

$$T_m^{\alpha\beta} = |b|^2 (u^{\alpha} u^{\beta} + \tfrac{1}{2} g^{\alpha\beta}) - b^{\alpha} b^{\beta}.$$

It is well known that the Maxwell equation (2.137) contains a "constraint" part and an "evolution" part. The constraint can be taken to be

$$F = \xi_{\beta} \nabla_{\alpha} \psi^{\alpha\beta} = 0 \tag{2.139}$$

and the evolution equations we can take to be

$$F^{\mu} = H^{\mu}_{\beta} \nabla_{\alpha} \psi^{\alpha\beta} = 0, \tag{2.140}$$

with

$$H^{\mu}_{\beta} = \delta^{\mu}_{\beta} + \xi^{\mu} \xi_{\beta}.$$

It is easy to prove that if $F=0$ holds on a hypersurface $\mathscr{F}$ transverse to ξ^μ, then $F=0$ holds in a neighborhood of $\mathscr{F}$ as a consequence of the "evolution" equations $F^\mu=0$. In fact, differentiating $F^\mu=0$ yields

$$\nabla_\mu F^\mu = \nabla_\beta \nabla_\alpha \psi^{\alpha\beta} + \xi^\mu \xi_\beta \nabla_\mu \nabla_\alpha \psi^{\alpha\beta} + \nabla_\mu(\xi^\mu \xi_\beta)\nabla_\alpha \psi^{\alpha\beta} = 0,$$

whence

$$\xi^\mu \nabla_\mu F + (\nabla_\mu \xi^\mu)F = 0,$$

which proves the statement.

Therefore, it would seem natural to replace, in the field equations, equation (2.137) with its evolution part (2.140). However, by doing so we lose the conservative nature of the field equations and the whole theory developed in Section 2.5 would not be applicable.

However, the theory can be applied in the case where the timelike vector field ξ^μ is hypersurface orthogonal, which occurs in most applications.

Then it is possible to introduce, at least locally, coordinates (x^μ) such that

$$\xi_\mu = \delta^0_\mu.$$

In these coordinates, equation (2.137) splits into

$$\Phi = \nabla_i \psi^{0i} = 0, \tag{2.141}$$

$$\Phi^i = \nabla_0 \psi^{0i} + \nabla_k \psi^{ki} = 0. \tag{2.142}$$

Then equation (2.141) is the constraint equation whereas equations (2.142) are the evolution equations.

We can then take as field equations equations (2.135), (2.136), and (2.142).

By introducing the column vector

$$F^\alpha = \begin{pmatrix} T^{\alpha\beta} \\ \rho u^\alpha \\ \psi^{\alpha k} \end{pmatrix}, \tag{2.143}$$

then the field equations can be written

$$\nabla_\alpha F^\alpha = 0 \tag{2.144}$$

together with the constraint

$$\nabla_\alpha \psi^{0\alpha} = 0 \tag{2.145}$$

and the supplementary conservation law

$$\nabla_\alpha h^\alpha = 0, \tag{2.146}$$

with

$$h^\alpha = -\rho S u^\alpha. \tag{2.147}$$

Since we are dealing with systems of conservation laws with constraints the theory developed in Section 2.5 is not directly applicable but requires a slight generalization. In this case it amounts to an obvious extension. Ruggeri and Strumia (1981b) have tackled this problem but they did not properly take into account the constraint (2.145).

The general case of an arbitrary system of conservation laws with constraints has been treated, in a noncovariant framework, by Boillat (1982a, b).

The compatibility relations arising from (2.144)–(2.146) are

$$w_\beta \,\mathrm{d}T^{\alpha\beta} + \psi\,\mathrm{d}(\rho u^\alpha) + \lambda_i\,\mathrm{d}\psi^{\alpha i} + k\,\mathrm{d}\psi^{0\alpha} = \mathrm{d}(-\rho S u^\alpha), \tag{2.148}$$

where w_β, ψ, λ_i are the components of the main field $\mathbf{U}'$ and k is the multiplier for the constraint.

By contracting equation (2.148) with $\xi_\alpha = (1, 0, 0, 0)$ we obtain

$$w_\beta^d(T^{0\beta}) + \psi\,\mathrm{d}(\rho u^0) + \lambda_i\,\mathrm{d}\psi^{0i} = -\mathrm{d}(\rho S u^0). \tag{2.149}$$

As in the previous section we have the following identity

$$T\,\mathrm{d}(-\rho S u^0) = u_\beta\,\mathrm{d}(T_f^{0\beta}) + (G+1)\,\mathrm{d}(\rho u^0)$$

and, therefore,

$$T\,\mathrm{d}(-\rho S u^0) = u_\beta\,\mathrm{d}(T^{0\beta}) + (G+1)\mathrm{d}(\rho u^0) - u_\beta\,\mathrm{d}(T_m^{0\beta}). \tag{2.150}$$

It is easy to see that

$$u_\beta\,\mathrm{d}(T_m^{0\beta}) = -b_i\,\mathrm{d}\psi^{0i},$$

hence

$$T\,\mathrm{d}(-\rho S u^0) = u_\beta\,\mathrm{d}(T^{0\beta}) + (G+1)\,\mathrm{d}(\rho u^0) + b_i\,\mathrm{d}\psi^{0i}. \tag{2.151}$$

It follows that the main field $\mathbf{U}'$ is given by

$$\mathbf{U}' = \frac{1}{T}\begin{pmatrix} u_\beta \\ G+1 \\ b_i \end{pmatrix}. \tag{2.152}$$

Now we prove that $h = h^\alpha \xi_\alpha = h^0$ is a convex function of $\mathbf{U}'$. By taking as the field $\mathbf{U} = F^\alpha \xi_\alpha = F^0$ and proceeding as in the previous section this amounts to proving that the quadratic form

$$Q = \delta U'^A \delta U^A$$

is positive definite, for all variations $\delta U'^A$.

Now,

$$Q = \delta\left(\frac{u_\beta}{T}\right)\delta\left(T_m^{0\beta} + T_f^{0\beta}\right) + \delta\left(\frac{G+1}{T}\right)\delta(\rho u^0) + \delta\left(\frac{b_i}{T}\right)\delta\psi^{0i}. \quad (2.153)$$

Let Q_f be the corresponding quadratic form for the fluid without a magnetic field ($|b| = 0$). Then

$$Q = Q_f + \frac{1}{T}(\delta u_\beta \delta T_m^{0\beta} + \delta b_k \delta\psi^{0k}). \quad (2.154)$$

After some calculations we find that

$$TQ = TQ_f + 2\delta u^0 b_k \delta b^k - 2b^0 \delta u_k \delta b^k + u^0 |b|^2 \delta u_k \delta u^k + u^0 \delta b_k \delta b^k + u^0 \delta b_0 \delta b^0 + u^0 |b|^2 \delta u^0 \delta u_0,$$

which can be rewritten in a covariant fashion as follows

$$T(Q - Q_f) = (u^\alpha \xi_\alpha)|b|^2 \delta u_\beta \delta u^\beta + (u^\alpha \xi_\alpha)\delta b_\beta \delta b^\beta + 2b_\beta(\delta u^\alpha \xi_\alpha)\delta b^\beta - 2(b^\alpha \xi_\alpha)\delta u_\beta \delta b^\beta.$$

This expression, being an invariant scalar, can be evaluated in the rest frame (2.84).

It follows, after some calculations, that

$$T(Q - Q_f) = \xi_0[|b|^2(\delta u^2)^2 + |b|^2(\delta u^3)^2 + (\delta b^1)^2 + (\delta b^2)^2 + (\delta b^3)^2] + 2|b|\xi_2 \delta u^2 \delta b^1 - 2|b|\xi_1 \delta u^3 \delta b^3.$$

Under the assumption $\xi_0 > 0$ one finds that the eigenvalues of the matrix of the quadratic form in the right-hand side of (2.155) are all nonnegative and, therefore, that $T(Q - Q_f)$ is semipositive definite. This, together with the positive definiteness of Q_f which can be proved as in the previous section, completes the proof. The case $\xi_0 < 0$ can be dealt with by replacing h^α by $-h^\alpha$.

Now we turn to the determination of the multiplier k.

From equation (2.148) and equation (2.152) we obtain

$$u_\beta \, \mathrm{d}T^{\alpha\beta} + (G+1)\,\mathrm{d}(\rho u^\alpha) + b_i \, \mathrm{d}\psi^{\alpha i} + kT\,\mathrm{d}\psi^{0\alpha} = T\,\mathrm{d}(-\rho S u^\alpha). \quad (2.155)$$

Subtracting from (2.155) its expression for $b^\alpha = 0$, $\mathrm{d}b^\alpha = 0$, yields

$$u_\beta \, \mathrm{d}T_m^{\alpha\beta} + b_i \, \mathrm{d}\psi^{\alpha i} + kT\,\mathrm{d}\psi^{0\alpha} = 0,$$

which yields

$$k = -\frac{b_0}{T}. \quad (2.156)$$

Now, at variance with the previous sections, we take into account the constraint by defining

$$h'^{\alpha} = U'^{A}F^{\alpha A} + k\psi^{0\alpha} - h^{\alpha}. \tag{2.157}$$

Then, by using the compatibility relations (2.148), we have

$$\frac{\partial h'^{\alpha}}{\partial U'^{B}} = F^{\alpha B} + \frac{\partial k}{\partial U'^{B}}\psi^{0\alpha},$$

whence

$$A'^{\alpha A}_{B} = \frac{\partial F^{\alpha A}}{\partial U'^{B}} = \frac{\partial^2 h'^{\alpha}}{\partial U'^{A}\partial U'^{B}} - \frac{\partial^2 k}{\partial U'^{A}\partial U'^{B}}\psi^{0\alpha} - \frac{\partial k}{\partial U'^{A}}\frac{\partial \psi^{0\alpha}}{\partial U'^{B}}.$$

It follows that the field equations (2.144) can be written in the symmetric form, as a consequence of equation (2.145),

$$\mathscr{M}^{\alpha}_{AB}\nabla_{\alpha}U'^{B} = 0, \tag{2.158}$$

with

$$\mathscr{M}^{\alpha}_{AB} = \frac{\partial^2 h'^{\alpha}}{\partial U'^{A}\partial U'^{B}} - \frac{\partial^2 k}{\partial U'^{A}\partial U'^{B}}\psi^{0\alpha}. \tag{2.159}$$

Obviously, one has

$$\mathscr{M}^{\alpha}_{AB}\xi_{\alpha} = \frac{\partial^2 h'}{\partial U'^{A}\partial U'^{B}}, \tag{2.160}$$

with

$$h' = h'^{\alpha}\xi_{\alpha} = U'^{A}U^{A} - h.$$

From the well-known properties of the Legendre transformation, h' is also a convex function of U' and therefore $\det(\mathscr{M}^{\alpha}_{AB}\xi_{\alpha}) \neq 0$. It follows that both conditions for the hyperbolicity (Definition 2.1) are satisfied.

An interesting problem which remains to be solved is that of extending the theory expounded in this section to the case when the timelike vector ξ^{α} is not hypersurface orthogonal. Such an extension would be required in order to treat magnetohydrodynamic accretion onto a rotating black hole.

3

Singular hypersurfaces in space-time

3.0. Introduction

In the previous chapter we proved that, under suitable restrictions on the state equation, the equations of nondissipative relativistic fluid dynamics and magneto-fluid dynamics can be cast in the form of quasi-linear hyperbolic systems. A distinguishing feature of hyperbolic systems is the occurrence of nonsmooth solutions and in particular the generation of discontinuities in initially smooth profiles (Jeffrey, 1976). The physical interpretation of this phenomenon is that, in the absence of dissipation, the nonlinearity of the field equations tends to steepen a given initial profile until (if some conditions are met; Whitham, 1974) a discontinuity in some derivative of the field variables (or in the fields themselves) appears. Therefore, we must expect solutions with discontinuities to be a common occurrence in relativistic fluid dynamics and magneto-fluid dynamics. Due to the hyperbolic character of the equations the discontinuities will propagate as surfaces, called singular surfaces. This chapter is devoted to the study of propagating singular surfaces in a covariant framework. This provides the groundwork for the investigation of weak discontinuity waves in Chapter 4 and of shock waves in Chapters 8, 9, and 10.

A singular surface is a propagating surface across which the field variables characterizing the medium, and their derivatives, suffer a jump discontinuity (Truesdell and Toupin, 1960). When the fields themselves are continuous but some of their derivatives are discontinuous, the surface is called a weak discontinuity (or ordinary discontinuity) surface (Truesdell and Toupin, 1960; Jeffrey, 1976).

A great merit of the concept of a singular surface is that it allows for a mathematically *exact treatment* of the evolution of a given propagating singular surface (at least in the case of a weak discontinuity), free of any approximation. This topic will be treated in detail in the next chapter.

On the other side, one of the drawbacks of the concept of a singular surface is that it is limited to impulsive waves (with time and length scales very short compared to the other scales present in the physical problem under consideration). Sometimes this idealization may be too extreme (particularly when the length scale of the impulsive wave is of order of some mean free path in the material medium) and consequently the results must be judged with great care.

In the theory of singular surfaces a key role is played by the geometric and kinematic compatibility relations (Hadamard, 1903; Thomas, 1957; Truesdell and Toupin, 1960; Chen, 1976; Kosinski; 1986).

Let us recall briefly the compatibility relations as they are expressed in the framework of nonrelativistic continuum mechanics.

Let Σ_t be a moving surface in the Euclidean space $\mathbb{R}^3$, described by the parametric equation

$$x^a = x^a(v^\Gamma, t), \quad a = 1, 2, 3, \quad \Gamma = 1, 2,$$

where x^a are Cartesian coordinates and v^Γ are curvilinear coordinates on Σ_t.

Let Ψ be a function that is regularly discontinuous across Σ_t [i.e., such that the limits $\Psi_\pm = \lim_{P \to \Sigma_t^\pm} \Psi(P)$ are finite and different, where $\Sigma_t^\pm$ denotes the two orientations of Σ_t; a precise definition will be given in the next section]. Then the *geometric compatibility relations of first order* are

$$[[\psi_{,b}]] = n_b[[\psi_{,a}n^a]] + x_{b;\Delta}a^{\Gamma\Delta}[[\psi]]_{;\Gamma}, \tag{3.1}$$

where for any regularly discontinuous function $[[f]] = f_- - f_+$, n_b is the unit normal to Σ_t, $a^{\Gamma\Delta}$ is the induced metric on Σ_t, and the semicolon denotes covariant differentiation with respect to the induced Riemannian connection on Σ_t.

Similarly, the *kinematic compatibility relations of first order* are

$$\left[\left[\frac{\partial\psi}{\partial t}\right]\right] = -V_\Sigma\left[\left[n^a\frac{\partial\psi}{\partial x^a}\right]\right] + \frac{\delta[[\psi]]}{\delta t}, \tag{3.2}$$

where V_Σ is the normal speed of displacement of Σ_t and $\frac{\delta}{\delta t}$ denotes the Thomas displacement derivative (Thomas, 1957), that is,

$$\frac{\delta}{\delta t} = \frac{\partial}{\partial t} + V_\Sigma n^a \frac{\partial}{\partial x^a}.$$

In a space-time formulation, as we shall see in the next section, the geometric and kinematic compatibility relations can be considered as the space and time components, respectively, of a single covariant compatibility relation, across a singular hypersurface Σ, representing the history of Σ_t in space-time.

In relativity the covariant compatibility relations have been obtained, in a local form, by Maugin (1976) for a nonnull hypersurface.

The case of a null hypersurface is not encompassed by Maugin's treatment because the projection operator cannot be defined. However, null hypersurfaces are very important in relativity as possible carriers of electromagnetic and gravitational discontinuity waves.

In nonrelativistic continuum mechanics, an approach different from the standard one, leading to the compatibility relations (3.1)–(3.2), is due to Cattaneo (1978). This approach, suitably extended to a relativistic framework, is capable of dealing with both null and nonnull hypersurfaces (Anile, 1982), and applies also to the space-time formulation of Newtonian physics.

The plan of the chapter is the following.

In Section 3.1, first we introduce the concept of the *inner covariant derivative* for nonnull hypersurfaces. This is done covariantly in a local form, that is, restricting oneself to local charts in the space-time manifold. We follow a different route than Maugin's in the sense that we do not use parametric coordinates on the hypersurface Σ. The results are, however, equivalent. Then we introduce the concept of the Thomas displacement derivative, in a covariant way, for a nonnull hypersurface in an arbitrary space-time.

Also, we recall some basic concepts from distribution theory, define the Dirac distribution for a nonnull hypersurface, and prove some basic results for the derivatives of regularly discontinuous tensor fields.

Following the method of Cattaneo (1978) we introduce a definition of inner covariant derivative which is appropriate for null hypersurfaces and we obtain the relevant compatibility relations.

Finally, following Friedlander (1975), we give a definition of the Dirac distribution for a hypersurface which applies also in the null case and obtain some basic formulas for the derivatives of regularly discontinuous tensor fields.

3.1. Regularly discontinuous tensor fields across a hypersurface

Let $\mathscr{M}$ be a space-time, $\Omega \subset \mathscr{M}$ an open subset of $\mathscr{M}$, and $\varphi : \Omega \to \mathbb{R}$ a differentiable function. The equation

$$\varphi(x^\alpha) = 0 \tag{3.3}$$

will define a hypersurface Σ in Ω.

First of all, we shall assume that Σ is a nonnull hypersurface, that is, $g^{\alpha\beta}\varphi_{,\alpha}\varphi_{,\beta} \neq 0$.

Then the unit normal to Σ can be defined by

$$n_\mu = \frac{\varphi_{,\mu}}{|g^{\alpha\beta}\varphi_{,\alpha}\varphi_{,\beta}|^{1/2}}, \tag{3.4}$$

with $n_\mu n^\mu = \pm 1$ according to whether Σ is timelike or spacelike.

The projection tensor $\tilde{h}^{\mu\nu}$ onto Σ is

$$\tilde{h}^\nu_\mu = \delta^\nu_\mu \mp n_\mu n^\nu. \tag{3.5}$$

Let $v^\mu(x^\alpha)$ be a smooth vector field of Ω defined on Σ; then we define the *inner covariant derivative* of $v^\mu(x^\alpha)$ by

$$\tilde{\nabla}_\nu v^\mu = \tilde{h}^\alpha_\nu \nabla_\alpha v^\mu. \tag{3.6}$$

Remark 1. This operation represents a directional derivative along directions lying on the hypersurface Σ. It is to be distinguished from the induced derivative, defined for vector fields which are tangent to Σ,

$$''\nabla_\nu v^\mu = \tilde{h}^\alpha_\nu \tilde{h}^\mu_\beta \nabla_\alpha v^\beta, \tag{3.7}$$

where the result of the inner covariant derivative is further projected onto Σ, in order to obtain a tensor which is tangent to Σ. Obviously, for scalars, $\tilde{\nabla}$ and $''\nabla$ coincide.

Remark 2. The definition (3.6) can easily be extended, by linearity, to arbitrary tensor fields. The extension to a field **U** consisting of tensor fields is straightforward.

Remark 3. When v^μ is defined only on points of Σ, the definition (3.6) still makes sense. In fact, let us consider local coordinates adapted to Σ, that is, $y^0 \equiv \varphi(x^\alpha)$ and y^i are three coordinates on Σ, and let $\hat{v}^\mu(y^\alpha)$ be any smooth extension of v^μ off Σ. Then

$$\tilde{\nabla}_\nu v^\mu = \tilde{h}^\alpha_\nu \nabla_\alpha v^\mu = \tilde{h}^\alpha_\nu \{\partial_\alpha \hat{v}^\mu + \Gamma^\mu_{\alpha\beta} v^\beta\}. \tag{3.8}$$

Now, in the coordinates (y^α),

$$\tilde{h}^\alpha_\nu \partial_\alpha \hat{v}^\mu = \tilde{h}^0_\nu \partial_0 \hat{v}^\mu + \tilde{h}^i_\nu \partial_i v^\mu = \tilde{h}^i_\nu \partial_i v^\mu$$

because $\tilde{h}^0_\nu = 0$ in these coordinates.

Therefore, $\tilde{\nabla}_\nu v^\mu$ does not depend on the particular extension $\hat{v}^\mu$.

Now we introduce the regularly discontinuous tensor fields and the compatibility relations.

DEFINITION 3.1. *Let (y^0, y^i) be coordinates adapted to Σ, $y^0 = \varphi(x^\alpha)$, y^i are* *three coordinates on Σ. Then a smooth tensor field T^{α}_{β} on $\Omega-\Sigma$ will be said to be regularly discontinuous across Σ if there exist two tensor fields $\tilde{T}^{\alpha\cdot\cdot}_{\pm\beta\cdot\cdot}$ which are smooth in the domains $\Omega_+ = \{(y^0, y^i), y^0 \geq 0\}$, $\Omega_- = \{(y^0, y^i), y^0 \leq 0\}$ and such that in the open subdomains $\overset{0}{\Omega}_+ = \{(y^0, y^i),\ y^0 > 0\}$, $\overset{0}{\Omega}_- = \{(y^0, y^i), y^0 < 0\}$ coincide with $T^{\alpha\cdot\cdot}_{\beta\cdot\cdot}$.*

The jump of T across Σ is defined as

$$[[T^{\alpha\cdot\cdot}_{\beta\cdot\cdot}(y^i)]] = \tilde{T}^{\alpha\cdot\cdot}_{-\beta\cdot\cdot}(0, y^i) - \tilde{T}^{\alpha\cdot\cdot}_{+\beta\cdot\cdot}(0, y^i).$$

The jump $[[T]]$ is a tensor field on Σ and we can apply to it the inner covariant derivative.

Obviously the jumps $[[\nabla_\mu T^{\alpha\cdot\cdot}_{\beta\cdot\cdot}]]$ are not all independent because, for derivatives along Σ, the jump and derivative operations commute. This can be made precise by the first order *compatibility relations*, which, in the case where Σ is nonnull, are

$$[[\nabla_\mu T^{\alpha\cdot\cdot}_{\beta\cdot\cdot}]] = \tilde{\nabla}_\mu[[T^{\alpha\cdot\cdot}_{\beta\cdot\cdot}]] + n_\mu[[n^\lambda \nabla_\lambda T^{\alpha\cdot\cdot}_{\beta\cdot\cdot}]]. \tag{3.9}$$

In order to prove equation (3.9) it is sufficient to consider only vector fields. The extension to tensor fields by linearity is trivial.

Let (y^0, y^i) be coordinates adapted to Σ, then

$$\varphi_\alpha = (1, 0, 0, 0), \quad g^{\alpha\beta}\varphi_\alpha\varphi_\beta = g^{00} \neq 0,$$

$$n_\alpha = \frac{1}{\sqrt{|g^{00}|}}(1, 0, 0, 0), \quad n^\alpha = \left(\pm\sqrt{|g^{00}|}, \frac{g^{i0}}{\sqrt{g^{00}}}\right).$$

Equation (3.9), in these coordinates, reads

$$[[\nabla_j v^\alpha]] = \tilde{\nabla}_j[[v^\alpha]] = \tilde{h}^\mu_j \nabla_\mu[[v^\alpha]] = \nabla_j[[v^\alpha]],$$

which holds because y^j are coordinates on Σ, and

$$[[\nabla_0 v^\alpha]] = \tilde{\nabla}_0[[v^\alpha]] + \frac{1}{\sqrt{|g^{00}|}}[[n^\mu \nabla_\mu v^\alpha]],$$

which holds because

$$\tilde{\nabla}_0[[v^\alpha]] = -\frac{g^{i0}}{|g^{00}|}\nabla_i[[v^\alpha]].$$

For a field **U** *consisting of tensor fields which are regularly discontinuous across a nonnull hypersurface Σ, the first order compatibility relations will* *obviously be*

$$[[\nabla_\alpha \mathbf{U}]] = \tilde{\nabla}_\alpha [[\mathbf{U}]] + n_\alpha [[\nabla_\mathbf{n} \mathbf{U}]], \tag{3.10}$$

where $\nabla_\mathbf{n} \mathbf{U} \equiv n^\nu \nabla_\nu \mathbf{U}$.

By iteration one can obtain compatibility relations of higher order for the jump $[[\nabla_{\alpha_1} \nabla_{\alpha_2} \cdots \nabla_{\alpha_m} \mathbf{U}]]$.

As an example we shall derive the second order compatibility relations. We shall prove them for vector fields. The extension to a field $\mathbf{U}$ consisting of tensor fields is straightforward.

Let v^μ be a vector field which is regularly discontinuous across Σ. By applying (3.9) to $\nabla_\alpha v^\mu$ we have

$$\begin{aligned}[[\nabla_\beta \nabla_\alpha v^\mu]] = {} & \tilde{\nabla}_\beta \tilde{\nabla}_\alpha [[v^\mu]] + \chi_{\alpha\beta} [[\nabla_\mathbf{n} v^\mu]] \\ & + n_\alpha \tilde{\nabla}_\beta [[\nabla_\mathbf{n} v^\mu]] + n_\beta [[n^\nu \nabla_\nu \nabla_\alpha v^\mu]],\end{aligned} \tag{3.11}$$

where

$$\chi_{\alpha\beta} = \tilde{\nabla}_\beta n_\alpha = \tilde{\nabla}_\alpha n_\beta \tag{3.12}$$

is the second fundamental form of Σ.

Now, a well-known result of differential geometry is that (Misner et al., 1973)

$$\nabla_\beta \nabla_\alpha v^\mu - \nabla_\alpha \nabla_\beta v^\mu = R^\mu_{\sigma\beta\alpha} v^\sigma, \tag{3.13}$$

where $R^\alpha_{\beta\gamma\delta}$ is the Riemann tensor of the Riemannian connection. Therefore, from equations (3.11)–(3.13) it follows that

$$\begin{aligned}[[\nabla_\alpha \nabla_\beta v^\mu]] + R^\mu_{\sigma\beta\alpha} [[v^\sigma]] = {} & \tilde{\nabla}_\beta \tilde{\nabla}_\alpha [[v^\mu]] + \chi_{\alpha\beta} [[\nabla_\mathbf{n} v^\mu]] \\ & + n_\alpha \tilde{\nabla}_\beta [[\nabla_\mathbf{n} v^\mu]] + n_\beta [[n^\nu \nabla_\nu \nabla_\alpha v^\mu]].\end{aligned} \tag{3.14}$$

It is easy to see that, for an arbitrary vector field w^μ, one has

$$\tilde{\nabla}_\alpha \tilde{\nabla}_\beta w^\mu - \tilde{\nabla}_\beta \tilde{\nabla}_\alpha w^\mu = \tilde{h}^\rho_\alpha \tilde{h}^\tau_\beta R^\mu_{\sigma\rho\tau} w^\sigma - n_\beta \chi^\rho_\alpha \tilde{\nabla}_\rho w^\mu + n_\alpha \chi^\rho_\beta \tilde{\nabla}_\rho w^\mu. \tag{3.15}$$

From the previous two equations it follows that

$$\begin{aligned}[[\nabla_\alpha \nabla_\beta v^\mu]] + R^\mu_{\sigma\beta\alpha} [[v^\sigma]] = {} & \tilde{\nabla}_\alpha \tilde{\nabla}_\beta [[v^\mu]] - \tilde{h}^\rho_\alpha \tilde{h}^\tau_\beta R^\mu_{\sigma\rho\tau} [[v^\sigma]] \\ & + n_\beta \chi^\rho_\alpha \tilde{\nabla}_\rho [[v^\mu]] - n_\alpha \chi^\rho_\beta \tilde{\nabla}_\rho [[v^\mu]] \\ & + \chi_{\alpha\beta} [[\nabla_\mathbf{n} v^\mu]] + n_\alpha \tilde{\nabla}_\beta [[\nabla_\mathbf{n} v^\mu]] \\ & + n_\beta [[n^\nu \nabla_\nu \nabla_\alpha v^\mu]].\end{aligned} \tag{3.16}$$

By contracting equation (3.16) with n^α we obtain

$$\begin{aligned}[[n^\nu \nabla_\nu \nabla_\alpha v^\mu]] = {} & -n^\nu R^\mu_{\sigma\alpha\nu} [[v^\sigma]] - \chi^\rho_\alpha \tilde{\nabla}_\rho [[v^\mu]] + \tilde{\nabla}_\alpha [[\nabla_\mathbf{n} v^\mu]] \\ & + n_\alpha [[n^\lambda n^\nu \nabla_\nu \nabla_\lambda v^\mu]],\end{aligned}$$

which after substitution into equation (3.11) yields

$$\begin{aligned}[[\nabla_\beta\nabla_\alpha v^\mu]] = {} & \tilde{\nabla}_\beta\tilde{\nabla}_\alpha[[v^\mu]] + \chi_{\alpha\beta}[[\nabla_{\mathbf{n}} v^\mu]] + n_\alpha\tilde{\nabla}_\beta[[\nabla_{\mathbf{n}} v^\mu]] \\ & + n_\beta\tilde{\nabla}_\alpha[[\nabla_{\mathbf{n}} v^\mu]] - n_\beta\chi^\rho_\alpha\tilde{\nabla}_\rho[[v^\mu]] \\ & + n_\alpha n_\beta[[n^\lambda n^\nu\nabla_\nu\nabla_\lambda v^\mu]] - n_\beta n^\nu R^\mu_{\sigma\alpha\nu}[[v^\sigma]]. \end{aligned} \tag{3.17}$$

Equations (3.17) *are the second order compatibility relations for a regularly discontinuous vector field.*

In order to treat the case of a field **U** consisting of tensor fields it is convenient to introduce a matrix $R_{\alpha\beta}$ with components $(R_{\alpha\beta})^B_A$, $A, B = 1, 2, \ldots, N$, such that the commutation relations (3.13) are suitably generalized (Pham Mau Quam, 1969), giving

$$\nabla_\alpha\nabla_\beta U^B - \nabla_\beta\nabla_\alpha U^B = (R_{\alpha\beta})^B_A U^A. \tag{3.18}$$

Then the *second order compatibility relations for a regularly discontinuous field* **U** are

$$\begin{aligned}[[\nabla_\beta\nabla_\alpha \mathbf{U}]] = {} & \tilde{\nabla}_\beta\tilde{\nabla}_\alpha[[\mathbf{U}]] + \chi_{\alpha\beta}[[\nabla_{\mathbf{n}}\mathbf{U}]] + n_\alpha\tilde{\nabla}_\beta[[\nabla_{\mathbf{n}}\mathbf{U}]] \\ & + n_\beta\tilde{\nabla}_\alpha[[\nabla_{\mathbf{n}}\mathbf{U}]] - n_\beta\chi^\rho_\alpha\tilde{\nabla}_\rho[[\mathbf{U}]] \\ & + n_\alpha n_\beta[[n^\lambda n^\nu\nabla_\lambda\nabla_\nu\mathbf{U}]] - n_\beta n^\nu R_{\alpha\nu}[[\mathbf{U}]]. \end{aligned} \tag{3.19}$$

Now we shall introduce the concept of the Thomas displacement derivative.

Let us consider first the case of Minkowski space $\mathscr{M} = \mathbb{R}^4$ with the metric, in inertial coordinates (t, x^i), given by

$$g_{\mu\nu} = \eta_{\mu\nu} = \mathrm{diag}(-1, 1, 1, 1)$$

and let f be a function which is regularly discontinuous across Σ, assumed to be a timelike hypersurface.

The first order compatibility relations are then

$$[[\partial_\alpha f]] = \tilde{\nabla}_\alpha[[f]] + n_\alpha[[n^\mu\partial_\mu f]]. \tag{3.20}$$

The normal speed of propagation of Σ with respect to the family of inertial observers, having

$$u^\mu = (1, 0, 0, 0),$$

is given by

$$V_\Sigma\Gamma_\Sigma = -n_0, \tag{3.21}$$

where $\Gamma_\Sigma = (1 - V^2_\Sigma)^{1/2}$ is the Lorentz factor of V_Σ.

Then, from the normalization condition $n_\mu n^\mu = 1$, we have

$$n_i = \Gamma_\Sigma v_i,$$

where $v_i v_i = 1$, v_i is a unit three-vector in Euclidean $\mathbb{R}^3$. Hence

$$n_\mu = (-\Gamma_\Sigma V_\Sigma, \Gamma_\Sigma v_i), \quad n^\mu = \Gamma_\Sigma (V_\Sigma, v_i), \tag{3.22}$$

and the components of the projection tensor $\tilde{h}^\mu_\nu = \delta^\mu_\nu - n_\mu n^\nu$ are

$$\tilde{h}^0_0 = \Gamma^2_\Sigma, \qquad \tilde{h}^i_0 = \Gamma^2_\Sigma V_\Sigma v_i,$$
$$\tilde{h}^0_i = \delta^j_i - \Gamma^2_\Sigma v_i v^j, \quad \tilde{h}^0_i = -\Gamma^2_\Sigma V_\Sigma v_i.$$

Then the compatibility relations (3.20) for $\alpha = 0$ yield

$$[[\partial_0 f]] = \mathscr{D}[[f]] - V_\Sigma[[v^i \partial_i f]], \tag{3.23}$$

where

$$\mathscr{D} = \partial_0 + V_\Sigma \partial_i \tag{3.24}$$

can then be interpreted as the *Thomas displacement derivative*, because (3.23) are the standard three dimensional kinematic compatibility relations (3.2).

Similarly, for $\alpha = i$ we obtain, after some manipulations,

$$[[\partial_i f]] = \pi^j_i \partial_j [[f]] + v_i [[v^j \partial_j f]] \tag{3.25}$$

with $\pi^j_i = \delta^j_i - v_i v^j$ the projection tensor onto the 2-surface of $\mathbb{R}^3$ given by $\Sigma \cap (t = \text{const.})$. Equations (3.25) are the standard three dimensional geometric compatibility relations, equations (3.1).

A *covariant definition of the Thomas displacement derivative* can be obtained in an arbitrary space-time as follows. Let Σ be timelike with unit normal n^μ.

Let u^μ be a timelike vector field, $u_\mu u^\mu = -1$, and introduce the following vectors

$$k^\mu = \frac{1}{\Gamma_\Sigma}(u^\mu + \Gamma_\Sigma V_\Sigma n^\mu), \tag{3.26}$$

$$q^\mu = \frac{1}{\Gamma_\Sigma}(n^\mu - \Gamma_\Sigma V_\Sigma u^\mu), \tag{3.27}$$

where V_Σ is the normal speed of propagation of Σ with respect to u^μ, $n^\mu u_\mu = -\Gamma_\Sigma V_\Sigma$.

We have $k^\mu n_\mu = 0$, $k^\mu k_\mu = -1$, $q^\mu u_\mu = 0$, and $q^\mu q_\mu = 1$. Then k^μ is a timelike unit vector on Σ and q^μ a spacelike unit vector in the 3-space orthogonal to u^μ.

The compatibility relations (3.8), contracted with u^μ, yield

$$[[u^\mu \nabla_\mu T^{\alpha\cdot\cdot}_{\beta\cdot\cdot}]] = u^\mu \tilde{\nabla}_\mu [[T^{\alpha\cdot\cdot}_{\beta\cdot\cdot}]] - V_\Sigma \Gamma_\Sigma [[n^\mu \nabla_\mu T^{\alpha\cdot\cdot}_{\beta\cdot\cdot}]],$$

which are equivalent to

$$[[u^\mu \nabla_\mu T^{\alpha\cdot\cdot}_{\beta\cdot\cdot}]] = \mathscr{D}[[T^{\alpha\cdot\cdot}_{\beta\cdot\cdot}]] - V_\Sigma [[q^\mu \nabla_\mu T^{\alpha\cdot\cdot}_{\beta\cdot\cdot}]], \tag{3.28}$$

where

$$\mathscr{D}[[T^{\alpha\cdot\cdot}_{\beta\cdot\cdot}]] = \frac{1}{\Gamma_\Sigma} k^\mu \nabla_\mu [[T^{\alpha\cdot\cdot}_{\beta\cdot\cdot}]]. \tag{3.29}$$

Equations (3.28) are a covariant formulation of the kinematic compatibility relations with respect to the timelike vector field u^μ, and therefore $\mathscr{D}$ *can be interpreted as the Thomas displacement derivative.*

Some properties of the Thomas derivative of the vector field k^μ are of noticeable interest and will be useful in what follows.

Obviously, $\mathscr{D}k^\mu$ is orthogonal to k^μ. Furthermore, let e_2^μ, e_3^μ be two spacelike unit vector fields, defined on Σ, mutually orthogonal and orthogonal to k^μ and q^μ. Then the following formula holds:

$$e_A^\mu \mathscr{D} k_\mu = -V_\Sigma \Gamma_\Sigma e_A^\mu \nabla_\mu V_\Sigma + \frac{V_\Sigma}{\Gamma_\Sigma^2} e_A^\mu n^\alpha (\nabla_\alpha u_\mu - \nabla_\mu u_\alpha)$$
$$+ \frac{1}{\Gamma_\Sigma^3} e_A^\mu u^\alpha \nabla_\alpha u_\mu, \quad A = 2, 3. \tag{3.30}$$

In fact, one has

$$e_A^\mu \mathscr{D} k_\mu = \frac{1}{\Gamma_\Sigma} e_A^\mu k^\alpha \nabla_\alpha k_\mu = \frac{1}{\Gamma_\Sigma} e_A^\mu k^\alpha \left(\frac{1}{\Gamma_\Sigma} \nabla_\alpha u_\mu + V_\Sigma \nabla_\alpha n_\mu \right)$$
$$= \frac{1}{\Gamma_\Sigma^2} e_A^\mu k^\alpha \nabla_\alpha u_\mu + \frac{V_\Sigma}{\Gamma_\Sigma} e_A^\mu k^\alpha \nabla_\mu n_\alpha$$
$$= \frac{1}{\Gamma_\Sigma^3} e_A^\mu u^\alpha \nabla_\alpha u_\mu - V_\Sigma \Gamma_\Sigma e_A^\mu \nabla_\mu V_\Sigma$$
$$+ \frac{V_\Sigma}{\Gamma_\Sigma^2} e_A^\mu n^\alpha (\nabla_\alpha u_\mu - \nabla_\mu u_\alpha).$$

In many applications one has to differentiate tensor fields which are regularly discontinuous across the hypersurface Σ. The mathematically sound way of doing this is through the use of distribution theory. Here we shall present some of the basic results which will be of use later.

First of all, we recall some basic results for distributions on a manifold (Friedlander, 1975).

DEFINITION 3.2. *Let Ω be an open subset of $\mathbb{R}^n$ and $C_0^\infty(\Omega)$ denote the space of C^∞ functions on Ω with compact support.*

Let $u: C_0^\infty(\Omega) \to \mathbb{R}$ be a linear form on $C_0^\infty(\Omega)$. Then u is said to be a distribution on Ω if, for any compact $K \subseteq \Omega$, there exist two constants C_K, N_K, such that

$$(u, \Phi) \leq C_K \sum_{|\alpha| \leq N_K} \sup |\partial^\alpha \Phi|, \tag{3.31}$$

$\forall \Phi \in C_0^\infty(\Omega)$ with $\operatorname{supp} \Phi \subseteq K$, *and where (u, ϕ) denotes the value of the form u on the function ϕ.*

Remark 4. If f is locally integrable in Ω, $f \in L^1_{\text{loc}}(\Omega)$, then f defines in a natural way a distribution $\hat{f}$ on Ω by

$$(f, \Phi) = \int_\Omega f \Phi \mathrm{d}x, \tag{3.32}$$

$\forall \Phi \in C_0^\infty(\Omega)$, with $\mathrm{d}x$ the Lebesgue measure in $\mathbb{R}^n$.

In fact, (3.31) is verified with $N_K = 0$ and $C_K = \int_\Omega |f| \mathrm{d}x$. The vector space of distributions on Ω is denoted by $\mathscr{D}'(\Omega)$.

Now let $\mathscr{M}$ be a space-time, and Ω be an open subset of $\mathscr{M}$.

DEFINITION 3.3. *Let $u: C_0^\infty(\Omega) \to \mathbb{R}$ be a linear form. Then u is said to be a distribution if in any local coordinates (x^α), ranging in an open subset Ω' of $\mathbb{R}^n$ there exists a distribution u' on Ω' such that*

$$(u, \Phi) = (u', \Phi(x^\alpha)|g(x^\alpha)|^{1/2})$$

for any differentiable function Φ with compact support in Ω, with $g = \det(g_{\mu\nu})$.

The vector space of distributions on Ω is denoted by $\mathscr{D}'(\Omega)$.

Let $\mathscr{D}^r_{0s}(\Omega)$ denote the vector space of differentiable tensor fields with compact support.

DEFINITION 3.4. *Let $T: \mathscr{D}^r_{0s}(\Omega) \to \mathbb{R}$ be a linear form. Then T is said to be a tensor distribution of type (r, s) if, in any local coordinates (x^α) ranging in an open subset Ω' of $\mathbb{R}^n$, there exist n^{r+s} distributions $T'^{\alpha 1 \alpha 2 \ldots \alpha r}_{\beta 1 \beta 2 \ldots \beta s}$ on Ω' such that*

$$(T, \Phi) = (T'^{\alpha 1 \alpha 2 \ldots \alpha r}_{\beta 1 \beta 2 \ldots \beta s}, \Phi^{\beta 1 \beta 2 \ldots \beta s}_{\alpha 1 \alpha 2 \ldots \alpha r}(x^\alpha)|g(x^\alpha)|^{1/2})$$

for any $\Phi \in \mathscr{D}^r_{0s}(\Omega)$.

The vector space of tensor distributions of type (r, s) is denoted by $\mathscr{D}'^{r}_{s}$.

Let T be a tensor field of type (r, s) defined in a bounded domain Ω of $\mathscr{M}$, which is locally integrable, that is, such that in any local coordinates its components $T^{\alpha 1\alpha 2\ldots\alpha r}_{\beta 1\beta 2\ldots\beta s}$ are integrable.

Then T defines a tensor distribution $\hat{T}$ of type (r, s) by

$$\hat{T}:\mathscr{D}^{s}_{0r}(\Omega)\to\mathbb{R},$$

$$\langle\hat{T},\Phi\rangle=\int_{\Omega}T^{\alpha 1\ldots\alpha r}_{\beta 1\ldots\beta s}\Phi^{\beta 1\ldots\beta s}_{\alpha 1\ldots\alpha r}\mu, \tag{3.33}$$

where μ is the invariant volume element of $\mathscr{M}$,

$$\mu=\sqrt{|g|}\,\mathrm{d}x^{0}\wedge\mathrm{d}x^{1}\wedge\mathrm{d}x^{2}\wedge\mathrm{d}x^{3}$$

in local coordinates.

The covariant derivative of a tensor distribution is defined as follows.

DEFINITION 3.5. *Let $T\in\mathscr{D}'^{r}_{s}$. Then the covariant derivative ∇T is the tensor distribution of type $(r, s+1)$, $\nabla T\in\mathscr{D}'^{r}_{s}+1$ such that*

$$(\nabla T,\Phi)=(T,\delta\Phi),\quad\forall\Phi\in\mathscr{D}^{s+1}_{0r}, \tag{3.34}$$

where

$$(\delta\Phi)^{\alpha 1\ldots\alpha s}_{\beta 1\ldots\beta r}=-\nabla_{\nu}\Phi^{\nu\alpha 1\ldots\alpha s}_{\beta 1\ldots\beta r}. \tag{3.35}$$

Now we introduce some important distributions which will be of use later.

Let $\phi:\Omega\to\mathbb{R}$ be a differentiable function and Σ be the hypersurface of Ω defined by $\phi(x^{\alpha})=0$.

Let $\Omega_{+}=(x\in\Omega:\phi(x)\geq 0)$, $\Omega_{-}=(x\in\Omega:\phi(x)\leq 0)$ and let χ_{+} and χ_{-} be the characteristic functions of $\overset{0}{\Omega}_{\pm}$. Obviously, $\chi_{\pm}$ are locally integrable and they define two distributions, also denoted by $\chi_{\pm}$.

Let Σ_{+} denote the hypersurface Σ with positive orientation and n^{α} its outward unit normal. We define the Dirac distribution δ_{Σ} relative to Σ by

$$(\delta_{\Sigma},f)=\int_{\Sigma}f|_{\Sigma}\mu_{\Sigma}, \tag{3.36}$$

where μ_{Σ} is the induced volume element of Σ, $f\in C^{\infty}_{0}(\Omega)$.

It is easy to show that

$$\nabla_{\alpha}\chi_{\pm}=\mp n_{\alpha}\delta_{\Sigma}. \tag{3.37}$$

In fact, for any smooth vector field f^α with compact support in Ω, one has

$$(\nabla_\alpha \chi_+, f^\alpha) = -(\chi_+, \nabla_\alpha f^\alpha) = -\int_{\overset{\circ}{\Omega}_+} \partial_\alpha(f^\alpha \sqrt{|g|}) \mathrm{d}x^0 \mathrm{d}x^1 \mathrm{d}x^2 \mathrm{d}x^3$$

$$= -\int_{\partial \overset{\circ}{\Omega}} f^\alpha n_\alpha \mu_\Sigma = -\int_{\Sigma_+} f^\alpha n_\alpha \mu_\Sigma = -(n_\alpha \delta_\Sigma, f^\alpha).$$

The induced volume element μ_Σ, because Σ is nonnull, is defined as follows. In local coordinates (y^0, y^i) adapted to Σ, such that $y^0 = \phi(x^\alpha)$, y^i are three coordinates on Σ having the same orientation as (x^α),

$$\mu_\Sigma = \sqrt{|\tilde{g}|}\, \mathrm{d}y^1 \mathrm{d}y^2 \mathrm{d}y^3,$$

with $\tilde{g}_{ij}$ the induced metric tensor on Σ. When Σ is null the definition of μ_Σ requires more care (Friedlander, 1975).

Now, let $T^{\alpha\cdots}_{\beta\cdots}$ be a tensor field in Ω, regularly discontinuous across Σ, and $\tilde{T}^{\alpha\cdots}_{\pm\beta\cdots}$ be the two tensor fields introduced in Definition 3.2. Then T defines a tensor distribution in Ω. In fact, let

$$\tilde{T}^{\alpha\cdots}_{\beta\cdots} = \chi_+ \tilde{T}^{\alpha\cdots}_{+\beta\cdots} + \chi_- \tilde{T}^{\alpha\cdots}_{-\beta\cdots}, \tag{3.38}$$

where $\chi_\pm$ represent the distributions associated with the characteristic functions. By using equation (3.37) we obtain

$$\nabla_\mu \hat{T}^{\alpha\cdots}_{\beta\cdots} = n_\mu [[T^{\alpha\cdots}_{\beta\cdots}]] \delta_\Sigma + (\nabla_\mu \widehat{T^{\alpha\cdots}_{\beta\cdots}}), \tag{3.39}$$

where $(\nabla_\mu \widehat{T^{\alpha\cdots}_{\beta\cdots}})$ denotes the tensor distribution defined by the regularly discontinuous tensor field $\nabla_\mu T^{\alpha\cdots}_{\beta\cdots}$.

The case when the hypersurface Σ is null requires special care because of the following two points:

(i) the definition (3.6) of *inner covariant derivative* does not apply since the projection tensor $\tilde{h}^\nu_\mu$ is not defined;
(ii) the definition of the Dirac distribution δ_Σ, equation (3.36), and hence also equation (3.37), must be changed because of the nonexistence of the unit normal.

In order to treat null hypersurfaces we utilize a method originally introduced by Cattaneo (1978) in a noncovariant framework, suitably modified to make it covariant. First of all, we assume that space-time $\mathscr{M}$ (or an open connected subset Ω of it) is time-orientable, that is, there exists a smooth unit timelike vector field ξ^α. Let $\phi: \Omega \to \mathbb{R}$ be a differentiable function and consider the hypersurface Σ given by

$$\phi(x^\alpha) = 0.$$

Let $l_\alpha = \nabla_\alpha \phi$ and assume that Σ is null, that is,

$$l^\alpha l_\alpha = 0.$$

We notice that $l^\mu \xi_\mu \neq 0$. In fact, at a given space-time point we can always use inertial coordinates such that $\xi^\mu = (1, 0, 0, 0)$ and the metric has the Minkowski form $\eta_{\mu\nu}$. Then

$$l_\mu \xi^\mu = l_0 \neq 0$$

because

$$l_\mu l^\mu = -l_0^2 + l_i l^i = 0.$$

Let $v^\mu(x^\alpha)$ be a smooth vector field of Ω defined on Σ. We define the inner covariant derivative of v^μ by

$$\tilde{\nabla}_\alpha v^\beta = \nabla_\alpha v^\beta - \frac{l_\alpha}{(l_\mu \xi^\mu)} \xi^\nu \nabla_\nu v^\beta. \tag{3.40}$$

Notice that

$$\xi^\alpha \tilde{\nabla}_\alpha v^\beta = 0$$

and also that if w^μ is tangent to Σ, that is, $w^\mu l_\mu = 0$, then

$$w^\mu \tilde{\nabla}_\mu v^\beta = w^\mu \nabla_\mu v^\beta.$$

The above definition can be extended in a trivial way to a tensor field. A coordinate-free formulation can be obtained very easily (Anile, 1982).

Proceeding as before, we obtain the compatibility relations in the case of a null hypersurface. They are

$$[[\nabla_\alpha v^\beta]] = \tilde{\nabla}_\alpha [[v^\beta]] + l_\alpha [[n^\nu \nabla_\nu v^\beta]], \tag{3.41}$$

where

$$n^\nu = \frac{\xi^\nu}{l^\mu \xi_\mu},$$

and can easily be extended to tensor fields.

The definition of the Dirac distribution δ_Σ must be modified when Σ is null because the induced volume element cannot be defined. Instead one utilizes the Leray form $\mu_{L\Sigma}$ defined by the requirement

$$\mu = \mathrm{d}\phi \wedge \mu_{L\Sigma}, \tag{3.42}$$

where

$$\mu = \sqrt{|g|}\, \mathrm{d}x^0 \wedge \mathrm{d}x^1 \wedge \mathrm{d}x^2 \wedge \mathrm{d}x^3$$

is the space-time volume four-form (Friedlander, 1975). Then the definition of δ_Σ becomes

$$(\delta_\Sigma, f) = \int_{\Sigma_+} f \mu_{L\Sigma} \tag{3.43}$$

and equation (3.37) is modified to

$$\nabla_\alpha \chi_\pm = \mp l_\alpha \delta_\Sigma. \tag{3.44}$$

Accordingly, equation (3.39) is modified to

$$\nabla_\mu \hat{T}^{\alpha\cdots}_{\beta\ldots} = l_\mu [[T^{\alpha\cdots}_{\beta\ldots}]] \delta_\Sigma + (\widehat{\nabla_\mu T^{\alpha\cdots}_{\beta\ldots}}). \tag{3.45}$$

Notice that the Leray form is well defined for any kind of hypersurface (whether Σ is null or nonnull) and hence equation (3.45) holds for any kind of hypersurface.

In the next chapter we will apply the method expounded in this section to the study of the propagation of weak discontinuity waves.

4
Propagation of weak discontinuities

4.0. Introduction

In the previous chapter we introduced the concept of propagating singular surfaces (or singular hypersurfaces in space-time) and derived the compatibility relations among the jumps of the field variables. When only derivatives of the field variables can be discontinuous across a propagating singular surface, we are dealing with *weak discontinuities*. When the fields themselves can be discontinuous we have *strong discontinuities* (among which *shock waves* are of paramount importance).

Whereas shock waves can be produced from an initially smooth solution as a consequence of nonlinear steepening and breaking, weak discontinuities can only be produced as a result of discontinuities in initial or boundary conditions. For instance, in gas dynamics, a jump in the derivatives of the velocity can appear in a flow along a solid obstacle with angles. Also, a jump in the derivatives of the pressure can appear (among other discontinuities like shocks and contact discontinuities) when the initial condition consists of two adjoining masses of gas compressed to different pressures (Landau and Lifshitz, 1959a). Although conceptually these two examples remain valid also for relativistic fluid motion, for relativistic flow in an astrophysical context only the latter is meaningful. For instance, the case of a cloud moving relativistically in a jet has been considered by Blandford and Königl (1979) in the context of models for the rapid variations in extragalactic radio sources' emissions.

An extremely important application of the concept of weak discontinuity waves is to the study of impulsive gravitational waves. In fact, the geometrical quantities corresponding to the physical degrees of freedom of the gravitational field are not the metric components or the Christoffel symbols but the Riemann tensor components (Misner, et al., 1973). Therefore a gravitational impulse could be modeled as a space-time where the metric and the Christoffel symbols are continuous but the second derivatives of the metric (and hence the Riemann tensor) are discontinuous across a hypersurface (Synge, 1960). In this case, by using the theory of weak discontinuities, one can prove that gravitational impulses propagate as waves with the speed of light and one can also define a local energy conservation for these waves, thereby providing an invariant definition of gravitational radiation, at least in this limiting case (Pirani, 1957).

Once they have been produced by given initial and boundary conditions, the propagation of weak discontinuities can be studied relatively easily (at variance with shock waves).

The evolution of a weak discontinuity can be studied exactly because, in some sense, its motion decouples from that of the smooth part of the solution. In fact, the motion of the wave separates from the solution behind the wavefront and its evolution can be determined, once the state ahead of the wavefront has been specified and the initial data for the wave parameters have been assigned. More precisely, one introduces a class of curves (rays), and the problem of determining the evolution of the wave is reduced to that of solving: (i) a partial differential equation (which is the characteristic equation); (ii) transport equations for the wave parameters (amplitude and polarization) along the rays (the transport equations are in the form of systems of ordinary differential equations of the Bernoulli type).

The theory of weak discontinuities in the context of Newtonian physics is rather old (Hadamard, 1903; Thomas, 1957; Truesdell and Toupin, 1960; Chen, 1976). In the past it has been applied mainly to fluid dynamics and elasticity theory (Truesdell and Toupin, 1960; Chen, 1976). An elegant and concise presentation of the theory, in a form suitable for general hyperbolic systems, has been given by Boillat (1965).

In a relativistic framework the propagation of weak discontinuities has been studied by several authors (Boillat, 1969, 1973; Cattaneo, 1970; Saini, 1976; Giambo', 1982, among others), and the theory can be presented in a concise and fully covariant formulation.

The theory of the propagation of weak discontinuities has the very interesting feature that it allows an exact analytical treatment of the phenomenon of nonlinear breaking of impulsive waves. Furthermore, this theory can be of great relevance also in connection with the testing of elaborate numerical codes for relativistic fluid dynamics and magneto-fluid dynamics (for instance in the context of accretion onto black holes; Hawley, Smarr, and Wilson, 1984a, b). In fact, numerical codes based on finite differencing have great difficulty in coping with discontinuities because they tend to be ignored or smeared out in a nonphysical way. Therefore, exact analytical results for the propagation of weak discontinuities could provide a benchmark against which to check various differencing methods. Discontinuities can be dealt with relatively easily by the method of characteristics (which however is applicable only in one space dimension; Whitham, 1974).

The plan of the chapter is the following.

In Section 4.1 we review the fundamental theory of characteristic surfaces and bicharacteristic curves. The treatment is modeled after that of Choquet-Bruhat (1968) and of Cattaneo (1981). In Section 4.2 we give a covariant formulation of the theory of weak discontinuities for general hyperbolic systems. In particular, we derive the transport equation, discuss the breaking of the discontinuity wave, and give a covariant formulation of the exceptionality condition. We follow the treatments of Boillat (1973), Cattaneo (1981), and Giambo' (1982). In Section 4.3 we treat in detail weak discontinuities in relativistic fluid dynamics. In particular the transport equation is derived explicitly and the exceptionality condition is analyzed following the work of Boillat (1973). In Section 4.4 we discuss weak discontinuities in relativistic magneto-fluid dynamics. We focus on the exceptionality condition for magnetoacoustic and Alfvén waves, following the work of Greco (1972). In Section 4.5 we discuss electromagnetic and gravitational discontinuities, following the work of Lichnerowicz (1960). We derive transport equations and conservation laws both in the electromagnetic and in the gravitational case.

4.1. Characteristic hypersurfaces

We recall some basic definitions and results.

Let us consider the quasi-linear hyperbolic system

$$A_B^{\alpha A} \nabla_\alpha U^B = f^A \tag{4.1}$$

with $A_B^{\alpha A}, f^A$ differentiable functions of $\mathbf{U} \in \mathbf{D} \subseteq \mathbb{R}^N$.

Let Σ be a hypersurface, and (y^0, y^i) be local coordinates adapted to Σ, in which the local equation defining Σ is $y^0 = 0$ and y^i are three coordinates on Σ. Then we can state the following definition.

DEFINITION 4.1. *Σ is called a characteristic hypersurface for the system* (4.1) *if, in local coordinates adapted to Σ, $\partial_0 U^A$ cannot be expressed in terms of $\partial_i U^A$.*

We have the following proposition.

PROPOSITION 4.1. *Let Σ be a characteristic hypersurface for the system* (4.1) *and let $\phi(x^\alpha) = 0$ be the local equation of Σ.*

Then ϕ satisfies the characteristic equation

$$\det(A_B^{\alpha A}\phi_\alpha)=0. \tag{4.2}$$

Proof. In local coordinates adapted to Σ equation (4.1) reads

$$A_B^{0A}\partial_0 U^B + A_B^{iA}\partial_i U^B + Y^A = f^A,$$

where Y^A does not depend on the derivatives $\partial_\alpha U^A$. Therefore $\partial_0 U^A$ cannot be expressed as a function of $\partial_i U^A$ iff $\det(A_B^{0A})=0$, a condition which in general coordinates coincides with equation (4.2). Q.E.D.

Now we shall study the characteristic equation (4.2). We shall limit ourselves to local considerations and therefore work in $\mathbb{R}^n$. We write equation (4.2) as

$$\Psi(x^\alpha,\phi_\alpha)=0, \quad x^\alpha\in\mathbb{R}^n, \tag{4.3}$$

where Ψ is homogeneous of degree N in ϕ_α.

The Cauchy problem for equation (4.3) can be formulated as follows. Let W_{n-1} be a submanifold of dimension $n-1$ of $\mathbb{R}^{n-1}$, and

$$f: W_{n-1}\to\mathbb{R}^{n+1}$$

be an immersion, that is, a map which in local coordinates can be expressed as

$$(u^i)\to f(u^i)=\left(\underset{0}{x}^\alpha(u^i),\underset{0}{z}(u^i)\right), \quad \alpha=1,\ldots,n, \quad i=1,\ldots,n-1, \tag{4.4}$$

where (u^i) are local coordinates in W_{n-1}, and (x^α,z) are in $\mathbb{R}^{n+1}$ and such that the Jacobian matrix has rank $n-1$ (Choquet-Bruhat, 1968). The Cauchy problem is to find a differentiable function $\phi(x^\alpha)$ such that it satisfies (4.3) and, furthermore,

$$\phi\left(\underset{0}{x}^\alpha(u^i)\right)=\underset{0}{z}(u^i). \tag{4.5}$$

Let $p_\alpha=\phi_\alpha$ and introduce coordinates (x^α,z,p_α) in $\mathbb{R}^{2n+1}$.

Then equation (4.3) is equivalent to the following exterior differential system in $\mathbb{R}^{2n+1}$,

$$\Psi(x^\alpha,p_\alpha)=0, \tag{4.6}$$

$$\omega\equiv \mathrm{d}z-p_\alpha \mathrm{d}x^\alpha=0. \tag{4.7}$$

An integral manifold of the differential system (4.6)–(4.7) is defined as follows. Let W_m be a manifold of dimension m,

$$\sigma: W_m \to \mathbb{R}^{2n+1}$$

a differentiable map given in local coordinates by

$$(w^a) \to (x^\alpha(w^a), z(w^a), p_\alpha(w^a)),$$

where (w^a) are local coordinates in W_m. Then the couple (W_m, σ) is an integral manifold of the system (4.6)–(4.7) if:

(i) $\Psi(\sigma(p)) = 0, \forall p \in W_m$;
(ii) $\sigma^*\omega(p) = 0$

[where σ^* denotes the pull-back operator (Choquet-Bruhat, 1968)], which in local coordinates read

(i) $\Psi(x^\alpha(w^a), p_\alpha(w^a)) = 0$;
(ii) $\dfrac{\partial z}{\partial w^a} - p_\alpha(w^a)\dfrac{\partial x^\alpha}{\partial w^a} = 0.$

Now we construct the *initial integral manifold* of the system (4.6)–(4.7). Let W_{n-1} be a submanifold of $\mathbb{R}^{n-1}$ of dimension $n-1$ and $f: W_{n-1} \to \mathbb{R}^{n+1}$ the previously defined immersion, equation (4.4). Define the differentiable map

$$\bar{f}: W_{n-1} \to \mathbb{R}^{2n+1}, \tag{4.8}$$

$$(u^i) \to \left(\underset{0}{x}^\alpha(u^i), \underset{0}{z}(u^i), \underset{0}{p}_\alpha(u^i) \right),$$

where $\underset{0}{x}^\alpha(u^i), \underset{0}{z}(u^i)$ are given by equation (4.4) and $\underset{0}{p}_\alpha(u^i)$ are obtained by solving the system

$$\Psi\left(\underset{0}{x}^\alpha(u^i), \underset{0}{p}_\alpha(u^i) \right) = 0, \tag{4.9}$$

$$\frac{\partial \underset{0}{z}}{\partial u^i} - \underset{0}{p}_\alpha(u^i)\frac{\partial \underset{0}{x}^\alpha}{\partial u^i} = 0. \tag{4.10}$$

This system has solutions for $\underset{0}{p}_\alpha(u^i)$ provided the Jacobian Δ is nonvanishing, where

$$\Delta = \det \begin{bmatrix} \dfrac{\partial \underset{0}{x}^1}{\partial u^1} & \cdots\cdots\cdots & \dfrac{\partial \underset{0}{x}^n}{\partial u^1} \\ \vdots & & \vdots \\ \dfrac{\partial \underset{0}{x}^1}{\partial u^{n-1}} & \cdots\cdots\cdots & \dfrac{\partial \underset{0}{x}^n}{\partial u^{n-1}} \\ \dfrac{\partial \Psi}{\partial \underset{0}{p}_1} & \cdots\cdots\cdots & \dfrac{\partial \Psi}{\partial \underset{0}{p}_n} \end{bmatrix}.$$

Now, since f is an immersion, rank $(\partial \underset{0}{x}^\alpha/\partial u^i) = n-1$. Then $\Delta \neq 0$ provided the derivatives $\partial \Psi/\partial \underset{0}{p}_\alpha$ do not vanish simultaneously. Henceforth we shall assume the latter condition to hold, and in this case W_{n-1} is said to be *not Cauchy-characteristic*. In general, $\Psi(x^\alpha, p_\alpha)$ is nonlinear in p_α and therefore the system (4.9)–(4.10) will have several solutions which will correspond to different initial integral manifolds. Now we can state the following theorem.

THEOREM 4.1. *If W_{n-1} is not Cauchy-characteristic, there exists an integral manifold W_n of the exterior differential system* (4.6)–(4.7), *passing through W_{n-1}, to which there corresponds a local solution of the Cauchy problem for equation* (4.3).

Proof. We consider the characteristic system associated with the exterior differential system (4.6)–(4.7), that is,

$$\frac{\mathrm{d}x^\alpha}{\mathrm{d}v} = \frac{\partial \Psi}{\partial p_\alpha}, \quad \frac{\mathrm{d}p_\alpha}{\mathrm{d}v} = -\frac{\partial \Psi}{\partial x^\alpha},$$

$$\frac{\mathrm{d}z}{\mathrm{d}v} = 0, \tag{4.11}$$

whose general integral can be written as

$$x^\alpha = x^\alpha\left(v, \underset{0}{x}^\alpha(u^i), \underset{0}{z}(u^i), \underset{0}{p}^b(u^i)\right),$$

$$p_\alpha = p_\alpha\left(v, \underset{0}{x}^\alpha(u^i), \underset{0}{z}(u^i), \underset{0}{p}^b(u^i)\right), \tag{4.12}$$

$$z = \underset{0}{z}(u^i),$$

where

$$x^\alpha\left(0, \underset{0}{x}^\beta(u^i), \underset{0}{z}(u^i), \underset{0}{p}^b(u^i)\right) = \underset{0}{x}^\alpha(u^i),$$

$$p_\alpha\left(0, \underset{0}{x}^\beta(u^i), \underset{0}{z}(u^i), \underset{0}{p}^b(u^i)\right) = \underset{0}{p}_\alpha(u^i)$$

are the initial data for $v = 0$.

Let $W_n = W_{n-1} \times \mathbb{R}$, and introduce local coordinates (u^i, v) in W_n. Then we define the differentiable map

$$F: W_n \to \mathbb{R}^{2n+1},$$

with F given in local coordinates by equation (4.12).

Obviously, for $v = 0$, W_n and W_{n-1} are diffeomorphic. Now we show that the couple (W_n, F) is an integral manifold of (4.6)–(4.7). First of all, we show that

$$\Psi(x^\alpha, p_\alpha) = 0,$$

where x^α, p_α are given by (4.12). In fact, from (4.11) we have

$$\frac{\mathrm{d}\Psi}{\mathrm{d}v} = 0,$$

hence

$$\Psi(x^\alpha, p_\alpha) = \Psi\left(\underset{0}{x}^\alpha, \underset{0}{p}_\alpha\right) = 0.$$

Then, from (4.11), we have

$$\frac{\partial z}{\partial v} - p_\alpha \frac{\partial x^\alpha}{\partial v} = -p_\alpha \frac{\partial \Psi}{\partial p_\alpha} = -N\Psi = 0$$

because Ψ is homogeneous of degree N. It remains to be shown that

$$\frac{\partial z}{\partial u^i} - p_\alpha \frac{\partial x^\alpha}{\partial u^i} = 0.$$

Let

$$H = \frac{\partial z}{\partial u^i} - \frac{\partial x^\alpha}{\partial u^i} p_\alpha.$$

Then

$$\frac{\partial H}{\partial v} = -\frac{\partial p_\alpha}{\partial v}\frac{\partial x^\alpha}{\partial u^i} - p_\alpha \frac{\partial^2 x^\alpha}{\partial v \partial u^i} = \frac{\partial \Psi}{\partial x^\alpha}\frac{\partial x^\alpha}{\partial u^i} - p_\alpha \frac{\partial^2 x^\alpha}{\partial v \partial u^i}.$$

But from $p_\alpha \frac{\partial x^\alpha}{\partial v} = 0$ it follows that

$$\frac{\partial p_\alpha}{\partial u^i}\frac{\partial x^\alpha}{\partial v} + p_\alpha \frac{\partial^2 x^\alpha}{\partial u^i \partial v} = 0,$$

hence

$$\frac{\partial H}{\partial v} = \frac{\partial \Psi}{\partial x^\alpha}\frac{\partial x^\alpha}{\partial u^i} + \frac{\partial \Psi}{\partial p_\alpha}\frac{\partial p_\alpha}{\partial u^i} = \frac{\partial \Psi}{\partial u^i} = 0.$$

Therefore,

$$H = H_0 = \frac{\partial \underset{0}{z}}{\partial u^i} - \underset{0}{p_\alpha}\frac{\partial \underset{0}{x^\alpha}}{\partial u^i} = 0.$$

A solution to the Cauchy problem can be obtained from (4.12) by inverting $u^i = u^i(x^\alpha)$ and substituting into $\underset{0}{z}(u^i)$, that is,

$$\phi(x^\alpha) = \underset{0}{z}(u^i(x^\alpha)).$$

This is possible in a neighborhood of the initial manifold, which corresponds to $v = 0$ because the Jacobian of the transformation

$$(u^i, v) \rightarrow (x^\alpha)$$

is

$$\det \begin{bmatrix} \frac{\partial x^1}{\partial u^1} & \cdots\cdots & \frac{\partial x^n}{\partial u^1} \\ \vdots & & \vdots \\ \frac{\partial x^1}{\partial u^{n-1}} & \cdots\cdots & \frac{\partial x^n}{\partial u^{n-1}} \\ \frac{\partial x^1}{\partial v} & \cdots\cdots & \frac{\partial x^n}{\partial v} \end{bmatrix},$$

which, for $v = 0$, coincides with Δ. Q.E.D.

Let W_{n-1} be a submanifold of $\mathbb{R}^n$ of dimension $n-1$.
It is easy to find the characteristic hypersurfaces Σ such that

$$\Sigma \cap W_{n-1} = \hat{\sigma},$$

where $\hat{\sigma}$ is a preassigned n–2-submanifold of $\mathbb{R}^n$ which is also a submanifold of W_{n-1} (Fig. 4.1).

In fact, let (u^i) be local coordinates in W_{n-1} and let

$$\underset{0}{z}(u^i) = 0$$

be the local equation of $\hat{\sigma}$ in W_{n-1}.

Then, provided that W_{n-1} is not Cauchy-characteristic, to the several solutions of (4.9)–(4.10) there correspond different characteristic hypersurfaces with local equations $\phi(x^\alpha) = 0$ constructed according to the previous theorem, such that on W_{n-1} one has

$$\phi(x^\alpha(u^i)) = \underset{0}{z}(u^i).$$

The construction of solutions of the eikonal equation will be a crucial ingredient in the study of the propagation of weak discontinuities and also in the theory of asymptotic waves. In the next section we shall derive a set of propagation equations for the amplitudes of weak discontinuity waves along the bicharacteristics. Therefore, it will be assumed that, for the problem under consideration, the relevant solutions of the characteristic equation have been found using the results of this section.

Fig. 4.1. Initial value problem for the characteristic equation in space-time. Σ is the characteristic manifold (simple root) which on the initial manifold W_{n-1} coincides with $\hat{\sigma}$.

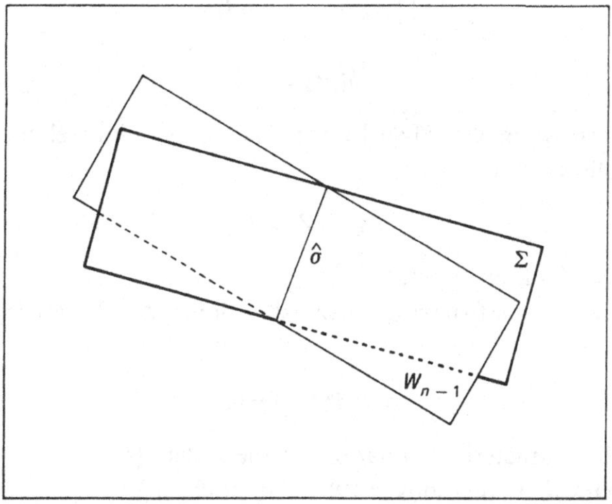

4.2. Weak discontinuities

Weak discontinuities are a mathematical idealization modeling impulsive waves. A precise definition is as follows.

DEFINITION 4.2. *The field* $\mathbf{U}$ *is said to have a weak discontinuity of order* m *across the hypersurface* Σ *if* $[[\mathbf{U}]]=0$, $[[\nabla_\alpha \mathbf{U}]]=0,\ldots,[[\nabla_{\alpha 1}\cdots\nabla_{\alpha m-1}\mathbf{U}]]=0$, *but, in general,* $[[\nabla_{\alpha 1}\cdots\nabla_{\alpha m}\mathbf{U}]]\neq 0$, *and* $\mathbf{U}$ *is differentiable elsewhere.*

Here we shall consider weak discontinuities of order 1, that is,

$$[[\mathbf{U}]]=0,$$

but, in general, $[[\nabla_\alpha \mathbf{U}]]\neq 0$.

Then the compatibility relations (3.9) and (3.41) reduce to

$$[[\nabla_\alpha U^A]]=Z^A n_\alpha \tag{4.13}$$

or

$$[[\nabla_\alpha U^A]]=Z^A l_\alpha \tag{4.14}$$

according to whether Σ is nonnull or null, with, obviously, a different meaning for Z^A. Now let us consider the following quasi-linear system

$$A^{\alpha A}_B \nabla_\alpha U^B = f^A, \tag{4.15}$$

with $A^{\alpha A}_B(U^C)$, $f^A(U^C)$ differentiable functions of $\mathbf{U}\in\mathbf{D}\subseteq\mathbb{R}^N$. Then

$$[[f^A]]=0 \tag{4.16}$$

and

$$[[A^{\alpha A}_B \nabla_\alpha U^B]]=0. \tag{4.17}$$

Let

$$\phi(x^\alpha)=0 \tag{4.18}$$

be the local equation of Σ. Then, by using the compatibility relations (4.13)–(4.14), it follows that

$$A^{\alpha A}_B \phi_\alpha Z^B=0 \tag{4.19}$$

in both cases (null or nonnull).

Therefore, in order to have a nonvanishing jump, $\mathbf{Z}\neq 0$, it is necessary for ϕ to obey

$$\det(A^{\alpha A}_B \phi_\alpha)=0, \tag{4.20}$$

which is the characteristic equation for the system (4.15).

We remark that equations (4.19)–(4.20) hold on Σ.

In the following we shall restrict ourselves to the case when Σ is timelike. The case of Σ null will be treated separately.

For the equations of relativistic fluid dynamics and magneto-fluid dynamics the case of Σ null corresponds to a particular limiting equation of state (the extreme stiff fluid, for which the speed of sound coincides with the speed of light). The results pertaining to the propagation of weak discontinuities in this case can be easily obtained from those presented in this section by a simple limiting procedure. As we shall see, for electromagnetic and gravitational discontinuities Σ is null and such a case will be treated separately in Section 4.5.

First of all, we consider the case when $\phi(x^\alpha)=0$ corresponds to *a simple root* of the characteristic equation (4.20).

Let

$$\mathbf{R}=(R^A)^{\mathrm{T}}, \quad \mathbf{L}=(L_A)$$

be the right and left eigenvectors of the matrix $A^{\alpha A}\phi_\alpha$ defined up to a factor, corresponding to the zero eigenvalue,

$$A^{\alpha A}_B\phi_\alpha R^B=0, \quad L_A A^{\alpha A}_B\phi_\alpha=0. \tag{4.21}$$

Then, from (4.19), we obtain

$$Z^A=\Pi(x^\alpha)R^A, (x^\alpha)\in\Sigma, \tag{4.22}$$

where the scalar Π is called the *amplitude of the discontinuity*. By differentiating the system (4.15), we obtain

$$A^{\alpha A}_B\nabla_\beta\nabla_\alpha U^B + A^{\alpha A}_{BC}(\nabla_\beta U^C)(\nabla_\alpha U^B)=f^A_C\nabla_\beta U^C, \tag{4.23}$$

where we have put

$$A^{\alpha A}_{BC}=\frac{\partial A^{\alpha A}_B}{\partial U^C}, \quad f^A_B=\frac{\partial f^A}{\partial U^B}. \tag{4.24}$$

By taking the jumps of equation (4.23) across Σ and using the second order compatibility relations (3.19), we obtain

$$\begin{aligned} &A^{\alpha A}_B(\chi_{\alpha\beta}Z^B+n_\alpha\tilde{\nabla}_\beta Z^B+n_\beta\tilde{\nabla}_\alpha Z^B+n_\alpha n_\beta W^B)\\ &\quad+A^{\alpha A}_{BC}(n_\alpha Z^B(\nabla_\beta U^C)_+ + n_\beta Z^C(\nabla_\alpha U^B)_+ + n_\alpha n_\beta Z^B Z^C) - f^A_B n_\beta Z^B=0, \end{aligned} \tag{4.25}$$

which holds on Σ and where $(\quad)_+$ signifies that the quantity within parentheses is evaluated on Σ_+.

By contracting with n^β, and multiplying by L_A, equation (4.25) yields

$$K^\alpha\tilde{\nabla}_\alpha\Pi+N\Pi^2+M\Pi=0, \tag{4.26}$$

where

$$K^\alpha = L_A A^{\alpha A}_B R^B, \tag{4.27}$$

$$N = L_A A^{\alpha A}_{BC} n_\alpha R^B R^C, \tag{4.28}$$

$$M = L_A(A^{\alpha A}_B \tilde{\nabla}_\alpha R^B - f^A_B R^B + A^{\alpha A}_{BC} n_\alpha n^\beta R^B (\nabla_\beta U^C)_+ + A^{\alpha A}_{BC} R^C (\nabla_\alpha U^B)_+). \tag{4.29}$$

It is easy to see that

$$K^\alpha n_\alpha = 0 \quad \text{on } \Sigma$$

and therefore K^α is tangent to Σ, hence

$$K^\alpha \tilde{\nabla}_\alpha = K^\alpha \nabla_\alpha.$$

By contracting equation (4.25) with $\tilde{h}^\mu_\beta$, where

$$\tilde{h}^\mu_\beta = \delta^\mu_\beta - n_\beta n^\mu,$$

yields

$$A^{\alpha A}_B R^B \chi_{\alpha\mu} + A^{\alpha A}_B n_\alpha \tilde{\nabla}_\mu R^B + A^{\alpha A}_{BC} n_\alpha R^B (\tilde{\nabla}_\mu U^C)_+ = 0.$$

This latter equation is an identity and follows from the equality

$$A^{\alpha A}_B n_\alpha R^B = 0$$

by differentiating along Σ.

A geometrical interpretation of the various terms appearing in equation (4.26) can be obtained with the help of the following propositions.

PROPOSITION 4.2. *Let $\bar{A} = \det(A^{\alpha A}_B \phi_\alpha)$. Then, on Σ, one has*

$$K^\alpha = k \frac{\partial \bar{A}}{\partial \phi_\alpha}, \tag{4.30}$$

with k a normalization factor.

Proof. Let $C^A_B = A^{\alpha A}_B \phi_\alpha$ and denote by $\hat{C}^B_A$ the adjoint of C^A_B,

$$C^A_B \hat{C}^B_A = \delta^A_B \bar{A}.$$

On Σ we have $\bar{A} = 0$ and therefore $\hat{C}^B_A$ is proportional both to R^B and to L_A,

$$\hat{C}^B_D = a R^B L_D,$$

with $a \neq 0$ (otherwise $\hat{C}^B_D = 0$ and the solution $\phi = 0$ would not correspond to a simple root of the characteristic equation).

Now, from the law for the derivation of a determinant,

$$\frac{\partial \bar{A}}{\partial \phi_\alpha} = \frac{\partial}{\partial \phi_\alpha}(A_B^{\mu A}\phi_\mu)\hat{C}_A^B = A_B^{\alpha A}\hat{C}_A^B = aL_A A_B^{\alpha A} R^B,$$

it follows that

$$K^\alpha = k\frac{\partial \bar{A}}{\partial \phi_\alpha}, k = \frac{1}{a}. \qquad \text{Q.E.D.}$$

PROPOSITION 4.3. *On Σ one has*

$$N = -\frac{k}{\sqrt{|G|}}\frac{\partial \bar{A}}{\partial U^C}R^C \tag{4.31}$$

with $G = g^{\mu\nu}\phi_\mu\phi_\nu$.

Proof.

$$\frac{\partial \bar{A}}{\partial U^C} = \frac{\partial}{\partial U^C}(A_B^{\alpha A}\phi_\alpha)\hat{C}_A^B = A_{BC}^{\alpha A}\phi_\alpha \frac{1}{k}R^B L_A,$$

hence

$$N = A_{BC}^{\alpha A}n_\alpha R^B R^C L_A = -\frac{k}{\sqrt{|G|}}\frac{\partial \bar{A}}{\partial U^C}R^C. \qquad \text{Q.E.D.}$$

Until now we have not assumed that the system (4.15) is hyperbolic. The only assumption we have made is that the hypersurface Σ corresponds to a solution of the characteristic equation of the system. In the case of a hyperbolic system we can obtain a suggestive expression for M. For this we need a few propositions.

PROPOSITION 4.4. *Let us consider the hyperbolic system*

$$A_B^{\alpha A}(\mathbf{U})\nabla_\alpha U^B = f^A(\mathbf{U}).$$

Then the right and left eigenvectors $\underset{(I)}{\mathbf{R}}, \underset{(J)}{\mathbf{L}}$ belonging to the eigenvalues $\underset{(I)}{\mu}$, $\underset{(J)}{\mu}$, obey

$$G_B^A \underset{(J)}{L_A} \underset{(I)}{R^B} = 0,$$

for $I \neq J$, with $G_B^A = A_B^{\alpha A}\xi_\alpha$.

Proof. An arbitrary ϕ_α (ϕ not a solution of the characteristic equation, in general) can always be decomposed as

$$\phi_\alpha = \zeta_\alpha - \mu \xi_\alpha.$$

Now $\underset{(I)}{\mathbf{R}}$ and $\underset{(I)}{\mu}$ satisfy

$$A_B^{\alpha A}\left(\zeta_\alpha - \underset{(I)}{\mu}\xi_\alpha\right)\underset{(I)}{R^B} = 0.$$

[there is no summation over (I)]. Then

$$A_B^{\alpha A}(\zeta_\alpha - \mu\xi_\alpha)\underset{(I)}{R^B} = -\left(\mu - \underset{(I)}{\mu}\right)A_B^{\alpha A}\xi_\alpha \underset{(I)}{R^B} = -\left(\mu - \underset{(I)}{\mu}\right)G_B^A \underset{(I)}{R^B}.$$

Similarly,

$$\underset{(J)}{L}A_B^{\alpha A}(\zeta_\alpha - \mu\xi_\alpha) = -\left(\mu - \underset{(J)}{\mu}\right)\underset{(J)}{L_A}G_B^A.$$

From these two equations it follows that

$$G_B^A \underset{(J)}{L_A}\underset{(I)}{R^B} = 0, \quad \text{for } I \neq J. \qquad \text{Q.E.D.}$$

Corollary.

$$\underset{(I)}{\hat{k}} = G_B^A \underset{(I)}{L_A}\underset{(I)}{R^B} \neq 0.$$

Proof. In fact, since $\det(G_B^A) \neq 0$, the vectors $G_B^A \underset{(I)}{L_A}$ are linearly independent and span $\mathbb{R}^N$, for $I = 1, \ldots, N$. Therefore, $\underset{(I)}{\mathbf{R}} \neq 0$ implies $G_B^A \underset{(I)}{L_A}\underset{(I)}{R^B} \neq 0$.

Q.E.D.

PROPOSITION 4.5. *On Σ one has the following formula*

$$L_A A_B^{\alpha A}\frac{\partial R^B}{\partial \phi_\beta}\phi_{\alpha\beta} = \hat{k}\frac{\partial k'}{\partial \phi_\alpha}\frac{\partial \bar{A}}{\partial \phi_\beta}\phi_{\alpha\beta} + k' L_A G^{AC}\frac{\partial R^C}{\partial \phi_\alpha}\frac{\partial \bar{A}}{\partial \phi_\beta}\phi_{\alpha\beta} + \frac{1}{2}k\frac{\partial^2 \bar{A}}{\partial \phi_\alpha \partial \phi_\beta}\phi_{\alpha\beta}, \tag{4.32}$$

where $k' = k/\hat{k}$ and the index (I) has been dropped for the sake of simplicity.

Proof. From

$$A_B^{\alpha A}\phi_\alpha \hat{C}_D^B = \delta_D^A \bar{A}$$

it follows, by multiplying by $R^C G_C^D$, that

$$A_B^{\alpha A}\phi_\alpha R^B = \frac{k}{\hat{k}} G_C^A R^C \bar{A}.$$

Differentiating twice with respect to ϕ_α and ϕ_β yields

$$\phi_{\alpha\beta} L_A\left(A_B^{\alpha A}\frac{\partial R^B}{\partial \phi_\beta} + A_B^{\alpha A}\frac{\partial R^B}{\partial \phi_\alpha}\right) = 2\hat{k}\frac{\partial k'}{\partial \phi_\alpha}\frac{\partial \bar{A}}{\partial \phi_\beta}\phi_{\alpha\beta} + 2k' L_A G^{AC}\frac{\partial R^C}{\partial \phi_\alpha}\frac{\partial \bar{A}}{\partial \phi_\beta}\phi_{\alpha\beta} + k\frac{\partial^2 \bar{A}}{\partial \phi_\alpha \partial \phi_\beta}\phi_{\alpha\beta}. \qquad \text{Q.E.D.}$$

Let $M_1 \equiv L_A A_B^{\alpha A}\tilde{\nabla}_\alpha R^B$ be the first term on the right-hand side of (4.29). Then

$$M_1 = L_A A_B^{\alpha A}\nabla_\alpha R^B = L_A A_B^{\alpha A}\frac{\partial R^B}{\partial \phi_\beta}\phi_{\alpha\beta} + L_A A_B^{\alpha A}\frac{\partial R^C}{\partial U^C}\nabla_\alpha U^C. \tag{4.33}$$

Furthermore, the bicharacteristic system for (4.20) is

$$\frac{dx^\mu}{ds} = \frac{\partial \bar{A}}{\partial \phi_\mu}, \quad \frac{d\phi_\mu}{ds} = -\frac{\partial \bar{A}}{\partial x^\mu},$$

hence

$$\frac{\partial \bar{A}}{\partial \phi_\beta}\phi_{\alpha\beta} = \frac{dx^\beta}{ds}\phi_{\alpha\beta} = \frac{d\phi_\alpha}{ds} = -\frac{\partial \bar{A}}{\partial x^\alpha}. \tag{4.34}$$

Now, let $M_2 \equiv L_A A_{BC}^{\alpha A} n_\alpha n^\beta R^B (\nabla_\beta U^C)_+$. From

$$A_{BC}^{\alpha A}\phi_\alpha R^B L_A = k\frac{\partial \bar{A}}{\partial U^C}$$

we easily obtain

$$M_2 = -\frac{k}{\sqrt{|G|}}\frac{\partial \bar{A}}{\partial U^C} n^\beta (\nabla_\beta U^C)_+.$$

Finally,

$$\begin{aligned} M = &-\hat{k}\frac{\partial k'}{\partial \phi_\alpha}\frac{\partial \bar{A}}{\partial x^\alpha} - k' L_A G^{AC}\frac{\partial R^C}{\partial \phi_\alpha}\frac{\partial \bar{A}}{\partial x^\alpha} + \frac{k}{2}\frac{\partial^2 \bar{A}}{\partial \phi_\alpha \partial \phi_\beta}\phi_{\alpha\beta} \\ &+ L_A\left(A_B^{\alpha A}\frac{\partial R^B}{\partial U^C}(\nabla_\alpha U^C)_+ - f_B^A R^B + A_{BC}^{\alpha A} R^C (\nabla_\alpha U^B)_+ \right) \\ &+ \frac{k}{\sqrt{|G|}}\frac{\partial \bar{A}}{\partial U^C} n^\beta (\nabla_\beta U^C)_+. \end{aligned} \tag{4.35}$$

The "rays" are introduced as the bicharacteristics of the characteristic equation (4.20), that is,

$$\frac{\mathrm{d}x^\mu}{\mathrm{d}s} = \frac{\partial \bar{A}}{\partial \phi_\mu} = \frac{1}{k} K^\mu. \tag{4.36}$$

Then equation (4.26) can be written as a transport equation along the rays,

$$k\frac{\mathrm{d}\Pi}{\mathrm{d}s} + N\Pi^2 + M\Pi = 0. \tag{4.37}$$

Let λ be a parameter along the rays related to s by

$$\frac{\mathrm{d}s}{\mathrm{d}\lambda} = k. \tag{4.38}$$

Then (4.37) becomes an equation of the Bernoulli type whose solution corresponding to the initial value $\Pi(0)$ at a given initial point on the ray is

$$\Pi(\lambda) = \frac{\Pi(0)}{\exp[Q(\lambda)]\left(1 + \Pi(0)\int_0^\lambda N(\lambda')\exp[-Q(\lambda')]\mathrm{d}\lambda'\right)}, \tag{4.39}$$

where

$$Q(\lambda) = \int_0^\lambda M(\lambda')\mathrm{d}\lambda'. \tag{4.40}$$

We see that the discontinuity amplitude $\Pi(\lambda)$ can become unbounded after a finite λ-time if either

(i) $\exp[-Q(\lambda)] \to \infty$ for $\lambda \to \lambda^*$; or

(ii) $\left(1 + \Pi(0)\int_0^\lambda N(\lambda')\exp[-Q(\lambda')]\mathrm{d}\lambda'\right) \to 0$ for $\lambda \to \lambda^*$.

The first case corresponds to the occurrence of caustics. Sometimes it is called a linear shock because it can also occur for linear equations (where $N = 0$). The second case corresponds to the gradient catastrophe and is interpreted as the formation of a shock wave (although a rigorous proof of this is still lacking in general) (Jeffrey, 1976). The finite parameter λ^* is called the critical time. A necessary condition for the existence of a critical time is

$$\Pi(0)N(\lambda) < 0.$$

The typically nonlinear case (ii) does not arise when the *exceptionality condition* $N=0$ holds, that is, when

$$\frac{\partial \bar{A}}{\partial U^C} R^C = 0. \tag{4.41}$$

An interesting formulation of the above condition, which is independent of the choice of the field **U**, is the following (Boillat, 1973).

For any function $f(\mathbf{U})$ define

$$\delta f = \frac{\partial f}{\partial U^C} R^C = \frac{\partial f}{\partial U^C} \delta U^C,$$

which is independent of the choice of **U**. Then equation (4.41) is equivalent to

$$\delta \bar{A} = 0. \tag{4.42}$$

For propagation into a constant state, $\underset{0}{\mathbf{U}}=$ constant, and in the absence of external sources, $\mathbf{f}=0$, equation (4.37) simplifies considerably, yielding

$$\frac{\mathrm{d}\Pi}{\mathrm{d}s} + \frac{1}{\sqrt{|G|}}(\delta \bar{A})\Pi^2 + \frac{\Pi}{2} \frac{\partial^2 \bar{A}}{\partial \phi_\alpha \partial \phi_\beta} \phi_{\alpha\beta} = 0. \tag{4.43}$$

Define

$$\bar{A}^\alpha = \frac{\partial \bar{A}}{\partial \phi_\alpha}.$$

Then when $N=0$ (which corresponds to the linear case or the exceptional case) equation (4.43) can be rewritten as

$$\nabla_\alpha(\Pi^2 \bar{A}^\alpha) = 0, \tag{4.44}$$

which is a conservation law for the amplitude discontinuity Π^2. Similar results are obtained in the case when $\phi(x^\mu)=0$ is a multiple root of the characteristic equation (4.20).

We shall limit ourselves to the case of hyperbolic systems. Let

$$\phi_\alpha = \zeta_\alpha - \mu\xi$$

and suppose that $\underset{(1)}{\mu}$ is a root with multiplicity r of the characteristic equation, the other roots being simple. Then

$$\bar{A}(\phi_\alpha) = H\left(\mu - \underset{(1)}{\mu}\right)^r \left(\mu - \underset{(2)}{\mu}\right) \cdots \left(\mu - \underset{(N-r)}{\mu}\right),$$

with H a proportionality factor. The roots $\underset{(I)}{\mu}$ must be homogeneous functions of ϕ_α of degree 1 and therefore

$$\mu - \underset{(1)}{\mu} = c^\alpha \phi_\alpha,$$

with c^α independent of ϕ_α. Let

$$\hat{A}(\phi_\alpha) = c^\alpha \phi_\alpha.$$

Then

$$\bar{A}(\phi_\alpha) = [\hat{A}(\phi_\alpha)]^r \tilde{A}(\phi_\alpha), \tag{4.45}$$

where $\hat{A}(\phi_\alpha)$ is of first degree in ϕ_α and $\tilde{A}(\phi_\alpha)$ does not contain $\hat{A}(\phi_\alpha)$. Let $\mathbf{R}_i$, $\mathbf{L}_i$, $i = 1, 2, \ldots, r$ be the right and left eigenvectors of

$$A_B^{\alpha A} \phi_\alpha$$

corresponding to the multiple root $\underset{(1)}{\mu}$.

Then from equation (4.19) we obtain

$$Z^A = \Pi_i R_i^A. \tag{4.46}$$

Proceeding as in the previous case yields

$$\begin{aligned} &L_{jA} A_B^{\alpha A} R_i^B \tilde{\nabla}_\alpha \Pi_i + L_{jA} A_{BC}^{\alpha A} n_\alpha R_i^B R_k^C \Pi_i \Pi_k \\ &+ \Pi_i L_{jA} (A_B^{\alpha A} \tilde{\nabla}_\alpha R_i^B + A_{BC}^{\alpha A} n_\alpha R_i^B n^\beta (\nabla_\beta U^C)_+ \\ &+ R_i^C (\nabla_\alpha U^C)_+ - f_B^A R_i^B) = 0. \end{aligned} \tag{4.47}$$

Now it can be proved that (Choquet-Bruhat, 1969)

$$L_{jA} A_B^{\alpha A} R_i^B = k_{ij} \frac{\partial \hat{A}}{\partial \phi_\alpha}, \tag{4.48}$$

where k_{ij} is a nonsingular matrix.

In fact, it is easy to see that

$$A_B^{\alpha A} \phi_\alpha R_i^B = \left(\mu - \underset{(1)}{\mu} \right) G_B^A R_i^B \tag{4.49}$$

and differentiating with respect to ϕ_α and multiplying by L_{Aj} yields, on the characteristic hypersurface,

$$L_{Aj} A_B^{\alpha A} R_i^B = k_{ij} \frac{\partial \left(\mu - \underset{(1)}{\mu} \right)}{\partial \phi_\alpha}.$$

Furthermore, from equation (4.49), differentiating with respect to U^C and multiplying by L_{Aj} and R_k^C gives, on the characteristic hypersurface,

$$L_{Aj}A^{\alpha A}_{BC}\phi_\alpha R_i^B R_k^C = \frac{\partial\left(\underset{(1)}{\mu} - \mu\right)}{\partial U^C} k_{ij} R_k^C,$$

which can be written as

$$L_{jA}A^{\alpha A}_{BC}n_\alpha R_i^B R_k^C = -\frac{k_{ij}}{\sqrt{|G|}}\frac{\partial \hat{A}}{\partial U^C} R_k^C. \tag{4.50}$$

Then, in this case the exceptionality condition becomes

$$\frac{\partial \hat{A}}{\partial U^C} R_k^C = 0.$$

4.3. Weak discontinuities in relativistic fluid dynamics

In this section we study the propagation of weak discontinuities in relativistic fluid dynamics. The characteristic equation, the propagation speeds, and the right and left eigenvectors have been studied in Section 2.3.

We have

$$\bar{A} = \det(A^{\alpha A}_B \phi_\alpha) = (e+p)^4 (u^\alpha \phi_\alpha)^4 [(u^\alpha \phi_\alpha)^2 - p'_e h^{\alpha\beta}\phi_\alpha\phi_\beta]. \tag{4.51}$$

The acoustic waves correspond to the roots of

$$\hat{A} = (u^\alpha \phi_\alpha)^2 - p'_e h^{\alpha\beta}\phi_\alpha\phi_\beta = 0. \tag{4.52}$$

The corresponding right and left eigenvectors are

$$\mathbf{R} = \begin{pmatrix} -h^{\nu\alpha}\phi_\alpha p'_e \\ (e+p)a \\ 0 \end{pmatrix}, \qquad \mathbf{L} = (a\phi_\mu, -a^2, -h^{\alpha\beta}\phi_\alpha\phi_\beta p'_s), \tag{4.53}$$

where $a = u^\alpha \phi_\alpha$.

Then the tangent vector to the rays, $K^\alpha = L_A A^{\alpha A}_B R^B$ is found to be

$$K^\alpha = -2a^2(e+p)(au^\alpha - p'_e h^{\mu\alpha}\phi_\mu). \tag{4.54}$$

From

$$\frac{\partial \bar{A}}{\partial \phi_\alpha} = 2(e+p)^4 a^4 (au^\alpha - p'_e h^{\mu\alpha}\phi_\mu)$$

it follows that

$$k = -\frac{1}{(e+p)^3 a^2}.$$

Also, it is easy to see that

$$N = \frac{a^5(e+p)}{\sqrt{|G|}}\left(2(1-p'_e) + \frac{(e+p)p''_e}{p'_e}\right). \tag{4.55}$$

The exceptionality condition is then

$$2(1-p'_e) + \frac{(e+p)p''_e}{p'_e} = 0. \tag{4.56}$$

If we assume $0 < p'_e \leq 1$ and $p''_e \geq 0$ then the only solution is $p'_e = 1$, which corresponds to the relativistic incompressible fluid (stiff matter).

Dropping the assumption $p''_e \geq 0$ we obtain the general integral of (4.56) (Boillat, 1973)

$$p = b - \frac{m^2}{(e+b)}, \tag{4.57}$$

with $m = m(S)$, $b = b(S)$ arbitrary functions, which does not seem to correspond to physically sensible fluids.

As an example we consider the special relativistic problem of a discontinuity wave propagating into a fluid at rest. Then, in the fluid ahead of the front

$$u^\mu = (1, 0, 0, 0)$$

and for the acoustic characteristic hypersurface we take

$$\phi = x^1 - c_s x^0,$$

with $c_s = \sqrt{p'_e}$ the speed of sound ahead of the front. It follows that

$$a = u^\mu \phi_\mu = -c_s.$$

From the definition (4.22) and the acoustic right eigenvector (4.53), the discontinuity amplitude is given by

$$\Pi = -\frac{(1-c_s^2)^{1/2}}{c_s^2}[[\partial_1 v]], \tag{4.58}$$

where v is the 1-component of the three-velocity.

For plane waves propagating into a constant state one has $M = 0$ and equation (4.26) reduces to

$$\frac{\partial}{\partial x^0}[[\partial_1 v]] + c_s[[\partial_1 v]] + \frac{W}{2}[[\partial_1 v]]^2 = 0, \tag{4.59}$$

where

$$W = \{2(1 - c_s^2) + \frac{e+p}{c_s^2}(p_e'')\}_+ \tag{4.60}$$

is the relativistic compressibility parameter (whose physical significance will be discussed in Chapters 8 and 9). From the latter equation it follows that

$$[[\partial_1 v]](x^0) = \frac{[[\partial_1 v]](0)}{1 + \dfrac{W}{2} x^0 [[\partial_1 v]](0)}. \tag{4.61}$$

Therefore, under the compressibility assumption $W > 0$, an initial negative slope discontinuity will break in a finite time x_B^0 given by

$$x_B^0 = -\frac{2}{W[[\partial_1 v]](0)}. \tag{4.62}$$

Notice that $x_B^0 \to \infty$ for the stiff fluid matter ($p_e' = 1$) and for the exceptional state equation (4.57).

The above exact law for the behavior of the discontinuity could be used in order to check the accuracy of numerical codes for relativistic fluid dynamics.

For material waves we have $\hat{A} = u^\alpha \phi_\alpha$, hence

$$\left.\frac{\partial \hat{A}}{\partial \phi_\alpha}\right|_\Sigma = u^\alpha.$$

Furthermore,

$$\frac{\partial \hat{A}}{\partial u^\nu} = \phi_\nu$$

and the exceptionality condition

$$\frac{\partial \hat{A}}{\partial U^C} R_i^C = 0$$

is easily verified. Therefore the material waves are exceptional.

Also,

$$\frac{\partial^2 \hat{A}}{\partial \phi_\alpha \partial \phi_\beta} = 0$$

and the transport equations reduce to

$$u^\alpha \nabla_\alpha \Pi_i = 0,$$

which states that the discontinuity is convected along the fluid flow lines.

4.4. Weak discontinuities in relativistic magneto-fluid dynamics

The characteristic equation, propagation speeds, and the right and left eigenvectors have been studied in Section 2.4. We recall that

$$\bar{A} = \det(A_B^{\alpha A}\phi_B) = Ea^2A^2N_4. \tag{4.63}$$

The magnetoacoustic waves correspond to the solutions of

$$N_4 = \eta(e'_p - 1)a^4 - (\eta + |b|^2)a^2G + B^2G = 0. \tag{4.64}$$

For magnetoacoustic waves the tangent to the rays is

$$K^\alpha = 4kEa^2A^2N^\alpha, \tag{4.65}$$

with

$$N^\alpha = \frac{1}{4}\frac{\partial N_4}{\partial \phi_\alpha}$$

and the nonlinear coefficient N of the transport equation given by

$$N = -\frac{k}{\sqrt{|G|}}Ea^2A^2\delta N_4 \tag{4.66}$$

using the definition of the operator δ given in Section 2.4. Now one has

$$\begin{aligned}\delta N_4 = {} & [4\eta(e'_p - 1)a^3 - 2(\eta + e'_p|b|^2)aG]\phi_\nu\delta u^\nu + [\eta a^4 e''_p \\ & + (e'_p - 1)a^4(e'_p + 1) - a^2G(e'_p + 1) - e''_p|b|^2a^2G]\delta p \\ & + [-2e'_p a^2Gb_\nu + 2BG\phi_\nu]\delta b^\nu.\end{aligned} \tag{4.67}$$

From (2.72) one has

$$\phi_\nu\delta u^\nu = -\frac{e'_p a}{\eta}\delta p. \tag{4.68}$$

From (2.74), after substituting into equation (2.70),

$$\nabla_\alpha b^\alpha + \frac{1}{(e+p)}b^\alpha\nabla_\alpha p = 0. \tag{4.69}$$

From (4.69) we obtain

$$\phi_\alpha\delta b^\alpha + \frac{B}{\eta}\delta p = 0. \tag{4.70}$$

From (2.75) we get

$$a\delta b^\beta - B\delta u^\beta + \frac{1}{\eta}(-e'_p ab^\beta + Bu^\beta)\delta p = 0 \tag{4.71}$$

and by contracting with b_β,

$$ab_\nu \delta b^\nu - Bb_\nu \delta u^\nu - \frac{e'_p}{\eta} a|b|^2 \delta p = 0. \tag{4.72}$$

From (2.74) one has

$$au_\nu \delta b^\nu - \frac{B}{\eta}\delta p = 0, \tag{4.72'}$$

and substituting into equation (4.71) finally gives

$$b_\nu \delta b^\nu = \left(\frac{e'_p}{\eta}|b|^2 - \frac{B^2}{\eta a^2}\right)\delta p. \tag{4.73}$$

Therefore, after some manipulations

$$\delta N_4 = a^2[a^2 K_1 + GK_2]\delta p, \tag{4.74}$$

where

$$K_1 = \eta e''_p + (e'_p - 1)(3 - 5e'_p), \tag{4.75}$$

$$K_2 = -e''_p|b|^2 + (e'_p - 1)\left(3 + \frac{2|b|^2}{\eta} e'_p\right). \tag{4.76}$$

The exceptionality condition for an arbitrary magnetic field gives

$$p''_e + \frac{2p'_e(1 - p'_e)}{e + p} = 0, \tag{4.77}$$

which corresponds to the stiff equation of state.

The Alfvén waves correspond to the solutions of

$$A = Ea^2 - B^2 = 0. \tag{4.78}$$

For the Alfvén waves $\delta p = \delta S = 0$, hence $\delta\eta = 0$. It follows that

$$\delta A = 2[a^2 b_\nu \delta b^\nu + Ea\phi_\nu \delta u^\nu - B\phi_\nu \delta b^\nu]. \tag{4.79}$$

From (2.91) one has

$$\delta b^\nu = \frac{B}{a}\delta u^\nu$$

and one can easily check that $\delta A = 0$ (Greco, 1972). Therefore the Alfvén waves are exceptional for any equation of state.

The material waves correspond to the double root $a^2 = 0$. Therefore, the rays coincide with the fluid streamlines and the exceptionality condition $\delta a = 0$ is easily verified from (2.89).

4.5. Electromagnetic and gravitational discontinuities

As an introduction to the gravitational case we shall first treat electromagnetic discontinuities (Lichnerowicz, 1960). Similar techniques will then be applied to the study of gravitational discontinuities. The unified treatment will show the great similarity between electromagnetic and gravitational waves.

We start with Maxwell's equations (2.46)–(2.47) and assume that only free currents and charges are present. Then the electromagnetic induction tensor $I^{\alpha\beta}$ obeys the constitutive law (2.60)

$$I^{\alpha\beta} = \frac{1}{\mu_0} F^{\alpha\beta}, \tag{4.80}$$

with μ_0 the vacuum magnetic permeability. Therefore the electromagnetic field equations become

$$[\partial_\alpha F_{\beta\gamma}] = 0, \tag{4.81}$$

$$\nabla_\beta F^{\alpha\beta} = 4\pi\mu_0 J^\alpha. \tag{4.82}$$

Our considerations will be local and we shall work in an open subset Ω of space-time $\mathscr{M}$, with local coordinates x^α. Let $\phi \in \mathscr{D}(\Omega)$ and consider the hypersurface Σ of Ω defined by

$$\phi(x^\alpha) = 0 \tag{4.83}$$

(where it has been assumed that $\mathrm{d}\phi \neq 0$ on Σ).

We shall study weak electromagnetic discontinuities of order 1, that is, *the electromagnetic field derivatives, $\partial_\gamma F_{\alpha\beta}$, suffer a jump discontinuity across* Σ. Let

$$l_\alpha = \partial_\alpha \phi.$$

Then the compatibility relations (4.13)–(4.14) can be written as

$$[[\partial_\gamma F_{\alpha\beta}]] = [[\nabla_\gamma F_{\alpha\beta}]] = \psi_{\alpha\beta} l_\gamma, \tag{4.84}$$

where $\psi_{\alpha\beta}$ is an antisymmetric tensor defined on Σ which has a different meaning according to whether Σ is null or nonnull. *We shall also assume that the current J^α is continuous across* Σ. Then, by taking the jumps of equations (4.81)–(4.82), one obtains

$$l_\alpha \psi_{\beta\gamma} + l_\beta \psi_{\gamma\alpha} + l_\gamma \psi_{\alpha\beta} = 0, \tag{4.85}$$

$$l^\alpha \psi_{\alpha\beta} = 0. \tag{4.86}$$

By contracting equation (4.85) with l^α and using (4.86), one has

$$l^\alpha l_\alpha \psi_{\beta\gamma} = 0,$$

whence, in order to have a nonvanishing discontinuity,

$$l^\alpha l_\alpha = 0 \tag{4.87}$$

and Σ is a null hypersurface.

Also,

$$l^\alpha \nabla_\alpha l_\beta = l^\alpha \nabla_\beta l_\alpha = 0 \tag{4.88}$$

and therefore l^α is a null geodesic vector field.

By suitably restricting Ω we can always introduce coordinates (x^α), adapted to Σ, in which the equation defining Σ is $x^0 = 0$. In these coordinates we have

$$\begin{aligned} l_\alpha &= (1, 0, 0, 0), \\ g^{00} &= 0, \\ l^\alpha &= (0, g^{0i}). \end{aligned} \tag{4.89}$$

Furthermore, in these coordinates, we also have

$$l_\alpha \nabla_\beta l^\alpha = l_0 \nabla_\beta l^0 + l_i \nabla_\beta l^i = 0,$$

hence

$$\nabla_\beta l^0 = 0. \tag{4.90}$$

Now we shall derive a transport equation for $[[\nabla_\sigma F_{\alpha\beta}]]$ along the null geodesics tangent to l^α (the rays).

It is easy to check that $[[\nabla_\sigma F_{\alpha\beta}]]$ obeys, on Σ, the relationships

$$l_\alpha [[\nabla_\sigma F_{\beta\gamma}]] + l_\beta [[\nabla_\sigma F_{\gamma\alpha}]] + l_\gamma [[\nabla_\sigma F_{\alpha\beta}]] = 0 \tag{4.91}$$

and

$$l_\alpha [[\nabla_\sigma F^{\alpha\beta}]] = 0. \tag{4.92}$$

In coordinates adapted to Σ, equations (4.91)–(4.92) read

$$[[\nabla_\sigma F_{ij}]] = 0, \tag{4.93}$$

$$[[\nabla_\sigma F^0_\beta]] = 0. \tag{4.94}$$

By differentiating equation (4.91) along Σ, putting $\alpha = i$, and summing on i, we obtain

$$\nabla_i (l^i [[\nabla_\sigma F_{\beta\gamma}]]) + \nabla_i (l_\beta [[\nabla_\sigma F^i_\gamma]]) + \nabla_i (l_\gamma [[\nabla_\sigma F^i_\beta]]) = 0,$$

which is, using $l^0 = 0$ and $\nabla_0 l^0 = 0$,

$$l^\rho \nabla_\rho [[\nabla_\sigma F_{\beta\gamma}]] + (\nabla_\rho l^\rho)[[\nabla_\sigma F_{\beta\gamma}]] + Q_{\sigma\beta\gamma} = 0, \tag{4.95}$$

with

$$Q_{\sigma\beta\gamma} = (\nabla_i l_\beta)[[\nabla_\sigma F^i_\gamma]] + (\nabla_i l_\gamma)[[\nabla_\sigma F^i_\beta]] + l_\beta[[\nabla_i \nabla_\sigma F^i_\gamma]] + l_\gamma[[\nabla_i \nabla_\sigma F^i_\beta]]. \tag{4.96}$$

Now we shall assume that the electromagnetic discontinuity propagates in vacuo, $J^\alpha = 0$. Then, by using (4.82), (4.94), and

$$[[\nabla_i \nabla_\sigma F^i_\gamma]] = [[\nabla_\sigma \nabla_i F^i_\gamma]],$$

$Q_{\sigma\beta\gamma}$ can be written in the form

$$Q_{\sigma\beta\gamma} = [[\nabla_\beta l_\rho \nabla_\sigma F^\rho_\gamma + \nabla_\gamma l_\rho \nabla_\sigma F^\rho_\beta]] - l_\beta[[\nabla_\sigma \nabla_0 F^0_\gamma]] - l_\gamma[[\nabla_\sigma \nabla_0 F^0_\beta]].$$

Also,

$$[[\nabla_\beta l_\rho \nabla_\sigma F^\rho_\gamma + \nabla_\gamma l_\rho \nabla_\sigma F^\rho_\beta]] = [[\nabla_\beta(l_\rho \nabla_\sigma F^\rho_\gamma) + \nabla_\gamma(l_\rho \nabla_\sigma F^\rho_\beta)]] - l^\rho[[\nabla_\sigma(\nabla_\beta F_{\gamma\rho} + \nabla_\gamma F_{\rho\beta})]]$$

and by using (4.81) we get

$$Q_{\sigma\beta\gamma} = l^\rho \nabla_\rho [[\nabla_\sigma F_{\beta\gamma}]] + [[\nabla_\beta(l_\rho \nabla_\sigma F^\rho_\gamma) + \nabla_\gamma(l_\rho \nabla_\sigma F^\rho_\beta)]] - l_\beta[[\nabla_0 \nabla_\sigma F^0_\gamma]] - l_\gamma[[\nabla_0 \nabla_\sigma F^0_\beta]]. \tag{4.97}$$

Now it is easy to see that, in adapted coordinates,

$$Q_{\sigma ij} = l^\rho \nabla_\rho [[\nabla_\sigma F_{ij}]],$$
$$Q_{\sigma 0} = l^\rho \nabla_\rho [[\nabla_\sigma F_0]]$$

and therefore equation (4.95) is rewritten as

$$2l^\rho \nabla_\rho [[\nabla_\gamma F_{\alpha\beta}]] + (\nabla_\rho l^\rho)[[\nabla_\gamma F_{\alpha\beta}]] = 0, \tag{4.98}$$

which is the transport equation for $[[\nabla_\gamma F_{\alpha\beta}]]$.

By substituting equation (4.84) into equation (4.98) we also obtain

$$2l^\rho \nabla_\rho \psi_{\alpha\beta} + (\nabla_\rho l^\rho)\psi_{\alpha\beta} = 0. \tag{4.99}$$

Now, let n^α be a vector field such that $n^\alpha l_\alpha \neq 0$ on Σ (by suitably restricting $\Omega \cap \Sigma$, n^α can be constructed using an orthonormal frame). By contracting equation (4.85) with n^α we get

$$\psi_{\alpha\beta} = l_\alpha \psi_\beta - l_\beta \psi_\alpha \tag{4.100}$$

with ψ_α defined on Σ and such that

$$\psi_\alpha l^\alpha = 0. \tag{4.101}$$

By substituting the representation (4.100) into equation (4.99), it follows that

$$2l^\rho \nabla_\rho \psi_\alpha + (\nabla_\rho l^\rho)\psi_\alpha = x l_\alpha \tag{4.102}$$

for some scalar x.

It is easy to see that ψ_α is a spacelike vector. In fact, at a given point, one can always choose an orthonormal frame such that $l^\alpha = (1, 1, 0, 0)$, hence $\psi_\alpha = (\psi_0, -\psi_0, \psi_2, \psi_3)$ and $\psi_\alpha \psi^\alpha = (\psi_2)^2 + (\psi_3)^2 > 0$. Contracting equation (4.102) with ψ^α yields the conservation law

$$\nabla_\rho (l^\rho |\psi|^2) = 0, \tag{4.103}$$

with

$$|\psi|^2 = \psi^\alpha \psi_\alpha.$$

Now we turn our attention to the discussion of gravitational discontinuities (Lichnerowicz, 1960). Until now we have assumed that space-time is a *differentiable manifold* $\mathcal{M}$ and this implies that the coordinate transformations $x^\alpha = x^\alpha(x'^\beta)$ must be C^∞ functions. Although this assumption is adequate for the rest of the applications treated in this book, it is overly restrictive in the case of gravitational discontinuities and gravitational shocks. In fact, a thorough analysis of the initial value problem for the Einstein field equations leads naturally only to the restriction that *the coordinate transformations* $x^\alpha = x^\alpha(x'^\beta)$ *be of class* C^2, *piecewise* C^3, C^4 (Lichnerowicz, 1955; Synge, 1960). These transformations will be called *admissible coordinate transformations.* Consistently, with the differentiability properties of the coordinate transformations we shall assume that the components of the metric tensor, $g_{\alpha\beta}$, are of class C^1, piecewise C^2, C^3.

Let Ω be an open subset of $\mathcal{M}$, $\phi \in D(\Omega)$, and Σ be the hypersurface given by

$$\phi(x^\alpha) = 0,$$

with $\mathrm{d}\phi \neq 0$ on Σ.

We shall study *gravitational discontinuities, in the sense that the second derivatives of the gravitational potential,* $g_{\alpha\beta,\mu\nu}$, *suffer a jump discontinuity across* Σ.

Therefore the Riemann tensor will be discontinuous and we are dealing with impulsive gravitational waves. From the geodesic deviation equation (Misner et al., 1973) one sees that these waves induce jumps in the acceleration of a test particle and in principle could be detected in this way.

The compatibility relations (4.13)–(4.14) applied to $g_{\alpha\beta,\mu\nu}$ yield

$$[[g_{\alpha\beta,\mu\nu}]] = h_{\alpha\beta} l_\mu l_\nu, \tag{4.104}$$

where

$$l_\mu = \phi_{,\mu}$$

and $h_{\alpha\beta}$ is a symmetric tensor defined on Σ which has a different meaning according to whether Σ is null or nonnull. Under an admissible coordinate transformation

$$x^\alpha = x^\alpha(x'^\beta),$$

one has

$$g_{\sigma'\tau'} = A^\alpha_{\sigma'} A^\beta_{\tau'} g_{\alpha\beta},$$

whence

$$\begin{aligned}[[\partial^2_{\nu'\rho'} g_{\sigma'\tau'}]] = A^\alpha_{\sigma'} A^\beta_{\tau'} A^\lambda_{\nu'} A^\mu_{\rho'} [[\partial^2_{\lambda\mu} g_{\alpha\beta}]] \\ + g_{\alpha\beta}([[\partial^2_{\nu'\rho'} A^\alpha_{\sigma'}]] A^\beta_{\tau'} \\ + A^\alpha_{\sigma'} [[\partial^2_{\nu'\mu'} A^\beta_{\tau'}]]).\end{aligned} \tag{4.105}$$

The compatibility relations applied to

$$\partial^2_{\nu'\rho'} A^\alpha_{\sigma'}$$

yield

$$[[\partial^2_{\nu'\rho'} A^\alpha_{\sigma'}]] = t^\alpha l_{\nu'} l_{\rho'} l_{\sigma'},$$

with t^α defined on Σ. Finally, we obtain that, under suitable coordinate transformations which are tangent on Σ to the identity transformation, $h_{\alpha\beta}$ transforms according to

$$h_{\alpha\beta} \to h_{\alpha\beta} + t_\alpha l_\beta + t_\beta l_\alpha, \tag{4.106}$$

also called *gauge transformations.*

If the transformations are C^3, then $h_{\alpha\beta}$ is invariant.

From the definition of the Riemann tensor (Misner et al., 1973)

$$R^\alpha_{\beta\lambda\mu} = \partial_\lambda \Gamma^\alpha_{\beta\mu} - \partial_\mu \Gamma^\alpha_{\beta\lambda} + \Gamma^\alpha_{\rho\lambda}\Gamma^\rho_{\beta\mu} - \Gamma^\alpha_{\rho\mu}\Gamma^\rho_{\beta\lambda},$$

where

$$\Gamma^\alpha_{\beta\mu}$$

are the Christoffel symbols

$$\Gamma^\alpha_{\beta\mu} = \tfrac{1}{2} g^{\alpha\sigma}(\partial_\beta g_{\sigma\mu} + \partial_\mu g_{\sigma\beta} - \partial_\sigma g_{\beta\mu}),$$

we have

$$[[R_{\alpha\beta\lambda\mu}]] = \tfrac{1}{2}(l_\lambda l_\beta h_{\alpha\mu} - l_\mu l_\beta h_{\alpha\lambda} + l_\alpha l_\mu h_{\beta\lambda} - l_\lambda l_\alpha h_{\beta\mu}), \tag{4.107}$$

and this expression is invariant under the transformation (4.106). From the definition of the Ricci tensor (Misner et al., 1973)

$$R_{\alpha\beta} = g^{\rho\sigma} R_{\rho\beta\sigma\alpha},$$

we have

$$[[R_{\alpha\lambda}]] \tfrac{1}{2} g^{\beta\mu}(l_\lambda l_\beta h_{\alpha\mu} + l_\alpha l_\mu h_{\beta\lambda} - l_\mu l_\beta h_{\alpha\lambda} - l_\lambda l_\alpha h_{\beta\mu}), \tag{4.108}$$

which is also invariant under the transformation (4.106).

Einstein's field equations are (Misner et al., 1973)

$$R_{\alpha\beta} - \tfrac{1}{2} g_{\alpha\beta} R = 8\pi\chi T_{\alpha\beta}, \tag{4.109}$$

where

$$R = g^{\mu\nu} R_{\mu\nu}$$

is the curvature scalar, $T_{\alpha\beta}$ is the matter energy-momentum tensor, and χ is the gravitational constant. We shall assume that the matter energy-momentum tensor $T_{\alpha\beta}$ is continuous across Σ. From equation (4.109) it follows that

$$-R = 8\pi\chi T,$$

hence

$$[[R]] = 0$$

and, therefore,

$$[[R_{\alpha\beta}]] = 0. \tag{4.110}$$

Now let us introduce coordinates adapted to Σ, for which the defining equations are $x^0 = 0$ and $l_\alpha = (1, 0, 0, 0)$. In these coordinates the only discontinuities in the second derivatives of $g_{\alpha\beta}$ can be in $\partial^2_{00} g_{\alpha\beta}$. We have

$$[[\partial^2_{00} g_{ij}]] = h_{ij}, \; [[\partial^2_{00} g_{0\alpha}]] = h_{0\alpha}.$$

From the transformation (4.106) one sees that h_{ij} is invariant whereas $h_{0\alpha}$ transforms according to

$$h_{0\alpha} \to h_{0\alpha} + t_0 l_\alpha + t_\alpha.$$

By a suitable choice of t_α it is possible to make $h_{0\alpha}$ vanish and for this reason $h_{0\alpha}$ is said to represent inessential discontinuities whereas h_{ij} *represents essential discontinuities*. In these coordinates, equation (4.110) reads

$$[[R_{ij}]] = -\tfrac{1}{2} g^{\beta\mu} l_\beta l_\mu h_{ij} = 0, \tag{4.111}$$

$$[[R_{0i}]] = \tfrac{1}{2} g^{\beta\mu}(l_0 l_\mu h_{i\beta} - l_\beta l_\mu h_{0i}) = 0, \tag{4.112}$$

$$[[R_{00}]] = \tfrac{1}{2} g^{\beta\mu}(l_\beta h_{0\mu} + l_\mu h_{0\beta} - l_\mu l_\beta h_{00} - h_{\mu\beta}) = 0. \tag{4.113}$$

We require nonvanishing essential discontinuities, therefore (4.111) yields

$$g^{\alpha\beta}l_\alpha l_\beta = 0 \tag{4.114}$$

and, as in the electromagnetic case, one sees that l_α is a null geodesic vector field. Then equation (4.112) gives

$$h_{ij}l^j = 0. \tag{4.115}$$

Finally, equation (4.113) yields

$$2h_{0\mu}l^\mu - h = 0, \tag{4.116}$$

where

$$h = g^{\alpha\beta}h_{\alpha\beta}.$$

Equations (4.115)–(4.116) are equivalent to the covariant equation

$$h_{\alpha\beta}l^\alpha - \tfrac{1}{2}hl_\beta = 0, \tag{4.117}$$

which is invariant under the transformation (4.106).

The propagation equation for the discontinuities can be obtained along similar lines as in the electromagnetic case. Starting from the explicit representation (4.107) for $[[R^\alpha_{\beta\lambda\mu}]]$ it is easy to show that

$$l_\lambda[[R^\alpha_{\beta\mu\nu}]] + l_\mu[[R^\alpha_{\beta\nu\lambda}]] + l_\nu[[R^\alpha_{\beta\lambda\mu}]] = 0, \tag{4.118}$$

$$l_\alpha[[R^\alpha_{\beta\lambda\mu}]] = 0. \tag{4.119}$$

In coordinates adapted to Σ, equation (4.118)–(4.119) are

$$[[R_{\alpha\beta ij}]] = 0, \tag{4.120}$$

$$[[R^0_{\alpha\beta\mu}]] = 0. \tag{4.121}$$

Differentiating equations (4.118) along Σ with respect to ∇_i, putting $\nu = i$, and summing yields

$$l^\rho\nabla_\rho[[R_{\alpha\beta\lambda\mu}]] + (\nabla_\rho l^\rho)[[R_{\alpha\beta\lambda\mu}]] + Q_{\alpha\beta\lambda\mu} = 0, \tag{4.122}$$

with

$$\begin{aligned} Q_{\alpha\beta\lambda\mu} = {} & (\nabla_\rho l_\lambda)[[R^\rho_{\alpha\beta\mu}]] + (\nabla_\rho l_\mu)[[R^\mu_{\alpha\beta\lambda}]] \\ & + l_\lambda[[\nabla_i R^i_{\alpha\beta\mu}]] + l_\mu[[\nabla_i R^i_{\alpha\beta\lambda}]] \end{aligned} \tag{4.123}$$

[where equation (4.121) has been used.]

Now we assume that the propagation occurs in vacuo,

$$R_{\alpha\beta} = 0.$$

Therefore, from the Bianchi identities (Misner et al., 1973)

$$\nabla_\rho R_{\alpha\beta\lambda\mu} + \nabla_\lambda R_{\alpha\beta\mu\rho} + \nabla_\mu R_{\alpha\beta\rho\lambda} = 0, \tag{4.124}$$

it follows that

$$\nabla_\rho R^\rho_{\alpha\beta\mu} = 0. \tag{4.125}$$

By using this latter equation we have

$$Q_{\alpha\beta\lambda\mu} = [[(\nabla_\lambda l_\rho)R^\rho_{\alpha\beta\mu} + (\nabla_\mu l_\rho)R^\rho_{\alpha\beta\lambda}]] - l_\lambda[[\nabla_0 R^0_{\alpha\beta\mu}]] \\ - l_\mu[[\nabla_0 R^0_{\alpha\beta\lambda}]].$$

Also, by using equation (4.124)

$$Q_{\alpha\beta\lambda\mu} = l^\rho[[\nabla_\rho R_{\alpha\beta\lambda\mu}]] + [[\nabla_\lambda(l_\rho R^\rho_{\alpha\beta\mu}) + \nabla_\mu(l_\rho R^\rho_{\alpha\beta\lambda})]] \\ - l_\lambda[[\nabla_0 R^0_{\alpha\beta\mu}]] - l_\mu[[\nabla_0 R^0_{\alpha\nu\lambda}]].$$

In adapted coordinates we have

$$Q_{\alpha\beta ij} = l^\rho \nabla_\rho[[R_{\alpha\beta ij}]],$$
$$Q_{\alpha\beta i0} = l^\rho \nabla_\rho[[R_{\alpha\beta i0}]],$$

and, therefore, equation (4.122) becomes

$$2l^\rho \nabla_\rho[[R_{\alpha\beta\lambda\mu}]] + (\nabla_\rho l^\rho)[[R_{\alpha\beta\lambda\mu}]] = 0. \tag{4.126}$$

By substituting the representation (4.107) into (4.126) we obtain

$$l_\lambda l_\beta c_{\alpha\mu} - l_\mu l_\beta c_{\alpha\lambda} + l_\alpha l_\mu c_{\beta\lambda} - l_\alpha l_\lambda c_{\beta\mu} = 0, \tag{4.127}$$

with

$$c_{\alpha\mu} = 2l^\rho \nabla_\rho h_{\alpha\mu} + (\nabla_\rho l^\rho)h_{\alpha\mu}. \tag{4.128}$$

Let n^α be a vector field on Σ such that $n^\alpha l_\alpha \neq 0$. Contracting equation (4.127) with $n^\lambda n^\beta$ gives

$$c_{\alpha\mu} = \Phi_\alpha l_\mu + \Phi_\mu l_\alpha - \Phi l_\alpha l_\mu, \tag{4.129}$$

with

$$\Phi_\alpha = n^\lambda c_{\alpha\lambda}, \quad \Phi = n^\lambda n^\alpha c_{\alpha\lambda}. \tag{4.130}$$

By defining

$$\hat{\Phi}_\alpha = \Phi_\alpha - \tfrac{1}{2}\Phi l_\alpha, \tag{4.131}$$

we obtain the transport equation for $h_{\alpha\mu}$

$$2l^\rho \nabla_\rho h_{\alpha\mu} + (\nabla_\rho l^\rho)h_{\alpha\mu} = \hat{\Phi}_\alpha l_\mu + \hat{\Phi}_\mu l_\alpha. \tag{4.132}$$

By a gauge transformation (4.106) we can always make $h = 0$, hence

$$h_{\alpha\beta} l^\beta = 0. \tag{4.133}$$

Therefore, by contracting equation (4.132) with $h^{\alpha\mu}$, the following conservation law is yielded

$$\nabla_\rho(\mathscr{E} l^\rho) = 0, \tag{4.134}$$

with

$$\mathscr{E} = h_{\alpha\mu}h^{\alpha\mu}.$$

Equation (4.134) can be interpreted as energy conservation for gravitational discontinuities.

In general relativity the concept of energy density for the gravitational field is rather elusive. However, it is remarkable that one can derive a local energy conservation law for impulsive gravitational waves.

5
Relativistic simple waves

5.0. Introduction

Simple waves are exact solutions of quasi-linear equations representing traveling waves. They are the most natural nonlinear analog of the plane traveling waves of the linear theory. Although they correspond to special initial conditions, they are still sufficiently general to be of physical interest (as witnessed by Friedrichs' theorem, which, loosely speaking, states that any one dimensional smooth solution neighboring a constant state must be a simple wave). Simple waves are exact analytical solutions which show clearly some of the main features of nonlinear wave propagation in general, such as steepening, breaking, and shock formation. They are also sufficiently complex to be useful as benchmarks against which to test numerical codes.

In classical fluid dynamics simple waves are of paramount importance for several reasons. Simple waves are the basic ingredients for constructing analytical solutions to the Riemann problem (or shock-tube problem: the evolution of an initial state corresponding to two adjacent fluids at different pressures). These solutions are among the standard tests for numerical hydrodynamical codes (Sod, 1978) because they comprise some of the key features of general hydrodynamical behavior (nonlinear steepening and the occurrence of discontinuities). Also, analytical solutions to the Riemann problem can be used in order to construct sophisticated numerical codes able to deal very accurately with shock front tracking (Plohr, Glimm, and McBryan, 1983). Simple waves are also used in order to construct approximate analytical solutions to the problem of weak shock decay (Landau and Lifshitz, 1959a; Whitham, 1974; Courant and Friedrichs, 1976) and in Whitham's theory of geometric shock dynamics (Whitham, 1974).

Given the importance of simple waves in classical fluid dynamics it is to be expected that simple waves solutions would play an analogous role in relativistic fluid dynamics.

Relativistic simple waves are relevant in a variety of astrophysical problems; here we shall mention only a few examples arising from the theory of extragalactic radio sources. Several other examples could be mentioned from relativistic astrophysics.

Jets seem to be characteristic features of extragalactic radio sources. They might be associated with collimated relativistic outflow, a characteristic property of active galactic nuclei (which are thought to power extragalactic radio sources). Jets may delineate a channel along which power from the nucleus is fed into the more extended radio structure. In some jets the light comes primarily from a regularly spaced series of bright knots. These knots could be associated with shocks behind which the relative energy of the flow can be dissipated in particle acceleration and synchrotron radiation. There are several hypothesis about the origin of these shocks. Rees (1978) suggested that the knots be attributed to the steepening of nonlinear acoustic waves in the jet. The acoustic waves would be generated by variations in the outflow velocity of a beam produced in the nucleus. An alternative hypothesis is that the knots be identified with dense blobs of gas which are compressed and swept outwards by the gas pressure of the jet (Blandford and Königl, 1979). Assuming the blobs are supersonic with respect to the jet, they would be followed by strong bow shocks. Obviously, other explanations could be envisaged for the origin of the knots.

The process of shock formation could be very important also in other astrophysical situations. For instance an interesting example is provided by the nonlinear evolution of adiabatic perturbations in the early universe ad this topic will be touched upon in Section 5.5.

As we shall see, a study of relativistic simple waves sheds light on the problem of relativistic shock formation due to nonlinear steepening.

Simple waves in relativistic fluid dynamics were first studied by Taub (1948) who introduced them through the Riemann invariants. Subsequently, they were analyzed in detail in several contexts by Liang (1977a) and Königl (1980), using Eulerian coordinates, and by Lanza, Miller, and Motta (1985) using Lagrangian coordinates. Relativistic simple waves have been used by Thompson (1986) in order to construct solutions to the relativistic shock-tube problem, a crucial benchmark for testing numerical relativistic hydrodynamical codes.

In many situations magnetic effects can be extremely important (for instance the magnetic field in a jet can be dynamically significant; Begelmann et al., 1984) and, therefore, one has to resort to relativistic magneto-fluid dynamics.

Simple waves for relativistic magneto-fluid dynamics have been considered in the plasma physics literature but have not been studied in great depth (Akhiezer et al., 1975). A detailed analysis was performed by Anile and Muscato (1983) for some cases, using Lagrangian coordinates. General qualitative properties can be found in Anile and Muscato (1988) and a numerical integration in several cases of physical interest has been performed by Muscato (1988b). These solutions are relevant for the relativistic magnetohydrodynamical shock-tube problem, which could be a very useful test for numerical relativistic MFD codes (Sloan and Smarr, 1986; where the tests are performed only by using the linearized magnetoacoustic and Alfvén wave modes).

The breaking of simple waves in relativistic fluid dynamics and magnetofluid dynamics has been studied in detail by Muscato (1988a).

The plan of the chapter is the following. In Section 5.1 we recall the basic definition of simple waves and Friedrich's theorem. In Section 5.2 we determine the general expression of the Riemann invariants for relativistic fluid dynamics. In Section 5.3 we write the equations of one dimensional relativistic fluid dynamics in Lagrangian coordinates, which will be useful in what follows, and in Section 5.4 we determine the simple wave solutions in these coordinates. In Section 5.5 we give explicit expressions for the Riemann invariants in the case of relativistic fluid dynamics for several state equations and we present some applications to the problem of the evolution of adiabatic perturbations in the early universe. These sections deal in detail with the nonlinear evolution of relativistic acoustic simple waves. Simple waves can also occur within supersonic flow (Courant and Friedrichs, 1976) and analytically are usually treated in the framework of stationary potential flow. Sections 5.6 to 5.8 aim at providing the relativistic extension of the classical theory of stationary potential flow. In particular, in Section 5.6 we introduce the basic definitions and properties of isentropic flow for relativistic fluid dynamics. In Section 5.7 we study, by using the hodograph method, the general properties of unsteady one dimensional and isentropic flow in special relativity (Lichnerowicz, 1967). In Section 5.8 we derive the equations describing stationary potential two dimensional flow in special relativity and determine the corresponding Riemann invariants and simple waves (Königl, 1980). Finally, in Section 5.9 we investigate the properties of simple waves in relativistic magneto-fluid dynamics and we study in detail their behavior for several state equations.

5.1. General formalism

In this section we consider homogeneous quasi-linear hyperbolic systems in the one dimensional case, that is, dependent only on two Cartesian coordinates x^0, x^1. They can be written as

$$\mathscr{A}^0(\mathbf{U})\frac{\partial \mathbf{U}}{\partial x^0} + \mathscr{A}^1(\mathbf{U})\frac{\partial \mathbf{U}}{\partial x^1} = 0. \tag{5.1}$$

We can state the following definition.

DEFINITION 5.1. *A simple wave for the system (5.1) is a smooth solution* $\mathbf{U}(x^0, x^1)$ *of the form*

$$\mathbf{U} = \mathbf{U}(\varphi(x^0, x^1)), \tag{5.2}$$

with $\varphi(x^0, x^1)$ *an appropriate* C^1 *function (called phase).*

By substituting equation (5.2) into (5.1) we obtain

$$(\mathscr{A}^0\varphi_0 + \mathscr{A}^1\varphi_1)\frac{\mathrm{d}\mathbf{U}}{\mathrm{d}\varphi} = 0,$$

whence, in order to have $\dfrac{\mathrm{d}\mathbf{U}}{\mathrm{d}\varphi} \neq 0$, it follows that

$$\det(\mathscr{A}^0\varphi_0 + \mathscr{A}^1\varphi_1) = 0. \tag{5.3}$$

Therefore, $\varphi(x^0, x^1)$ must obey the characteristic equation. It is convenient to introduce

$$\lambda = -\frac{\varphi_0}{\varphi_1}. \tag{5.4}$$

Then (5.3) becomes

$$\det(\mathscr{A}^1 - \lambda\mathscr{A}^0) = 0. \tag{5.5}$$

Let $\lambda^{(k)}$ be a simple eigenvalue and $\overset{(k)}{\mathbf{d}}$ be the corresponding (normalized) right eigenvector. Then $\dfrac{\mathrm{d}\mathbf{U}}{\mathrm{d}\varphi}$ must be proportional to $\overset{(k)}{\mathbf{d}}$,

$$\frac{\mathrm{d}\mathbf{U}}{\mathrm{d}\varphi} = \pi\overset{(k)}{\mathbf{d}}, \tag{5.6}$$

with π a proportionality factor.

Let $\mathbf{U} = (u^1, \ldots, u^N)^{\mathrm{T}}$. Then it is possible to write (5.6) in the form of an exterior differential system

$$\frac{du^1}{d_1^{(k)}} = \cdots = \frac{du^N}{d_N^{(k)}}. \tag{5.7}$$

One takes $u^1 = u^1(\varphi)$ as an arbitrary function of φ, then $u^2, \ldots, u^N$ as functions of u^1 are obtained by solving the above system.

The solution of the system (5.7) is equivalent to the determination of $N-1$ first integrals $\overset{(k)}{J}_1(u^1, \ldots, u^N)$, $\overset{(k)}{J}_2(u^1, \ldots, u^N), \ldots, \overset{(k)}{J}_{N-1}(u^1, \ldots, u^N)$, which are called the Riemann invariants corresponding to the eigenvalue $\lambda^{(k)}$.

It is easy to prove the following properties:

PROPOSITION 5.1. *The solution vector* $\mathbf{U}$ *in* S *is constant along the family of* $C^{(k)}$ *characteristics, which are straight lines.*

Proof. In fact, $\varphi = \text{const.}$ along a $\overset{(k)}{C}$ characteristic, therefore $\mathbf{U} = \mathbf{U}(\varphi)$ is also constant on $\overset{(k)}{C}$. It also follows that $\lambda^{(k)} = \lambda^{(k)}(\mathbf{U})$ is constant on $\overset{(k)}{C}$, hence the characteristic

$$\frac{\mathrm{d}x^1}{\mathrm{d}x^0} = \lambda^{(k)}(\mathbf{U})$$

is a straight line. Q.E.D.

Let S be a $\lambda^{(k)}$-simple wave region, and $\overset{(k)}{J}_1(\mathbf{U}), \ldots, \overset{(k)}{J}_{N-1}(\mathbf{U})$ be the Riemann invariants. In S one has

$$\overset{(k)}{J}_1(\mathbf{U}) = \cdots = \overset{(k)}{J}_{N-1}(\mathbf{U}) = \text{const.},$$

which defines a submanifold $\mathbb{V}$ of the $\mathbf{U}$-space $\mathscr{D} \subseteq \mathbb{R}^N$. It follows that, on $\mathbb{V}$,

$$\frac{\partial \overset{(k)}{\underset{(i)}{J}}}{\partial U^A}\mathrm{d}U^A = 0, \quad i = 1, \ldots, N-1$$

and, by (5.6),

$$\frac{\partial \overset{(k)}{\underset{i}{J}}}{\partial U^A} \overset{(k)A}{d} = 0, \quad i = 1, \ldots, N-1. \tag{5.8}$$

We can now state Friedrich's theorem.

THEOREM 5. *Let $\mathscr{C}$ be a smooth nonintersecting curve of the plane given by $\psi(x^0, x^1) = 0$ and $\Omega_\pm = ((x^0, x^1): \psi(x^0, x^1) \geq 0)$. Let $\mathbf{U}_0$ be a constant solution of the system (5.1) in Ω_+ and $\mathbf{U}_1$ be another solution which is smooth in an open subset Ω' of Ω_- whose boundary γ intersects $\mathscr{C}$ and such that it joins continuously $\mathbf{U}_0$ across $\gamma \cap \mathscr{C}$. Then $\mathbf{U}_1$ is a simple wave in an open subset of Ω'.*

The proof can be found in Jeffrey (1976) or Cabannes (1970).

In a simple wave region corresponding to the eigenvalue $\lambda^{(k)}$ the $\lambda^{(k)}$-characteristics are the straight lines

$$x^1 = \xi + \lambda^{(k)} x^0,$$

with

$$x^1(x^0 = 0) = \xi.$$

Let $\varphi(0, \xi) = f(\xi)$ be the initial value for φ. Then, because φ is constant along the $\lambda^{(k)}$-characteristics, we have

$$\begin{aligned} \varphi &= f(\xi) \\ \xi &= x^1 - \lambda^{(k)}(\mathbf{U}(f(\xi)))x^0. \end{aligned} \tag{5.9}$$

The breaking occurs when two characteristics intersect and this implies (Whitham, 1974)

$$1 + \frac{d\lambda^{(k)}}{d\xi} x^0 = 0.$$

The critical time t_B for breaking is then

$$t_B = \inf_\xi - \frac{1}{\dfrac{d\lambda^{(k)}}{d\xi}}, \tag{5.10}$$

$$\lambda^{(k)} = \lambda^{(k)}(\mathbf{U}(f(\xi))).$$

In the examples given in what follows we shall consider sinusoidal initial values for the phase, of the kind

$$f(\xi) = a \sin \frac{2\pi}{d} \xi. \tag{5.10'}$$

In the next section we shall apply the above formalism to relativistic fluid dynamics in Eulerian coordinates.

5.2. Simple waves in relativistic fluid dynamics: Eulerian treatment

The field vector **U** is

$$\mathbf{U} \equiv (U^A) = \begin{bmatrix} u^\nu \\ e \\ S \end{bmatrix}. \tag{5.11}$$

We recall the basic results of Section 2.3. The acoustic waves correspond to the roots of

$$(u^\alpha \varphi_\alpha)^2 - p'_e h^{\alpha\beta} \varphi_\alpha \varphi_\beta = 0 \tag{5.12}$$

and the corresponding right eigenvectors are

$$\mathbf{d} = \begin{bmatrix} -h^{\nu\alpha}\varphi_\alpha p'_e \\ (e+p)a \\ 0 \end{bmatrix}, \quad a = u^\alpha \varphi_\alpha. \tag{5.13}$$

The acoustic simple waves are solutions of the differential system

$$\frac{\mathrm{d}\mathbf{U}}{\mathrm{d}\varphi} = \pi \mathbf{d}. \tag{5.14}$$

Then it is immediately possible to obtain the following Riemann invariant

$$S = \text{const.} = S_0, \tag{5.15}$$

and, therefore, acoustic simple waves are isentropic.

Let us introduce Minkowski coordinates (t, x^1, x^2, x^3) in Minkowski space $\mathscr{M}$. Then, because we are in the one dimensional case, $\varphi = \varphi(t, x)$, having set $x = x^1$.

We can always take

$$\varphi = x - \lambda t. \tag{5.16}$$

Let $u^\mu = \Gamma(1, v_x, v_y, v_z)$ in Minkowski coordinates. Then $a = \Gamma(v_x - \lambda)$ and the characteristic equation (5.12) is written

$$(1 - c_s^2)\Gamma^2(v_x - \lambda)^2 - c_s^2(1 - \lambda^2) = 0, \tag{5.17}$$

where $c_s^2 = p'_e$, $c_s > 0$ being the local speed of sound (corresponding to $v_x = v_y = v_z = 0$ in the characteristic equation).

The right eigenvectors read explicitly

$$\mathbf{d} = \begin{bmatrix} -c_s^2(\lambda + \Gamma a) \\ c_s^2(1 + \Gamma v_x a) \\ -c_s^2 \Gamma v_y a \\ -c_s^2 \Gamma v_z a \\ (e+p)a \end{bmatrix}. \tag{5.18}$$

Hence, from (5.14),

$$\frac{d(\Gamma v_y)}{d\varphi} = -\pi c_s^2 a \Gamma v_y, \tag{5.19}$$

$$\frac{d(\Gamma v_z)}{d\varphi} = -\pi c_s^2 a \Gamma v_z, \tag{5.20}$$

whence we obtain the Riemann invariant

$$\frac{v_y}{v_z} = \text{constant}. \tag{5.21}$$

If in the fluid flow there is a point with vanishing speed, $v_x^* = v_y^* = v_z^* = 0$, then from (5.19)–(5.20) we obtain

$$v_y = v_z = 0 \tag{5.22}$$

throughout.

Since we can always assume the existence of a zero speed point for a wave propagating into a uniform constant state (by a suitable Lorentz transformation), which is the case of physical interest, henceforth we shall assume (5.22) to hold.

Then the characteristic equation (5.17) yields

$$\lambda = \frac{v_x(1 - c_s^2)\Gamma^2 \pm c_s}{c_s^2 + (1 - c_s^2)\Gamma^2} = \frac{v_x \pm c_s}{1 \pm v_x c_s} \equiv \lambda_\pm, \tag{5.23}$$

which is the usual relativistic addition formula for velocities. The $C_\pm$ characteristics are those lines with slopes $\lambda_\pm$, respectively.

From (5.18) it follows that

$$\frac{dv_x}{d\varphi} = -\pi \frac{c_s^2}{1 \pm v_x c_s} \frac{1}{\Gamma^3}, \tag{5.24}$$

$$\frac{de}{d\varphi} = \mp \pi \frac{(e+p)c_s}{\Gamma} \frac{1}{1 \pm v_x c_s}, \tag{5.25}$$

hence

$$\frac{dv_x}{de} = \mp \frac{c_s}{(e+p)} \frac{1}{\Gamma^2}, \tag{5.26}$$

which integrates immediately yielding

$$J_\pm \equiv \tfrac{1}{2}\ln\left(\frac{1+v_x}{1-v_x}\right) \pm \int \frac{c_s}{(e+p)} de = \text{constant}, \tag{5.27}$$

where $J_\pm$ are Riemann invariants ($J_\pm$ associated with the $C_\mp$ characteristics) (Taub, 1948).

A right-propagating simple wave has

$$J_- = \text{constant} \tag{5.28}$$

throughout space-time. It can be obtained by assigning v_x as an arbitrary function of $\varphi = x - \dfrac{v_x + c_s}{1 + v_x c_s} t$,

$$v_x = f\left(x - \frac{v_x + c_s}{1 + v_x c_s} t\right), \tag{5.29}$$

where f is an arbitrary smooth function specifying the initial wave profile. Then the relationship between e and v_x is obtained by $J_- = \text{const.}$, that is, by

$$\int \frac{c_s}{(e+p)} de = \tfrac{1}{2}\ln\left(\frac{1+v_x}{1-v_x}\right) + \text{const.}, \tag{5.30}$$

with $p = p(e, S_0)$.

Similarly, a left-propagating simple wave has

$$J_+ = \text{constant} \tag{5.31}$$

throughout space-time. The solution can be obtained as

$$v_x = f\left(x - \frac{v_x - c_s}{1 - v_x c_s} t\right) \tag{5.32}$$

and the relationship between e and v is given by $J_+ = \text{const.}$, that is,

$$\int \frac{c_s}{(e+p)} de = -\tfrac{1}{2}\ln\left(\frac{1+v_x}{1-v_x}\right) + \text{const.}, \tag{5.33}$$

with

$$p = p(e, S_0).$$

It is interesting to notice that the Riemann invariants are not Lorentz scalars. In fact, under a Lorentz transformation with x-velocity β, one has

$$v'_x = \frac{v_x + \beta}{1 + v_x\beta},$$

hence

$$J'_{\pm} = \tfrac{1}{2}\ln\left(\frac{1+v'_x}{1-v'_x}\right) \pm \int \frac{c_s}{(e+p)}\,\mathrm{d}e = \tfrac{1}{2}\ln\left(\frac{1+v_x}{1-v_x}\right)$$
$$+ \tfrac{1}{2}\ln\left(\frac{1+\beta}{1-\beta}\right) \pm \int \frac{c_s}{(e+p)}\,\mathrm{d}e,$$

that is,

$$J'_{\pm} = J_{\pm} + \tfrac{1}{2}\ln\left(\frac{1+\beta}{1-\beta}\right). \tag{5.34}$$

However, the simple wave condition $J_{\pm} = \text{const.}$ is a definition of a Lorentz invariant because $\beta = \text{const.}$

In the next section we shall rederive the Riemann invariants in the Lagrangian framework.

5.3. One dimensional relativistic fluid dynamics in Lagrangian coordinates

Let $\mathbf{u}$ be the fluid's four-velocity vector, assumed to be a smooth vector field in a domain Ω of space-time $\mathscr{M}$. The integral curves of $\mathbf{u}$ represent the world lines of the fluid particles and their differential equations are

$$\frac{\mathrm{d}x^{\mu}}{\mathrm{d}s} = u^{\mu}(x^{\alpha}) \tag{5.35}$$

in an arbitrary coordinate system (x^{α}), where s is the proper time along these world lines. The solutions of (5.35) (which will exist provided that Ω is sufficiently small) can be written as

$$x^{\mu} = x^{\mu}(\xi^i, s), \tag{5.36}$$

where ξ^i, $i = 1, 2, 3$ are three parameters such that

$$x^{\mu} = x^{\mu}(\xi^i, 0)$$

are the parametric equations of a hypersurface Σ, the initial hypersurface on which the particles are located at proper time $s = 0$.

The four coordinates (ξ^i, s) form a coordinate system in the neighborhood of the initial hypersurface Σ (these coordinates cannot be extended to cover the whole manifold Ω because singularities would develop) and they are like the Lagrangian coordinates of Newtonian continuum mechanics. In relativity they are an example of the more general comoving coordinates. Under the transformation $x'^i = \xi^i$, $x'^0 = x'^0(\xi^i, s)$ the equation of the initial hypersurface $s = 0$ becomes

$$x'^0 = x'^0(x'^i, 0).$$

The x'^μ are general comoving coordinates. In these coordinates the line element of space-time is written

$$ds^2 = g'_{\mu\nu}\, dx'^\mu\, dx'^\nu. \tag{5.37}$$

Along a world line $x'^i = \text{const.}$ we have $ds^2 = g'_{00}\,(dx'^0)^2$ and therefore the components of the four-vector $\mathbf{u}$ in these coordinates are

$$u'^\mu = \frac{1}{\sqrt{|g'_{00}|}}\,\delta^\mu_0. \tag{5.38}$$

In the following we shall consider two differential sets of coordinates in Minkowski space: the Eulerian coordinates of the inertial frame which we denote by $X^\alpha \equiv (T, X, Y, Z)$ and the comoving coordinates $x^\alpha \equiv (t, x, y, z)$. We shall assume that at the initial time $T = 0$ and $t = 0$ and the two coordinates coincide, that is,

$$\begin{aligned} T(0, x, y, z) &= 0, \\ X(0, x, y, z) &= x, \\ Y(0, x, y, z) &= y, \\ Z(0, x, y, z) &= z. \end{aligned} \tag{5.39}$$

Let $g_{\mu\nu}$ be the components of the metric tensor in the comoving coordinates. From the transformation formulas

$$g_{\mu\nu} = \frac{\partial X^\alpha}{\partial x^\mu}\frac{\partial X^\beta}{\partial x^\nu}\eta_{\alpha\beta}$$

we obtain explicitly

$$\begin{aligned} -T_t^2 + X_t^2 + Y_t^2 + Z_t^2 &= g_{00}, \\ -T_x^2 + X_x^2 + Y_x^2 + Z_x^2 &= g_{11}, \\ -T_y^2 + X_y^2 + Y_y^2 + Z_y^2 &= g_{22}, \\ -T_z^2 + X_z^2 + Y_z^2 + Z_z^2 &= g_{33}, \\ -T_tT_x + X_tX_x + Y_tY_x + Z_tZ_x &= g_{01}, \\ -T_tT_y + X_tX_y + X_tY_y + Z_tZ_y &= g_{02}, \end{aligned}$$

$$-T_tT_z+X_tX_z+Y_tY_z+Z_tZ_z=g_{03},$$
$$-T_xT_y+X_xX_y+Y_xY_y+Z_xZ_y=g_{12},$$
$$-T_xT_z+X_xX_z+Y_xY_z+Z_xZ_z=g_{13},$$
$$-T_yT_z+X_yX_z+Y_yY_z+Z_yZ_z=g_{23}. \tag{5.40}$$

Let $u^\mu=\Gamma(1,v_x,v_y,v_z)$ be the components of the fluid's four-velocity in the inertial frame, with $\Gamma=(1-v_x^2-v_y^2-v_z^2)^{-1/2}$. From the definition we have

$$v_x=\frac{\mathrm{d}X}{\mathrm{d}T}=\frac{X_t}{T_t},$$

$$v_y=\frac{\mathrm{d}Y}{\mathrm{d}T}=\frac{Y_t}{T_t},$$

$$v_z=\frac{\mathrm{d}Z}{\mathrm{d}T}=\frac{Z_t}{T_t} \tag{5.41}$$

because x, y, z are constant for a given fluid particle.

Henceforth we shall assume one dimensional flow according to the following definition: *all the physical variables and the components of the metric are functions of the comoving coordinates* (x, t).

From (5.40) we have

$$T_t=\Gamma\sqrt{|g_{00}|}=f(x,t)$$

and, by differentiating with respect to y,

$$T_{ty}=T_{yt}=0,$$

hence

$$\frac{\partial}{\partial t}T_y=0 \quad \text{with the initial condition}$$

$$T_y(0,x,y,z)=0 \quad \text{which then yields}$$
$$T_y(t,x,y,z)=0.$$

Similarly we obtain $T_z(t,x,y,z)=0$, whence

$$T=\tilde{T}(t,x). \tag{5.42}$$

By analogous reasoning we obtain

$$T=\tilde{T}(t,x),$$
$$X=\tilde{X}(t,x),$$

$$Y = \tilde{Y}(t, x) + y,$$
$$Z = \tilde{Z}(t, x) + z, \tag{5.43}$$

with $\tilde{T}(0, x) = 0$, $\tilde{X}(0, x) = x$, $\tilde{Y}(0, x) = \tilde{Z}(0, x) = 0$.

From (5.40) we then get

$$g_{22} = g_{33} = 1, \quad g_{02} = \tilde{Y}_t, \quad g_{03} = \tilde{Z}_t,$$
$$g_{12} = \tilde{Y}_x, \quad g_{13} = \tilde{Z}_x, \quad g_{23} = 0.$$

Let us write $g_{01} = \sqrt{|g_{00}|}\, h(x, t)$. Then the metric reads

$$\mathrm{d}s^2 = (\sqrt{|g_{00}|}\,\mathrm{d}t + h\,\mathrm{d}x)^2 + (g_{11} - h^2)\,\mathrm{d}x^2 + \mathrm{d}y^2 + \mathrm{d}z^2 + 2g_{02}\,\mathrm{d}y\,\mathrm{d}t$$
$$+ 2g_{03}\,\mathrm{d}z\,\mathrm{d}t + 2g_{12}\,\mathrm{d}x\,\mathrm{d}y + 2g_{13}\,\mathrm{d}x\,\mathrm{d}z. \tag{5.44}$$

By a well-known theorem on differential forms, the differential form $\sqrt{|g_{00}|}\,\mathrm{d}t + h\,\mathrm{d}x$ has an integrating factor, that is, there exists a differentiable function $\Lambda(x, t) \neq 0$, such that

$$\sqrt{|g_{00}|}\,\mathrm{d}t + h\,\mathrm{d}x = \Lambda(x, t)\,\mathrm{d}t,$$

with t' a new time variable, $t' = t'(x, t)$.

It is easy to check that the transformation $(x, y, z, t) \Leftrightarrow (x, y, z, t')$ is invertible and that the new coordinates (x, y, z, t') are still comoving.

By relabeling the coordinates and the metric coefficients, finally we obtain for the line element

$$\mathrm{d}s^2 = g_{00}\,\mathrm{d}t^2 + g_{11}\,\mathrm{d}x^2 + \mathrm{d}y^2 + \mathrm{d}z^2 + 2g_{02}\,\mathrm{d}t\,\mathrm{d}y + 2g_{03}\,\mathrm{d}t\,\mathrm{d}z$$
$$+ 2g_{12}\,\mathrm{d}x\,\mathrm{d}y + 2g_{13}\,\mathrm{d}x\,\mathrm{d}z. \tag{5.45}$$

Now we shall restrict ourselves to one dimensional flow along the x axis, that is, $v_y = v_z = 0$. It follows that $\tilde{Y}_t = \tilde{Z}_t = 0$, hence $g_{02} = g_{03} = 0$. Also, from $g_{12} = \tilde{Y}_x$, we have $\partial_t g_{12} = \partial_t \tilde{Y}_x = \partial_x \tilde{Y}_t = 0$. Then, from the initial condition $g_{12}(0, x) = 0$ (in the reference state $t = 0$ the three-metric is Euclidean) it follows that $g_{12}(t, x) = 0$.

Similarly, $g_{13}(t, x) = 0$. With these restrictions the line element becomes

$$\mathrm{d}s^2 = g_{00}\,\mathrm{d}t^2 + g_{11}\,\mathrm{d}x^2 + \mathrm{d}y^2 + \mathrm{d}z^2. \tag{5.46}$$

The transformations from Eulerian to comoving coordinates reduce to

$$T = \tilde{T}(x, t),$$
$$X = \tilde{X}(x, t),$$
$$Y = y,$$
$$Z = z. \tag{5.47}$$

Let us consider the equation of mass conservation

$$\nabla_\alpha(\rho u^\alpha) = \frac{1}{\sqrt{|g|}}\partial_\alpha(\sqrt{|g|}\,\rho u^\alpha) = 0, \tag{5.48}$$

which in comoving coordinates is written

$$\partial_t\left(\sqrt{|g|}\frac{\rho}{\sqrt{|g_{00}|}}\right) = 0,$$

whence

$$\sqrt{|g|}\frac{\rho}{\sqrt{|g_{00}|}} = f(x). \tag{5.49}$$

It follows that

$$\rho\sqrt{|g_{11}|} = (\sqrt{|g_{11}|}\rho)_{t=0} = f(x).$$

Because, at $t = 0$, the two coordinate systems coincide, $g_{11}(x, 0) = 1$ hence $f(x) = \rho_0(x)$, the initial value for the mass density. Then

$$\sqrt{|g_{11}|}\,\rho = \rho_0(x). \tag{5.50}$$

By introducing a mass coordinate μ instead of x, defined by

$$\mu = \int_0^x \rho_0(x')\,\mathrm{d}x' \quad (\rho_0 \geq 0), \tag{5.51}$$

we can write the line element in the form

$$\mathrm{d}s^2 = a^2\,\mathrm{d}t^2 + \frac{1}{\rho^2}\mathrm{d}\mu^2 + \mathrm{d}y^2 + \mathrm{d}z^2, \tag{5.52}$$

where we have set $a^2 = -g_{00}$ (for continuity $g_{00} < 0$ because at the initial time $t = 0$, $g_{00} = -1$).

The choice of the mass coordinate μ is especially useful for numerical calculations.

Then from (5.40) we obtain

$$\begin{aligned} \tilde{T}_t &= a\Gamma, \quad & \tilde{T}_\mu &= \frac{\Gamma v_x}{\rho}, \\ \tilde{X}_t &= a\Gamma v, \quad & \tilde{X}_\mu &= \frac{\Gamma}{\rho} \end{aligned} \tag{5.53}$$

together with the inverse transformation

$$\begin{aligned} t_T &= \frac{\Gamma}{a}, \quad & t_X &= -\frac{\Gamma v_x}{\rho}, \\ \mu_T &= -\rho\Gamma v_x, \quad & \mu_X &= \rho\Gamma. \end{aligned} \tag{5.54}$$

The mass conservation equation (5.48) written in inertial coordinates (X^α) is

$$\frac{\partial}{\partial T}(\rho\Gamma) + \frac{\partial}{\partial X}(\rho U) = 0, \tag{5.55}$$

where

$$U = \Gamma v_x, \quad \Gamma^2 = 1 + U^2. \tag{5.56}$$

After some manipulations, using (5.53), equation (5.55) yields

$$\rho_t = -a\rho^2 \frac{U_\mu}{\Gamma}. \tag{5.57}$$

Because the coordinate transformations (5.47) are assumed to be sufficiently smooth (at least of class C^2), we have

$$X_{\mu t} = X_{t\mu},$$

and from equation (5.53) we obtain

$$U_t = \rho\Gamma a_\mu. \tag{5.58}$$

The energy-momentum conservation equations

$$\nabla_\alpha T^{\alpha\beta} = 0 \tag{5.59}$$

are explicitly written as

$$\frac{1}{\sqrt{|g|}} \frac{\partial(\sqrt{|g|}\, T^\beta_\alpha)}{\partial x^\beta} - \frac{1}{2}\frac{\partial g_{\beta\mu}}{\partial x^\alpha} T^{\beta\mu} = 0. \tag{5.60}$$

In Lagrangian coordinates, for $\beta = 0$ we obtain

$$\varepsilon_t + p\left(\frac{1}{\rho}\right)_t = 0 \tag{5.61}$$

and for $\beta = 1$,

$$a_\mu = -\frac{a p_\mu}{\rho f}. \tag{5.62}$$

It is easily checked that the Riemann tensor of the metric (5.52) reduces to one component, R_{1010} (apart from those obtained from it by symmetry), which vanishes as a consequence of equation (5.57) and (5.58).

5.4. Simple waves in Lagrangian coordinates

The field equations can be taken to be, by suitably arranging equations (5.57)–(5.62),

$$U_t + \frac{a\Gamma p_\mu}{f} = 0, \tag{5.63}$$

$$\rho_t + a\rho^2 \frac{U_\mu}{\Gamma} = 0, \tag{5.64}$$

$$\varepsilon_t + p\left(\frac{1}{\rho}\right)_t = 0. \tag{5.65}$$

Equation (5.61) is not an evolution equation and decouples from the others.

By assuming an equation of state

$$p = p(\rho, \varepsilon),$$

equations (5.63)–(5.65) are, in matrix form,

$$\mathscr{A}^0 \mathbf{Y}_t + \mathscr{A}^1 Y_\mu = 0,$$

where

$$Y = \begin{pmatrix} v \\ \rho \\ \varepsilon \end{pmatrix}, \quad \text{and} \tag{5.66}$$

$$\mathscr{A}^0 = \begin{bmatrix} 1 & 0 & 0 \\ 0 & 1 & 0 \\ 0 & -\dfrac{p}{\rho^2} & 1 \end{bmatrix}, \mathscr{A}^1 = \begin{bmatrix} 0 & \dfrac{a\Gamma p_\rho}{f} & \dfrac{a\Gamma p_\varepsilon}{f} \\ \dfrac{a\rho^2}{\Gamma} & 0 & 0 \\ 0 & 0 & 0 \end{bmatrix}. \tag{5.67}$$

The system can also be written in normal form

$$\mathbf{Y}_t + \mathscr{A}\mathbf{Y}_\mu = 0, \tag{5.68}$$

where

$$\mathscr{A} = \begin{bmatrix} 0 & \dfrac{a\Gamma p_\rho}{f} & \dfrac{a\Gamma p_\varepsilon}{f} \\ \dfrac{a\rho^2}{\Gamma} & 0 & 0 \\ \dfrac{a\rho}{\Gamma} & 0 & 0 \end{bmatrix}. \tag{5.69}$$

We look for solutions in the form of simple waves $\mathbf{Y} = \mathbf{Y}(\phi)$. Then (5.68)–(5.69) becomes

$$(\mathscr{A} - \lambda I)\mathbf{Y}' = 0, \tag{5.70}$$

where $\lambda = -\frac{\phi_t}{\phi_\mu}$. The eigenvalues are

$$\lambda_1 = 0, \quad \lambda_{2,3} = \pm a\sqrt{\frac{p_\rho \rho^2 + p_\varepsilon \rho}{f}} \tag{5.71}$$

and the corresponding right eigenvectors are, with α a factor,

$$\mathbf{d}_1 = \alpha \begin{pmatrix} 0 \\ -p_\varepsilon \\ p_\rho \end{pmatrix}, \quad \mathbf{d}_{2,3} = \alpha \begin{bmatrix} 1 \\ \pm\frac{\rho^2}{\Gamma}\sqrt{\frac{f}{p_\rho \rho^2 + p_\varepsilon \rho}} \\ \pm\frac{\rho}{\Gamma}\sqrt{\frac{f}{p_\rho \rho^2 + p_\varepsilon \rho}} \end{bmatrix}. \tag{5.72}$$

The case $\lambda = 0$ represents the material waves (where p and U are constant). The cases $\lambda \neq 0$ correspond to the acoustic waves. The simple waves are the solutions of the system

$$\frac{\mathrm{d}U}{\mathrm{d}\phi} = \alpha, \tag{5.73}$$

$$\frac{\mathrm{d}\rho}{\mathrm{d}\phi} = \pm\frac{\alpha\rho^2}{\Gamma}\left(\frac{f}{p_\rho \rho^2 + p_\varepsilon \rho}\right)^{1/2}, \tag{5.74}$$

$$\frac{\mathrm{d}\varepsilon}{\mathrm{d}\phi} = \pm\frac{\alpha\rho}{\Gamma}\left(\frac{f}{p_\rho \rho^2 + p_\varepsilon \rho}\right)^{1/2}. \tag{5.75}$$

From (5.74)–(5.75) it follows that

$$\frac{\mathrm{d}\varepsilon}{\mathrm{d}\phi} + p\frac{\mathrm{d}\left(\frac{1}{\rho}\right)}{\mathrm{d}\phi} = 0,$$

hence, from the first law of thermodynamics,

$$\frac{\mathrm{d}S}{\mathrm{d}\phi} = 0 \tag{5.76}$$

and, therefore, simple waves are isentropic. Then one can write $p = p(e)$, and it follows that

$$p_\rho = c_s^2(1 + \varepsilon),$$
$$p_\varepsilon = c_s^2 \rho,$$

with $c_s^2 = p'_e$. Substituting back into (5.76)–(5.77) yields

$$\frac{\mathrm{d}\rho}{\mathrm{d}\phi} = \pm \frac{\alpha\rho}{\Gamma c_s} \tag{5.77}$$

and

$$\frac{\mathrm{d}\varepsilon}{\mathrm{d}\phi} = \pm \frac{\alpha p}{\Gamma c_s \rho}. \tag{5.78}$$

Then, from (5.73)–(5.78) one obtains

$$\frac{\mathrm{d}U}{\mathrm{d}\rho} = \pm \frac{\Gamma c_s}{\rho}, \tag{5.79}$$

which, upon integrating, gives

$$\ln(\Gamma + U) \pm \int \frac{c_s \,\mathrm{d}\rho}{\rho} = J_\pm = \text{const.} \tag{5.80}$$

These $J_\pm$ are easily seen to coincide with the Riemann invariants introduced in Section 2, equation (5.27) (Lanza et al., 1985).

5.5. Evaluation of the Riemann invariants for relativistic acoustic simple waves

We shall calculate the Riemann invariants for the various state equations discussed in Section 2.2.

In the case of a barotropic fluid with state equation

$$p = (\gamma - 1)e, \tag{5.81}$$

with $1 \leq \gamma \leq 2$, we have for the speed of sound

$$c_s^2 = \gamma - 1. \tag{5.82}$$

Then

$$J_\pm = \tfrac{1}{2}\ln\left(\frac{1+v_x}{1-v_x}\right) \pm \frac{\sqrt{\gamma-1}}{\gamma}\ln e$$

and, therefore, right- and left-propagating simple waves are characterized, respectively, by

$$e = e_0\left(\frac{1+v_x}{1-v_x}\right)^{\pm\gamma/(2\sqrt{\gamma-1})}, \tag{5.83}$$

where $e = e_0$ at $v_x = 0$ (Fig. 5.1).

Notice that, for right (left)-propagating simple waves, $e \to \infty$ as $v_x \to 1$ ($v_x \to -1$).

The solutions we have found provide an interesting way to understand the nonlinear development of acoustic waves in the early universe. The content of the universe consists of three components: ordinary matter (baryon component), radiation (photon component), and possible dark matter (whose precise form is still under debate). Considering only the matter-radiation fluid, it is necessary to distinguish two kinds of perturbations.

Under the first kind, both the radiation and matter can be perturbed together in such a way that the ratio of the photon number density to the baryon density within the perturbation is the same as in the unperturbed medium. These perturbations are referred to as the *adiabatic modes* because in this case the entropy per baryon remains constant. For the second kind, the matter distribution can be perturbed, leaving the photon density unchanged. The temperature within such a perturbation is the same as in the unperturbed medium, and therefore these perturbations are referred to as *isothermal modes* or entropy fluctuations.

Prior to recombination the speed of sound in the matter-radiation mixture is essentially that appropriate for a photon gas, that is, $c_s \cong c/\sqrt{3}$, and the corresponding Jeans length is only slightly smaller than the cosmic horizon distance (Peebles, 1980). After recombination the speed of sound is that appropriate for a cool hydrogen gas and, therefore, the Jeans

Fig. 5.1. Progressive simple waves in a barotropic fluid with $\gamma = 4/3$ (radiation gas).

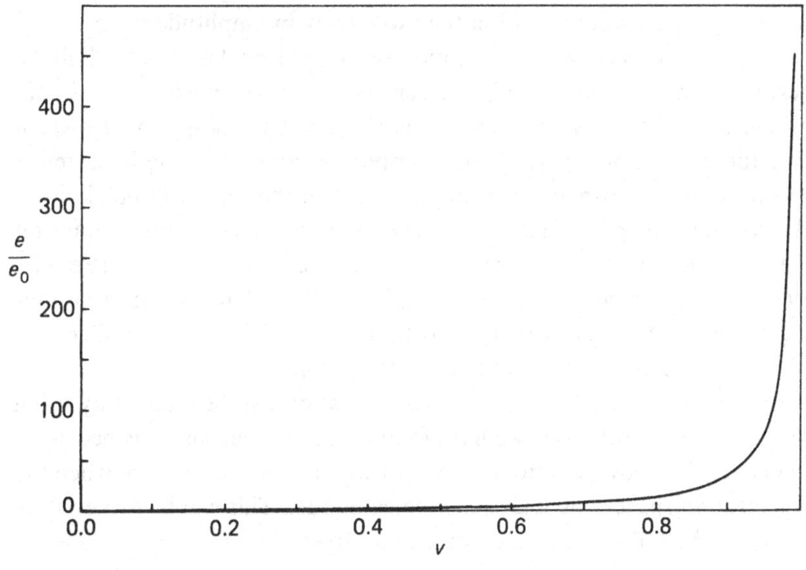

length scale drops by several orders of magnitude from its value before recombination.

The evolution of an adiabatic recombination, in the linear regime, can be depicted as follows. Let λ be the wavelength of the perturbation. We start following the evolution of the perturbation from the time when it first enters the horizon size, $\lambda \ll \lambda_H$ (otherwise, in general relativity, ambiguities may arise; Peebles, 1980). The perturbation will grow in amplitude as long as $\lambda > \lambda_J$. Adiabatic perturbations with wavelengths smaller than the Jeans length immediately prior to recombination will start to oscillate when $\lambda < \lambda_J$ and will continue to do so until t_{rec}. If they survive after recombination they will begin to grow because they would once again satisfy the condition $\lambda > \lambda_J$.

Adiabatic density fluctuations on sufficiently small scales are damped by the action of viscosity and heat conduction. If the damping time scale is shorter than the cosmic expansion time scale, the wave will be effectively damped before the universe has time to expand by an appreciable factor. This process selects a preferred damping scale λ_D as the shortest wavelength whose damping time scale equals the cosmic expansion time scale. The exact value of λ_D is determined by Thomson scattering of radiation off free electrons (Peebles, 1971). Only adiabatic fluctuations with $\lambda > \lambda_D$ can therefore survive after recombination.

The linear evolution of isothermal fluctuations is much simpler. During the prerecombination period the amplitude of isothermal waves remains constant since their growth is inhibited by the drag due to Thomson scattering. After recombination they will grow in amplitude if $\lambda > \lambda_J$.

Nonlinear effects could be important in the evolution of adiabatic perturbations. A weakly nonlinear density perturbation with a sinusoidal profile will tend to break and form shocks. If the breaking time t_b (which is a function of the perturbation amplitude and wavelength) is much shorter than the cosmic expansion time t, then the wave will break, form shocks, and dissipate. This nonlinear process imposes an upper limit on the amplitude of adiabatic perturbations at the equipartition time (Peebles, 1980). Precise calculations using relativistic fluid dynamics in a cosmological model have been performed by Liang (1977b), Anile, Miller, and Motta (1983), and by Carioli and Motta (1984).

In some cases the process of shock formation can be studied by using the simple wave solutions we have found. For this purpose it is necessary to extend the above solutions from special relativity to the case when the background space-time is a cosmological model. This can be achieved by applying the following result (Anile and Greco, 1978).

PROPOSITION 5.2. *Let $T^{\mu\nu}$ be a traceless symmetric tensor satisfying*

$$\nabla_\alpha T^{\alpha\beta} = 0$$

with respect to the conformally flat metric

$$g_{\alpha\beta} = \exp(\psi)\eta_{\alpha\beta}$$

(where $\eta_{\alpha\beta}$ denotes the Minkowski metric).

If one defines

$$\bar{T}^{\alpha\beta} = \exp(3\psi) T^{\alpha\beta}$$

then $T^{\alpha\beta}$ satisfies

$$\partial_\alpha \bar{T}^{\alpha\beta} = 0.$$

Proof. By direct computation, one has for the Christoffel symbols of $g_{\alpha\beta}$,

$$\Gamma^{\beta}_{\alpha\mu} = \tfrac{1}{2}(\psi_{,\mu}\delta^{\beta}_{\alpha} + \psi_{,\alpha}\delta^{\beta}_{\mu} - \eta_{\alpha\mu}\eta^{\beta\nu}\psi_{,\nu}),$$

hence

$$\nabla_\alpha T^{\alpha\beta} = \partial_\alpha T^{\alpha\beta} + 3\psi_{,\alpha} T^{\alpha\beta}. \qquad \text{Q.E.D.}$$

This proposition can be applied to a radiation fluid in a spatially flat Robertson–Walker Universe with the metric

$$\mathrm{d}s^2 = -\mathrm{d}t^2 + R^2(t)\delta_{ij}\,\mathrm{d}x^i\,\mathrm{d}x^i, \tag{5.84}$$

where t is the cosmic time and $R(t)$ is the universe expansion factor. By introducing the conformal time τ defined by

$$\frac{\mathrm{d}\tau}{\mathrm{d}t} = \frac{1}{R(t)}, \tag{5.85}$$

the above metric can be written in the conformally flat form

$$\mathrm{d}s^2 = \exp(\psi)(-d\tau^2 + \delta_{ij}\,\mathrm{d}x^i\,\mathrm{d}x^j),$$

with

$$\exp\left(\frac{\psi}{2}\right) = R.$$

For a test-radiation fluid with energy-momentum tensor

$$T^{\alpha\beta} = \tfrac{4}{3} e u^\alpha u^\beta + \frac{e}{3} g^{\alpha\beta}$$

the transformation $\bar{T}^{\alpha\beta} = \exp(3\psi)T^{\alpha\beta}$ amounts to

$$\bar{e} = R^4 e, \quad \bar{u}^\alpha = R u^\alpha. \tag{5.86}$$

Let

$$\bar{u}^\alpha = \bar{\Gamma}(1, \bar{v}, 0, 0), \quad \bar{\Gamma} = (1 - \bar{v}^2)^{-1/2}.$$

Then the simple wave solution (5.83) with $\gamma = (4/3)$ corresponds to

$$\bar{e} = \bar{e}_0 \left(\frac{1 + \bar{v}}{1 - \bar{v}} \right)^{\pm(2/\sqrt{3})}, \quad \bar{v} = f\left(x - \frac{\bar{v} \pm 1/\sqrt{3}}{1 \pm \bar{v}/\sqrt{3}} \tau \right). \tag{5.87}$$

We assume that the test fluid described by $T^{\alpha\beta}$ moves only in the x direction. The peculiar velocity v of the test fluid with respect to the fundamental observers of the Robertson–Walker Universes [which in the coordinates (τ, x^i) have four-velocity $\overset{(0)}{u}{}^\alpha = (R, 0, 0, 0)$] is defined by

$$\Gamma = -u_\alpha \overset{(0)}{u}{}^\alpha, \quad \Gamma = (1 - v^2)^{-1/2}.$$

Hence

$$u^\alpha = \frac{\Gamma}{R}(1, v, 0, 0)$$

and it follows that $v = \bar{v}$.

Therefore, the above solutions can be written, in terms of quantities physically defined in the Robertson–Walker Universe, as

$$e = e_0 \left(\frac{1 + v}{1 - v} \right)^{\pm(2\sqrt{3}/3)}, \quad v = f\left(x - \frac{v + 1/\sqrt{3}}{1 + v/\sqrt{3}} \tau \right), \tag{5.88}$$

where $e_0 = \dfrac{\bar{e}_0}{R^4}$ ($\bar{e}_0 = \text{constant}$) is the energy density of the background cosmological fluid.

These equations are the same as in special relativity. The only difference is that τ is the conformal time and x is a comoving coordinate. These solutions had already been found by Liang (1977a) by an ad hoc procedure, not as a special case of the general result shown above.

Analogous solutions may be found for test-radiation fluid dynamics in a nonspatially flat Robertson–Walker model.

These solutions would necessarily be more complicated because the transformations needed in order to put the nonspatially flat Robertson–Walker Universes into a conformally flat form are quite involved (Hawking and Ellis, 1973).

In the case of a *polytropic fluid* with state equation

$$p = k(S)\rho^\gamma, \tag{5.89}$$

we have for the local speed of sound

$$c_s^2 = \frac{\gamma p}{p+e}. \tag{5.90}$$

Then from the adiabaticity condition $u^\alpha \partial_\alpha S = 0$ and the first law of thermodynamics it follows that, along the flow lines (by a suitable choice of the additive constant in the internal energy)

$$p = (\gamma - 1)\rho\varepsilon, \tag{5.91}$$

whence

$$c_s^2 = \frac{\gamma p(\gamma - 1)}{(\gamma - 1)\rho + \gamma p}. \tag{5.92}$$

From this we see that

$$c_s^2 \leq \gamma - 1.$$

A straightforward calculation shows that

$$\int \frac{c_s^2}{(e+p)} \mathrm{d}e = 2\int \frac{\mathrm{d}c_s}{(\gamma-1) - c_s^2} = \frac{1}{\sqrt{\gamma - 1}} \ln \left(\frac{1 + \dfrac{c_s}{\sqrt{\gamma-1}}}{1 - \dfrac{c_s}{\sqrt{\gamma-1}}} \right),$$

Fig. 5.2. Progressive simple waves in a polytropic fluid with $\gamma = 4/3$; upper curve, $\bar{c}_s = 0.3$, lower curve $\bar{c}_s = 0.01$.

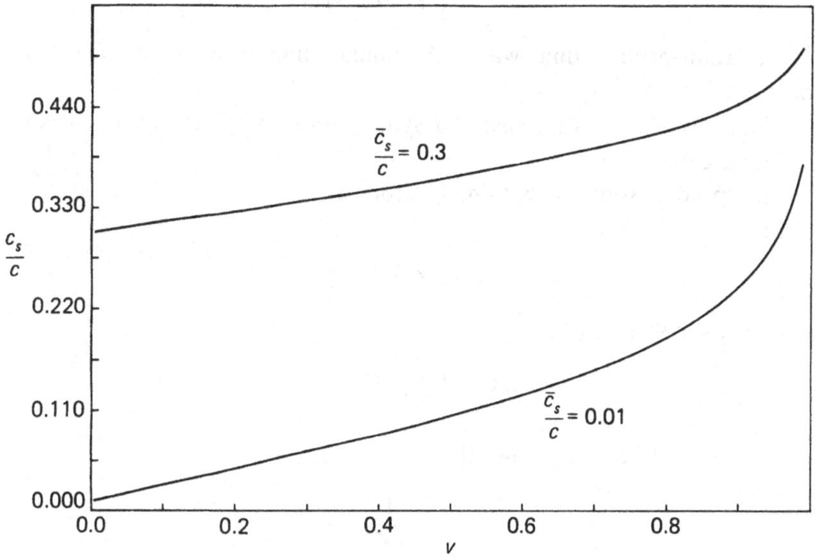

hence

$$J_{\pm} = \tfrac{1}{2}\ln\left(\frac{1+v_x}{1-v_x}\right) \pm \frac{1}{\sqrt{\gamma-1}}\ln\left(\frac{1+\dfrac{c_s}{\sqrt{\gamma-1}}}{1-\dfrac{c_s}{\sqrt{\gamma-1}}}\right).$$

Therefore, right(left)-propagating simple waves are characterized by

$$c_s = (\gamma-1)^{1/2}\frac{\dfrac{(\gamma-1)^{1/2}+\bar{c}_s}{(\gamma-1)^{1/2}-\bar{c}_s}\left(\dfrac{1+v_x}{1-v_x}\right)^{\pm\frac{\sqrt{\gamma-1}}{2}}-1}{\dfrac{(\gamma-1)^{1/2}+\bar{c}_s}{(\gamma-1)^{1/2}-\bar{c}_s}\left(\dfrac{1+v_x}{1-v_x}\right)^{\pm\frac{\sqrt{\gamma-1}}{2}}+1}, \tag{5.93}$$

where $\bar{c}_s = c_s(v_x = 0)$. Notice that $c_s \to \sqrt{\gamma-1}$ as $v_x \to \pm 1$ for right- and left-propagating simple waves, respectively (Fig. 5.2).

We remark that, like the situation for acoustic simple waves in Newtonian fluid dynamics, from (5.93) and the condition $c_s \geq 0$, we can draw the following limitation on the fluid's velocity in the direction opposite that of the wave propagation.

$$-v_x \leq \frac{1-\left(\dfrac{\sqrt{\gamma-1}-\bar{c}_s}{\sqrt{\gamma-1}+\bar{c}_s}\right)^{2/\sqrt{\gamma-1}}}{1+\left(\dfrac{\sqrt{\gamma-1}-\bar{c}_s}{\sqrt{\gamma+1}+\bar{c}_s}\right)^{2/\sqrt{\gamma-1}}} \tag{5.94}$$

for a right-propagating wave. A similar limitation holds for left-propagating waves.

Finally, we consider the case of a Synge gas given by the state equations (2.22)–(2.23).

The speed of sound is obtained as follows.

From

$$e = p[zG(z) - 1],$$

with $z = z(p, S)$, one has

$$e'_p = zG - 1 + p[G + zG']z'_p.$$

Now, from (2.23), it follows that

$$z'_p = \frac{1}{pzG'},$$

hence

$$e'_p = \frac{zG}{G'}\left(G + \frac{1}{z^2}\right).$$

Finally, one obtains

$$c_s^2 = \frac{G'/G}{z\left(G' + \frac{1}{z^2}\right)},$$

where $z = \dfrac{m}{k_B T}$ and $G(z) = \dfrac{K_3(z)}{K_2(z)}$ (Fig. 5.3).

Because simple waves are isentropic, we obtain

$$\frac{d\rho}{\rho} = z\left(G' + \frac{1}{z^2}\right)\mathrm{d}z$$

and, therefore, the Riemann invariants are

$$J_\pm = \tfrac{1}{2}\ln\left(\frac{1+v_x}{1-v_x}\right) \pm \int\left(\frac{z\left(G' + \frac{1}{z^2}\right)G'}{G}\right)^{1/2}\mathrm{d}z.$$

Fig. 5.3. Speed of sound in a relativistic gas (Synge's state equation) as a function of the parameter $z = m/k_B T$ in the region $0.1 < z < 1$.

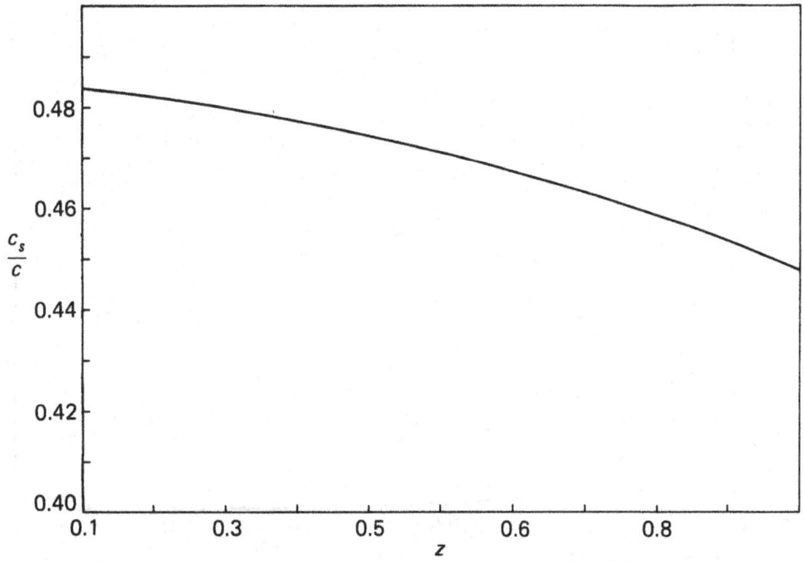

In this case it is not possible to analytically invert the condition $J_{\pm} =$ const. and therefore one must turn to numerical integration (Lanza et al., 1985). A simple wave is determined by first specifying an appropriate profile for the three-velocity v_x and then obtaining the corresponding values of z by inverting $J_{\pm} =$ const. From the numerical viewpoint it is more convenient to solve equation (5.26) for v_x as a function of z (Fig. 5.4).

The critical time for breaking can be obtained from (5.10) for an initially sinusoidal profile (5.10′)

$$v = v_0 \sin \frac{2\pi}{d} x.$$

For a barotropic fluid one obtains (Muscato, 1988a)

$$t_B = \frac{d}{2\pi} \frac{\sqrt{\gamma - 1}}{\{2[1 + 8v_0^2(\gamma - 1)]^{1/2} - 4v_0^2(\gamma - 1) - 2\}^{1/2}} \frac{\{3 - [1 + 8v_0^2(\gamma - 1)]^{1/2}\}^2}{2 - \gamma}.$$

Notice that $t_B \to \infty$ as $\gamma \to 2$ (Fig. 5.5). For a polytropic gas one must evaluate t_B numerically and one obtains (Muscato, 1988a) the results given in Figs. 5.6–5.7.

The next section will be devoted to the relativistic theory of isentropic

Fig. 5.4. Progressive simple waves in a Synge gas (v_x as a function of the parameter $z = m/k_B T$ in the range $0.1 < z < 1$).

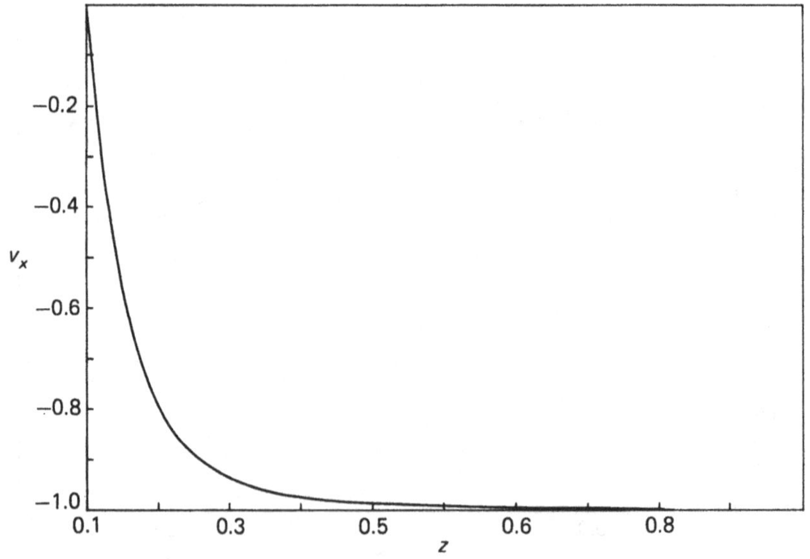

Fig. 5.5. Critical time in a barotropic fluid. Plot of $(t) = 2\pi t_B c_{s0}/d$ versus $(v) = v_0/c_{s0}$ for several values of γ. The lower curve corresponds to $\gamma = 4/3$, the upper curve to $\gamma = 2$, and the intermediate one to $\gamma = 1.9$.

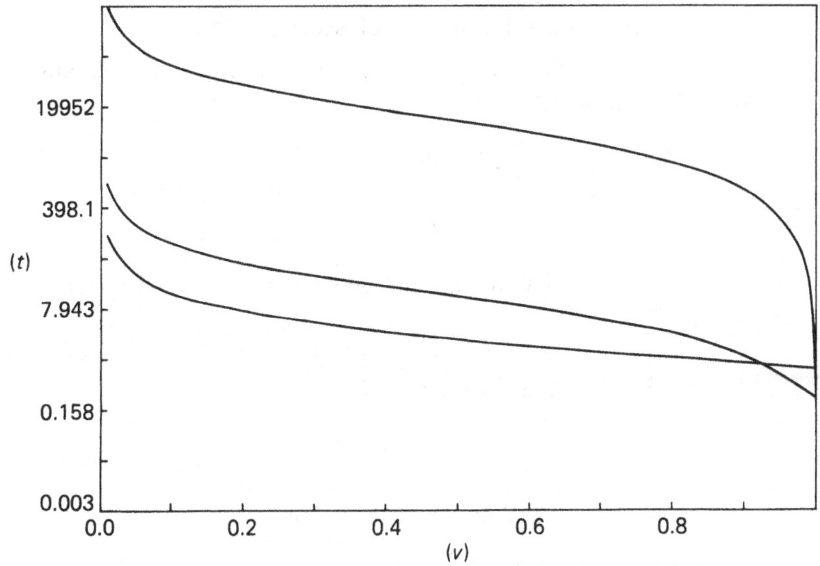

Fig. 5.6. Critical time for a polytropic fluid with $\gamma = 5/3$ in both classical and relativistic fluid dynamics. Quantities defined as in Fig. 5.5. The lower curve corresponds to the nonrelativistic case, for which

$$t_B = \frac{d}{\pi v_0(\gamma + 1)}$$

(Jeffrey, 1976), the intermediate one corresponds to $c_0 = 0.4$, and the upper one corresponds to $c_0 = 0.8$.

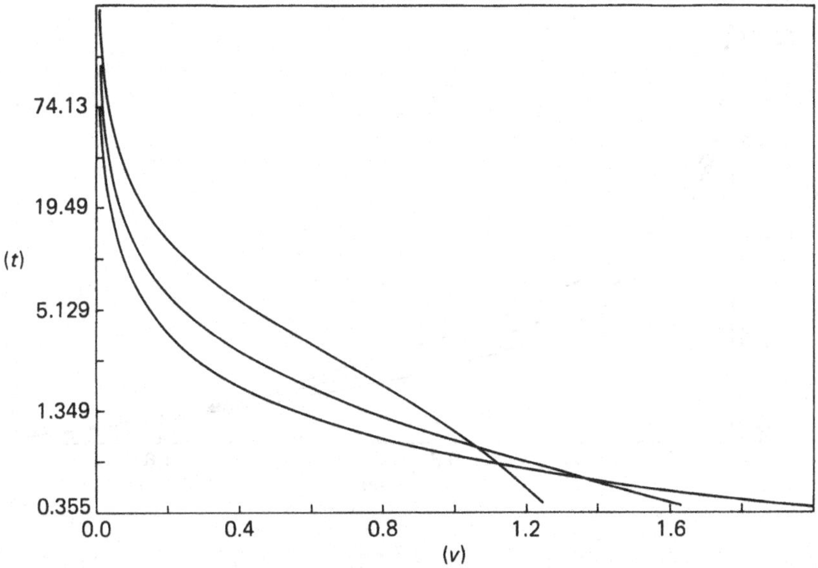

flow, a preliminary step for developing the relativistic theory of steady supersonic flow.

5.6. General properties of isentropic flow

In the case of isentropic flow, $S = \text{const.}$, the momentum conservation equation (2.10) can be written as

$$u^\alpha \nabla_\alpha u^\mu + h^{\alpha\mu} \nabla_\alpha \ln f = 0 \tag{5.95}$$

since

$$\mathrm{d}f = T\,\mathrm{d}S + \frac{\mathrm{d}p}{\rho}.$$

Equation (5.95) admits an elegant geometric interpretation (Lichnerowicz, 1955).

PROPOSITION 5.3. *The flow lines defined by*

$$\frac{\mathrm{d}x^\mu}{\mathrm{d}s} = u^\mu$$

$$u^\alpha \nabla_\alpha u^\mu + h^{\alpha\mu} \nabla_\alpha \ln f = 0$$

are the geodesics of the metric $\bar{g}_{\mu\nu} = f^2 g_{\mu\nu}$.

Fig. 5.7. Critical time for a polytropic fluid with $\gamma = 4/3$ in relativistic fluid dynamics. The lower curve corresponds to $c_0 = 0.4$; the intermediate one to $c_0 = 0.8$. The upper curve corresponds to a barotropic fluid with $\gamma = 4/3$. The quantities plotted are defined as in Fig. 5.5.

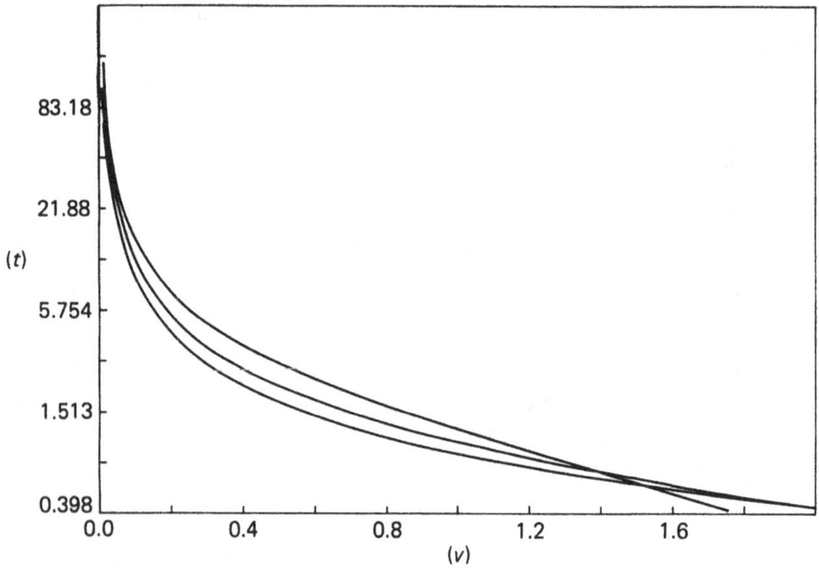

Furthermore equation (5.95) is equivalent to

$$c^\alpha \bar{\nabla}_\alpha c_\beta = 0, \tag{5.96}$$

where $c_\alpha = f u_\alpha$ and $\bar{\nabla}$ is the Riemannian connection of the metric $\bar{g}_{\mu\nu}$.

Proof. The Christoffel symbols of the metric $\bar{g}_{\mu\nu} = f^2 g_{\mu\nu}$ are

$$\bar{\Gamma}^\alpha_{\beta\gamma} = \Gamma^\alpha_{\beta\gamma} + \frac{1}{f}(\delta^\alpha_\beta f_{,\gamma} + \delta^\alpha_\gamma f_{,\beta} - g_{\beta\gamma} g^{\alpha\mu} f_{,\mu}),$$

where $\Gamma^\alpha_{\beta\gamma}$ are the Christoffel symbols of $g_{\mu\nu}$.

Therefore,

$$u^\alpha \bar{\nabla}_\alpha u_\beta = -k u_\beta, \quad k = u^\alpha (\ln f)_{,\alpha}$$

and the flow lines are geodesics.

Furthermore,

$$\begin{aligned} c^\alpha \bar{\nabla}_\alpha c_\beta &= f u^\alpha \{\bar{\nabla}_\alpha (f u_\beta)\} \\ &= f u^\alpha \{f \bar{\nabla}_\alpha u_\beta + u_\beta \bar{\nabla}_\alpha f\} \\ &= f^2 u^\alpha \bar{\nabla}_\alpha u_\beta + f u^\alpha u_\beta f_{,\alpha} \\ &= -f^2 \frac{u^\alpha}{f} f_{,\alpha} u_\beta + f u^\alpha u_\beta f_{,\alpha} = 0. \end{aligned}$$

Q.E.D.

The relativistic vorticity tensor $\Omega_{\alpha\beta}$ is defined by

$$\Omega_{\alpha\beta} = \bar{\nabla}_\alpha c_\beta - \bar{\nabla}_\beta c_\alpha. \tag{5.97}$$

Let ω be the 1-form

$$\omega = c_\alpha \mathrm{d}x^\alpha.$$

Then $\Omega_{\alpha\beta}$ can also be written as

$$\Omega = \mathrm{d}\omega.$$

In the Newtonian limit, $\Omega_{\alpha\beta}$ reduces to the usual vorticity tensor. In fact, $f = 1 + \varepsilon + p/\rho$ and in the Newtonian limit we can set $f = 1$.

Since $\bar{g}_{\mu\nu} = f^2 g_{\mu\nu}$, one has

$$\bar{g}^{\mu\nu} = f^{-2} g^{\mu\nu}$$

and therefore

$$\bar{g}^{\alpha\beta} c_\alpha c_\beta = -1. \tag{5.98}$$

It follows that

$$c^\alpha \bar{\nabla}_\beta c_\alpha = 0. \tag{5.99}$$

Then, from (5.96), one has

$$c^\alpha \Omega_{\alpha\beta} = 0. \tag{5.100}$$

It can be proved that $\Omega_{\alpha\beta}$ satisfies the relativistic analog of the Helmoltz theorem (Lichnerowicz, 1955, 1967), which in local coordinates is written,

$$c^\mu \nabla_\mu \Omega_{\alpha\beta} + (\nabla_\alpha c^\mu)\Omega_{\mu\beta} + (\nabla_\beta c^\mu)\Omega_{\alpha\mu} = 0.$$

The flow is said to be irrotational if $\Omega = 0$. In this case, in a simply connected domain W of $\mathscr{M}$, there exists a function $\sigma \in \mathscr{D}(W)$, the potential function, such that

$$c_\alpha = \frac{\partial \sigma}{\partial x^\alpha}.$$

In this case equation (5.98) can be written

$$\frac{\partial \sigma}{\partial x^\alpha}\frac{\partial \sigma}{\partial x^\beta} g^{\alpha\beta} + f^2 = 0. \tag{5.101}$$

Furthermore, the continuity equation

$$\nabla_\alpha(\rho u^\alpha) = 0$$

gives

$$\nabla^\alpha\left(\frac{f}{\rho}\sigma_{,\alpha}\right) = 0,$$

with f given by (5.101) and ρ as a function of f through the state equation.

In the next section we shall consider a particular example of isentropic and irrotational flow: one dimensional unsteady isentropic flow.

5.7. Unsteady one dimensional and isentropic flow in special relativity

For one dimensional motion we have, in Minkowski coordinates (t, x, y, z),

$$u^\mu = \Gamma(1, v, 0, 0),$$

and the motion occurs on the x-axis.

It is easy to see that $\Omega_{\alpha\beta} = 0$. In fact, $\Omega_{02} = \Omega_{03} = \Omega_{23} = \Omega_{12} = \Omega_{13} = 0$. Furthermore, from (5.100) it follows that

$$u^0 \Omega_{0\beta} + u^1 \Omega_{1\beta} = 0,$$

whence $\Omega_{01} = 0$. Therefore the flow is irrotational.

Equation (5.101) then can be written

$$f^2 - \sigma_t^2 + \sigma_x^2 = 0, \tag{5.102}$$

where

$$\sigma_t = -f\Gamma, \quad \sigma_x = f\Gamma v.$$

Now we apply the hodograph method (Liang, 1977a).

We perform a Legendre transformation by introducing the function χ

$$\chi = \sigma - f x v \Gamma + f t \Gamma. \tag{5.103}$$

We have

$$\mathrm{d}\chi = \Gamma(t - xv)\,\mathrm{d}f + \Gamma^3 f(tv - x)\,\mathrm{d}v.$$

By introducing the new variable

$$\tau = \tfrac{1}{2}\ln\left(\frac{1+v}{1-v}\right), \tag{5.104}$$

$\mathrm{d}\tau = \Gamma^2\,\mathrm{d}v$, $v = \tanh\tau$, $\Gamma = \cosh\tau$, hence

$$\mathrm{d}\chi = (t\cosh\tau - x\sinh\tau)\,\mathrm{d}f + f(t\sinh\tau - x\cosh\tau)\,\mathrm{d}\tau.$$

Therefore

$$t = \frac{1}{f}(\chi_f f\cosh\tau - \chi_\tau\sinh\tau),$$

$$x = \frac{1}{f}(\chi_f\sinh\tau - \chi_\tau\cosh\tau).$$

The mass conservation equation is

$$\partial_t(\rho\Gamma) + \partial_x(\rho\Gamma v) = 0. \tag{5.105}$$

By expressing it in terms of the variables (f, τ) we finally obtain

$$c_s^2 f^2 \chi_{ff} + f\chi_f - \chi_{\tau\tau} = 0, \tag{5.106}$$

which is a linear equation.

Sometimes it is more convenient to use the variable $\nu = \ln f$. Then equation (5.106) reads

$$c_s^2\chi_{\nu\nu} + (1 - c_s^2)\chi_\nu - \chi_{\tau\tau} = 0. \tag{5.107}$$

Note that for a barotropic fluid c_s = constant and the latter equation is very simple to solve (Liang, 1977a).

In this way the problem of determining the isentropic one dimensional flow corresponding to given initial and boundary conditions is reduced to an initial-boundary value problem for a linear equation. However, the applicability of the method is limited by the fact that expressing the physical initial-boundary conditions in terms of the variables ν, τ, and χ is not an easy task.

In the next section we shall treat another example of isentropic irrotational flow of great physical interest, the case of stationary potential flow in two dimensions.

5.8. Stationary potential two dimensional flow in special relativity

A flow is said to be *potential* if it is isentropic, $S =$ constant, and irrotational, $\Omega_{\alpha\beta} = 0$.

In special relativity a flow is stationary if, in an inertial system (x^i, t), all the physical quantities are independent of time t.

Let $u^\mu = \Gamma(1, v_1, v_2, v_3)$ be the four-velocity in this coordinate system. Then for a steady potential flow the requirement $\Omega_{\alpha\beta} = 0$ yields

$$f\Gamma = f_0 = \text{const.}, \tag{5.108}$$

with f_0 the value of f at the stagnation point ($\mathbf{v} = 0$) and

$$\partial_j v_i - \partial_i v_j = 0, \tag{5.109}$$

which states that the vector field $\mathbf{v} \equiv (v_1, v_2, v_3)$ is irrotational in $\mathbb{R}^3$. Equation (5.108) represents the relativistic Bernoulli theorem. In the Newtonian limit it is easily seen to reduce to

$$\tfrac{1}{2}v^2 + \varepsilon + \frac{p}{\rho} = \text{const.}$$

For stationary flow equation (5.95) reads

$$\Gamma v_i \partial_i(\Gamma v_j) + \partial_j \ln f + \Gamma^2 v_i v_j \partial_i \ln f = 0.$$

By contracting with v_j, using the thermodynamic relationship

$$\mathrm{d}f = \frac{f c_s^2}{\rho}\,\mathrm{d}\rho,$$

which holds for isentropic flow and the mass conservation equation, yields

$$\frac{1}{c_s^2} v_i(v_j \partial_j(\Gamma v_i)) = \partial_i(\Gamma v_i). \tag{5.110}$$

In the case of two dimensional flow, $v_3 = 0$, $v_1 = v_x$, $v_2 = v_y$, equations

(5.109) and (5.110) yield, after some manipulations,

$$(\mathcal{M}_x^2 - 1)\frac{\partial v_x}{\partial x} + \mathcal{M}_x \mathcal{M}_y \frac{\partial v_x}{\partial y} + \mathcal{M}_x \mathcal{M}_y \frac{\partial v_y}{\partial x} + (\mathcal{M}_y^2 - 1)\frac{\partial v_y}{\partial y} = 0, \tag{5.111a}$$

$$-\frac{\partial v_x}{\partial y} + \frac{\partial v_y}{\partial x} = 0, \tag{5.111b}$$

where

$$\mathcal{M}_x = \frac{\Gamma v_x}{\Gamma_{c_s} c_s}, \quad \mathcal{M}_y = \frac{\Gamma v_y}{\Gamma_{c_s} c_s}$$

are the proper Mach numbers for flows in the x and y directions, with $\Gamma_{c_s} = (1 - c_s^2)^{-1/2}$.

The system (5.111a)–(5.111b) can be written in the form (Königl, 1980)

$$\mathcal{A}^0 \partial_x \mathbf{U} + \mathcal{A}^1 \partial_y \mathbf{U} = 0,$$

where

$$\mathbf{U} = \begin{pmatrix} v_x \\ v_y \end{pmatrix}$$

is the field vector and $\mathcal{A}^0$, $\mathcal{A}^1$ are given by

$$\mathcal{A}^0 = \begin{pmatrix} \mathcal{M}_x^2 - 1, & \mathcal{M}_x \mathcal{M}_y \\ 0 & 1 \end{pmatrix} \quad \mathcal{A}^1 = \begin{pmatrix} \mathcal{M}_x \mathcal{M}_y, & \mathcal{M}_y^2 - 1 \\ -1 & 0 \end{pmatrix}.$$

It is easily seen that for supersonic flow

$$\mathcal{M}^2 = \mathcal{M}_x^2 + \mathcal{M}_y^2 > 1$$

the system (5.111) is hyperbolic.

The characteristic curves $\mathscr{C}_\pm$, are given by

$$\frac{\mathrm{d}y}{\mathrm{d}x} = \lambda_\pm, \tag{5.112}$$

where

$$\lambda_\pm = \frac{-\mathcal{M}_x \mathcal{M}_y \mp (\mathcal{M}^2 - 1)^{1/2}}{1 - \mathcal{M}_x^2}. \tag{5.113}$$

Then the system (5.11) can be written in characteristic form as

$$\left(\frac{\mathrm{d}v_y}{\mathrm{d}v_x}\right)_{\bar{c}_\pm} = \frac{\mathcal{M}_x \mathcal{M}_y \mp (\mathcal{M}^2 - 1)^{1/2}}{1 - \mathcal{M}_y^2}, \tag{5.114}$$

where $\bar{c}_\pm$ are the characteristics in the hodograph plane (v_x, v_y).

It is easy to check the orthogonality relations between the $\mathscr{C}_+$ and $\bar{c}_-$ and between the $\mathscr{C}_-$ and $\bar{c}_+$ families of characteristics,

$$\left(\frac{dy}{dx}\right)_{\mathscr{C}_\pm}\left(\frac{dv_y}{dv_x}\right)_{\bar{c}_\mp} = -1. \tag{5.115}$$

By introducing polar coordinates in the hodograph space (v_x, v_y); $v_x = v\cos\vartheta$, $v_y = v\sin\vartheta$, and also the *relativistic mach angle* ν defined by

$$\sin\nu = \frac{1}{\mathscr{M}}, \quad 0<\nu<\mathscr{M}, \tag{5.116}$$

equation (5.112) becomes

$$\left(\frac{dy}{dx}\right)_{\mathscr{C}_\pm} = \tan(\vartheta \pm \nu). \tag{5.117}$$

From equations (5.117) and (5.115) it follows that the normal to the $\bar{c}_-(\bar{c}_+)$ characteristic lies at an angle $\nu(-\nu)$ with respect to the velocity vector $\mathbf{v}$.

Also, equation (5.114) can be written

$$(d\vartheta)_{\bar{c}_\pm} = \pm\frac{dv}{v}(\mathscr{M}^2-1)^{1/2}, \tag{5.118}$$

which allows a ready determination of the Riemann invariants, because the relationship between $\mathscr{M}$ and v is provided by the Bernoulli equation (5.108) and the state equation.

For a relativistic gas we must use the Synge state equations (2.22)–(2.23). Then it is convenient to express f and c_s as a function of the variable $z = m/k_BT$ and use equation (5.108) in order to express v as a function of z. However, the final integral must be computed numerically. It is possible to obtain analytical results in the nonrelativistic and ultrarelativistic limits. In the ultrarelativistic limit we have

$$c_s^2 = \tfrac{1}{3}$$

and equation (5.118) yields for the Riemann invariants

$$J_\pm = \vartheta \mp E(\mathscr{M}) = \text{const.}, \tag{5.119}$$

where

$$E(\mathscr{M}) = \sqrt{3}\arctan\left(\frac{(\mathscr{M}^2-1)^{1/2}}{3}\right) - \arctan(\mathscr{M}^2-1)^{1/2}. \tag{5.120}$$

In the nonrelativistic limit,

$$f = G(x) \simeq 1 + \frac{5}{2z}$$

and from (5.108) it follows that

$$c_s^2 = \bar{c}_s^2 - \tfrac{1}{3}v^2,$$

where $\bar{c}_s$ is the speed of sound at the stagnation point ($v = 0$), and equation (5.120) then yields

$$J_\pm = \vartheta \mp E(\mathscr{M}) = \text{const.}, \tag{5.121}$$

where

$$E(\mathscr{M}) = 2\arctan\left(\frac{(\mathscr{M}^2 - 1)}{2}\right)^{1/2} - \arctan(\mathscr{M}^2 - 1)^{1/2}. \tag{5.122}$$

By employing standard methods of classical fluid dynamics (Landau and Lifshitz, 1959a) one can derive the relativistic analogs of the Chaplygin and the Euler–Tricomi equations (Königl, 1980).

The theory developed in this section could be used in order to analyze shock formation from simple waves in relativistic steady supersonic flow. Applications could be envisaged to the problem of head-on collisions of heavy ions (Sobel et al., 1975; Clare and Strottman, 1986) and to several astrophysical problems where relativistic supersonic flow might occur (stellar winds, accretion onto neutron stars and black holes, jets in extragalactic radio sources).

In many situations magnetic fields cannot be neglected and one needs a theory of relativistic magnetohydrodynamical simple waves. This will be the subject of the next section.

5.9. Simple waves in relativistic magneto-fluid dynamics

We introduce Minkowski coordinates (x^1, x^2, x^3, t) in Minkowski space $\mathscr{M}$.

Then the Maxwell equations

$$\nabla_\alpha(u^\alpha b^\beta - b^\alpha u^\beta) = 0$$

yield, for one dimensional flow, $x \equiv x^1$,

$$\partial_x(u^1 b^0 - u^0 b^1) = 0,$$
$$\partial_t(u^0 b^1 - u^1 b^0) = 0,$$

whence

$$u^0 b^1 - u^1 b^0 = J_1 = \text{const.}, \tag{5.123}$$

which is a first integral of one dimensional flow.

By writing

$$u^\mu = \Gamma(1, v_x, v_y, v_z), \tag{5.124}$$

from $u_\alpha b^\alpha = 0$ it follows that

$$b^0 = v_x b_x + v_y b_y + v_z b_z$$

and J_1 can be written as

$$J_1 = \Gamma\{(1-(v_x)^2)b_x - v_x v_y b_y - v_x v_z b_z\}.$$

In the nonrelativistic limit $J_1 \simeq b_x$, which is an integral of motion for the relativistic one dimensional flow.

In order to discuss simple waves we recall that the field vector is

$$\mathbf{U} = \begin{pmatrix} u^\nu \\ b^\nu \\ p \\ S \end{pmatrix}$$

and the characteristic equation is

$$a^2 A^2 N_4 = 0, \tag{5.125}$$

where

$$a = u^\alpha \varphi_\alpha, \quad A = Ea^2 - B^2,$$
$$N_4 = \eta(e_p' - 1)a^2 - (\eta + e_p'|b|^2)a^2 G + B^2 G,$$
$$G = g^{\mu\nu}\varphi_\mu \varphi_\nu, \quad B = b^\mu \varphi_\mu.$$

For the phase φ we have $\varphi = \varphi(t, x)$, hence

$$a = \Gamma(v_x - \lambda), \quad B = b_x - \lambda b^0, \quad G = 1 - \lambda^2, \tag{5.126}$$

where $\varphi_0 = -\lambda$, $\varphi_1 = 1$.

First of all, we treat the case of magnetoacoustic simple waves. We distinguish two cases:

(i) $e_p' > 1$.

A material wave can coincide with an Alfvén wave iff $a = 0$ is a solution to $A = 0$, which implies, because $E > 0$, $B = 0$. Similarly, a material wave can coincide with a magnetoacoustic wave iff $B^2 G = 0$, which implies (since $G > 0$ under the assumption $e_p' < 1$) $B = 0$.

In the following (except when stated otherwise) we shall assume that $B \neq 0$, and in this way we avoid the cases when a material wave can coincide with an Alfvén or a magnetoacoustic wave.

Also, an Alfvén wave can coincide with a magnetoacoustic wave iff $A = 0$ is a solution to $N_4 = 0$, which implies

$$\Lambda \equiv a^2 - G|b|^2 = 0.$$

Therefore, in order to exclude this case, we shall also assume that $a^2 - G|b|^2 \neq 0$.

We shall impose the conditions $B \neq 0$, $a^2 - G|b|^2 \neq 0$ at some given initial point and by continuity they will hold in a neighorhood of such a point. The extent of this neighborhood can be ascertained only by the integration of the full set of equations.

Under the above assumptions it has been shown (Section 2.4) that the equation $N_4 = 0$ admits four real and distinct roots for $\lambda(|\lambda| < 1)$.

The corresponding right eigenvectors are given by equation (2.96).

(ii) $e'_p = 1$.

In this case $N_4 = -AG$ and since we exclude Alfvén waves we must have $G = 0$. The corresponding right eigenvectors are formally the same as in the previous case.

In both cases one has the obvious Riemann invariant

$$J_0 = S = \text{const.}$$

and by taking p as an independent variable the equations defining magnetoacoustic simple waves are

$$\frac{du^\alpha}{dp} = \frac{d^\alpha}{Ea^2 A}, \tag{5.127}$$

$$\frac{db^\alpha}{dp} = \frac{d^{\alpha+4}}{Ea^2 A}. \tag{5.128}$$

From the above equations the following results of general character can be obtained (Anile and Muscato, 1988).

PROPOSITION 5.4. *Under the assumption $B \neq 0$, $\Lambda \neq 0$, and $e'_p > 1$ the quantity $|b|^2$ is an increasing function of p for fast magnetoacoustic simple waves, whereas it is a decreasing function of p for slow ones.*

Proof. It is easy to show, from equations (5.127)–(5.128), that

$$\frac{1}{2}\frac{d|b|^2}{dp} = e'_p|b|^2/\eta - B^2/\eta a^2. \tag{5.129}$$

Let v_Σ be the normal speed of propagation, defined as in equation (2.44). Then (5.129) can be written as:

$$\frac{1}{2}\frac{d|b|^2}{dp} = (e'_p v_\Sigma^2 - 1)/(1 - v_\Sigma^2). \tag{5.130}$$

Under the above assumptions, we have $0 \leq v_{\Sigma s} < (e'_p)^{-1/2} < v_{\Sigma f} < 1$, where $v_{\Sigma s}$, $v_{\Sigma f}$ are the slow and fast magnetoacoustic speeds, respectively, as defined in Proposition 2.4. Hence the statement follows. Q.E.D.

Remark 1. For $e'_p = 1$, for the root corresponding to $v_{\Sigma f} = 1$, we obtain $|b|^2 - 2p = \text{constant}$.

Remark 2. We notice that for a fast magnetoacoustic wave $A > 0$, whereas for a slow one $A < 0$. In fact,

$$A = E(a^2 + G)(v_\Sigma^2 - v_{\Sigma A}^2),$$

where $v_{\Sigma A}$ is the local Alfvén speed, satisfying $v_{\Sigma s} < v_{\Sigma A} < v_{\Sigma f}$.

Remark 3. Let $q = p + \frac{1}{2}|b|^2$ be the total pressure (gas pressure + magnetic pressure). Then, from equation (5.130),

$$\frac{\mathrm{d}q}{\mathrm{d}p} = \frac{v_\Sigma^2(e'_p - 1)}{1 - v_\Sigma^2}$$

and for $e'_p > 1$, $\mathrm{d}q/\mathrm{d}p > 0$; hence q is always a monotonically increasing function of p.

Under the assumptions $a \neq 0$, $B \neq 0$, we shall consider solutions with a stagnation point, where $v_x = v_y = v_z = 0$ at a given pressure p_0. Hence, $a(p_0) = -\lambda(p_0)$, $B(p_0) = b_x(p_0)$, and therefore a and B will maintain their initial signs.

One can always choose the reference frame such that $b_x(p_0) > 0$ and consider only progressive waves, for which $\lambda > 0$. With these choices, one always has $a < 0$, $B > 0$.

Another important result of general nature can be obtained concerning the behavior of the eigenvalue λ along the solution $\lambda = \lambda(p)$.

By substituting

$$\varphi_\alpha = (-\lambda, 1, 0, 0)$$

into $N_4 = 0$, we obtain

$$N_4(\mathbf{U}, \lambda) = 0,$$

which holds identically for a chosen root $\lambda = \lambda(p)$. Hence

$$\frac{\partial N_4}{\partial u^A}\frac{\mathrm{d}u^A}{\mathrm{d}p} + \frac{\partial N_4}{\partial \lambda}\frac{\mathrm{d}\lambda}{\mathrm{d}p} = 0,$$

which can be written as

$$\frac{\delta N_4}{\delta p} + \frac{\partial N_4}{\partial \lambda}\frac{\mathrm{d}\lambda}{\mathrm{d}p} = 0, \tag{5.131}$$

with

$$\frac{\delta N_4}{\delta p} = \left(\frac{\partial N_4}{\partial u^\alpha} d^\alpha + \frac{\partial N_4}{\partial b^\alpha} d^{\alpha+4}\right)\frac{1}{Ea^2 A} + \frac{\partial N_4}{\partial p}. \tag{5.132}$$

In Section 4.4 it has been shown that [equation (4.74)]

$$\delta N_4/\delta p = a^2(a^2 K_1 + G K_2), \tag{5.133}$$

where

$$K_1 = \eta e''_p + (e'_p - 1)(3 - 5e'_p),$$
$$K_2 = -e''_p|b|^2 + (e'_p - 1)(3 + 2e'_p|b|^2/\eta).$$

We proceed with the following propositions.

PROPOSITION 5.5. *Under assumptions $B \neq 0$, $a^2 - G|b|^2 \neq 0$, $e'_p > 1$ and the compressibility hypothesis*

$$W = -e''_p + 2e'_p(e'_p - 1)/\eta > 0 \quad (\textit{Weyl condition}),$$

one has

$$\delta N_4/\delta p \neq 0.$$

Proof. In equation (5.133) we substitute $a^2 = G v_\Sigma^2/(1 - v_\Sigma^2)$, hence

$$\frac{\delta N_4}{\delta p} = \frac{G}{1 - v_\Sigma^2}\{-v_\Sigma^2[WE + 3e'_p(e'_p - 1)] + W|b|^2 + 3(e'_p - 1)\}.$$

If $\delta N_4/\delta p = 0$ at some point, then v_Σ^2 must satisfy $N_4 = 0$. Now

$$N_4 = G^2 P(v_\Sigma^2)/(1 - v_\Sigma^2),$$

where

$$P(v_\Sigma^2) = \eta(e'_p - 1)v_\Sigma^4 - (\eta + e'_p|b|^2 - b_n^2)v_\Sigma^2(1 - v_\Sigma^2) + b_n^2(1 - v_\Sigma^2).$$

The compatibility between $N_4 = 0$ and $\delta N_4/\delta p = 0$ is then the equation:

$$Y = X_1|b|^4 + X_2|b|^2 + X_3 = 0,$$

with

$$X_1 = -\eta W^2 - 3e'_p W(e'_p - 1)^2,$$
$$X_2 = -\eta^2 W^2 - 3\eta W(e'_p - 1) - 9e'_p(e'_p - 1)^3 + b_n^2[\eta W^2 + 3W(e'_p - 1)],$$
$$X_3 = -3\eta^2 W(e'_p - 1) + b_n^2[\eta W + 3e'_p(e'_p - 1)][\eta W + 3(e'_p - 1)^2].$$

From assumption $B \neq 0$, $a^2 - G|b|^2 \neq 0$ it follows that $b_n^2 < |b|^2$ and by using this inequality, after some manipulations, we get

$$Y < -3(e'_p - 1)W[(e'_p - 1)|b|^2 - 2]^2 < 0. \qquad \text{Q.E.D.}$$

PROPOSITION 5.6. *Under assumptions $B \neq 0$, $a^2 - G|b|^2 \neq 0$, $e'_p > 1$ and the compressibility hypothesis $W > 0$, one has*

$$\frac{d\lambda}{dp} \gtrless 0$$

for progressive and retrograde waves, respectively.

Proof. Under our hypothesis the roots of $N_4 = 0$ are all distinct, hence $\partial N_4/\partial\lambda \neq 0$.

It follows that $\partial N_4/\partial\lambda$ has the same sign as at the stagnation point $p_0, v_x = v_y = v_z = 0$. A simple calculation shows that:

$$\left.\frac{\partial N_4}{\partial \lambda}\right|_{p_0} = \pm 2\lambda_0[(\eta + e'_p|b|^2 + B^2) - 4e'_pEB^2]^{1/2}_{p_0}$$

where

$$\lambda_0^2 = \left.\frac{(\eta + e'_p|b|^2 + B^2) \pm [(\eta - e'_p|b|^2 + B^2) - 4Ee'_pB^2]^{1/2}}{2Ee'_p}\right|_{p_0}.$$

The choice of the sign $\pm$ corresponds to fast and slow magnetoacoustic waves, respectively. Furthermore, since we shall deal with progressive waves, $\lambda_0 > 0$.

It follows that $\partial N_4/\partial\lambda$ is positive for fast magnetoacoustic waves, and negative for slow ones. Because $\partial N_4/\partial p \neq 0$, it follows that $\partial N_4/\delta p$ has the same sign at the stagnation point.

One finds, after lengthy calculations, that

$$\frac{1}{\lambda_0^2}\left(\frac{\delta N_4}{\delta p}\right)_{p_0} = \{W|b|^2 + 3(e'_p - 1) - \lambda^2[EW + 3e'_p(e'_p - 1)]\}_{p_0}.$$

It can be seen that the sign of $(\delta N_4/\delta p)_{p_0}$ is negative for the fast magnetoacoustic waves and positive for the slow ones. Then the statement follows from

$$d\lambda/dp = -(\delta N_4/\delta p)/(\partial N_4/\partial\lambda) > 0. \qquad \text{Q.E.D.}$$

Remarks. For $e'_p = 1$, from (5.133), one has $\delta N_4/\delta p = 0$, which corresponds to the exceptional case.

Equations (5.127)–(5.128) can be written explicitly as follows. In the case $e'_p > 1$

$$\frac{d(\Gamma v_x)}{dp} = \alpha_1 \Gamma v_x + \alpha_1/a + \alpha_2 b_x, \tag{5.134a}$$

$$\frac{d(\Gamma v_y)}{dp} = \alpha_1 \Gamma v_y + \alpha_2 b_y, \tag{5.134b}$$

$$\frac{d(\Gamma v_z)}{dp} = \alpha_1 \Gamma v_z + \alpha_2 b_z, \tag{5.134c}$$

$$\frac{db_x}{dp} = \beta_1 \Gamma v_x + \beta_2 b_x - (e'_p - 1)Ba^2/GA, \tag{5.134d}$$

$$\frac{db_y}{dp} = \beta_1 \Gamma v_y + \beta_2 b_y, \tag{5.134e}$$

$$\frac{db_z}{dp} = \beta_1 \Gamma v_z + \beta_2 b_z, \tag{5.134f}$$

where

$$\alpha_1 = -a^4(e'_p - 1)/AG, \quad \alpha_2 = aB(e'_p - 1)/\eta A,$$
$$\beta_1 = B(\alpha_1 - 1/\eta)/a, \quad \beta_2 = e'_p/\eta + B\alpha_2/a.$$

In the case $e'_p = 1$, under assumption $B \neq 0$, $a^2 - G|b|^2 \neq 0$, one has two roots $\lambda = \pm 1$ and equations (5.127)–(5.128) admit the following invariants, besides J_1,

$$\hat{J}_2 = v_y/v_z,$$
$$\hat{J}_3 = v_z/(v_x \mp 1),$$
$$\hat{J}_4 = (-J_2 b_z + b_y)p^{1/2},$$
$$\hat{J}_5 = [b_x \mp (v_x b_x + v_y b_y + v_z b_z)]^2 p,$$

$$\hat{J}_6 = \frac{v_x - K_1}{v_y - K_2}\frac{1}{p} \quad \text{with } K_1, K_2 \text{ constants.}$$

Now we turn our attention to the case $e'_p > 1$. From equations (5.134) we obtain

$$\frac{d}{dp}[\Gamma(v_y b_z - v_z b_y)] = (e'_p - 1)\frac{a^2}{A}[\Gamma(v_y b_z - v_z b_y)]. \tag{5.135}$$

At the stagnation point p_0 we have:

$$[\Gamma(v_y b_z - v_z b_y)]_{p_0} = 0,$$

hence, by the uniqueness theorem for the initial value problem, equation (5.135) yields the following invariant:

$$J_2 = v_y b_z - v_z b_y = 0. \tag{5.136}$$

Similarly, it can be seen that

$$\frac{\mathrm{d}}{\mathrm{d}p}\left(\frac{b_y}{b_z}\right)=\frac{\beta_1\Gamma}{b_z}\left(v_y-b_y\frac{v_z}{b_z}\right),$$

whence another invariant,

$$J_3=b_y/b_z. \tag{5.137}$$

In general, equations (5.134) are too complicated to be integrated analytically. In the following we shall treat some special cases which can be brought to quadratures.

First of all, we treat the case of a longitudinal magnetic field,

$$b_y=b_z=0.$$

From equations (5.134) we then have

$$\mathrm{v}_y=v_z=0.$$

The characteristic equation $N_4=0$ admits the following four solutions:

$$\lambda_m=(v_x+\bar{\lambda}_m)/(1+v_x\bar{\lambda}_m)\quad m=1,2,3,4,$$

where

$$(\bar{\lambda}_{1,2})=\pm(e'_p)^{-1/2},\quad(\bar{\lambda}_{3,4})=\pm(|b|^2/E)^{1/2}.$$

The roots $\bar{\lambda}_{3,4}$ coincide with the Alfvén waves and shall not be considered here. The simple waves for the acoustic speed are given by the following invariants:

$$J_1=b_x/\Gamma, \tag{5.138}$$

$$J_\pm=\ln\left(\frac{1+v_x}{1-v_x}\right)^{1/2}\mp\int\frac{[e'_p]^{1/2}}{\eta}\mathrm{d}p.$$

The invariants $J_\pm$ coincide with those of equation (5.27) obtained in the case of fluid dynamics.

The other case we shall consider is when the fluid's motion is purely longitudinal

$$v_y=v_z=0$$

and the magnetic field at the stagnation point is purely transverse

$$b_x(\mathbf{v}=0)=0.$$

Then from the invariant J_1 it follows that $b_x=0$ throughout the flow, hence $b^0=0$ and $B=0$. In this case we consider the following simple root of $N_4 = 0$, corresponding to the fast magnetoacoustic waves,

$$\lambda = (v_x + \bar{\lambda})/(1 + v_x\bar{\lambda}),$$

with

$$\bar{\lambda} = \pm[(\eta + e'_p|b|^2)/Ee'_p]^{1/2}.$$

From (5.134e)–(5.134f) we obtain the following invariants

$$J_4 = \ln b_y - \int \frac{e'_p}{\eta}\,dp, \quad J_5 = \ln b_z - \int \frac{e'_p}{\eta}\,dp. \tag{5.139}$$

Finally, equation (5.134a) gives the invariants:

$$J_\pm = \tfrac{1}{2}\ln[(1 + v_x)/(1 - v_x)] \mp \int \frac{1}{\eta}[e'_p(\eta + e'_p|b|^2)/E]^{1/2}\,dp.$$

For a fluid for which $p = p(\rho, S)$

$$\int \frac{e'_p}{\eta}\,dp = \ln \rho$$

and equation (5.139) gives

$$b_y/\rho = \text{const.}, \quad b_z/\rho = \text{const.},$$

which are analogous to the corresponding nonrelativistic invariants (Cabannes, 1970).

Remark 5. In the case $b_x = b_y = b_z = 0$ equations (5.134) reduce to the purely fluid equations:

$$\frac{dv_x}{dp} = \frac{\lambda v_x - 1}{\Gamma^2\eta(v_x - \lambda)},$$

$$\frac{dv_y}{dp} = \frac{\lambda v_y}{\Gamma^2\eta(v_x - \lambda)},$$

$$\frac{dv_z}{dp} = \frac{\lambda v_z}{\Gamma^2\eta(v_x - \lambda)},$$

$$\lambda = \frac{(e'_p -)\Gamma^2 v_x \pm [(e'_p - 1)\Gamma^2(1 - v_x^2) + 1]^{1/2}}{1 + (e'_p - 1)\Gamma^2}.$$

Another case of interest is when, at the stagnation point, $(b_x)_0 = 0$. It follows that $J_1 = 0$.

For plane polarized waves we can choose $b_y = v_y = 0$, hence from the invariant J_1 we obtain

$$b_x = v_x v_z b_z/\delta, \quad \delta = 1 - v_x^2.$$

For the magnetoacoustic waves we have

$$a = \frac{\dfrac{v_x \Delta}{\Gamma} \pm \left[\dfrac{\Delta^2}{\Gamma^2} + \delta\Delta\eta(e'_p - 1)\right]^{1/2}}{\eta(e'_p - 1) + \Delta/\Gamma^2},$$

with

$$\Delta = \eta + e'_p|b|^2 - (b^2)^2/\Gamma^2.$$

Finally, equations (5.134) yield, in this case,

$$\frac{dv_x}{dp} = \frac{(e'_p - 1)a^3}{A\Gamma G}(\lambda v_x - 1), \tag{5.140a}$$

$$\frac{dv_z}{dp} = \frac{(e'_p - 1)a^3\lambda v_z}{A\Gamma G} + \frac{(e'_p - 1)aB}{\eta A\Gamma}(b_z - b^0 v_z), \tag{5.140b}$$

$$\frac{db_z}{dp} = -\frac{B\Gamma v_z}{a}\left[\frac{(e'_p - 1)a^4}{AG} + \frac{1}{\eta}\right] + \left[\frac{e'_p}{\eta} + \frac{(e'_p - 1)B^2}{\eta A}\right]b_z.$$

From (5.140a), for the fast wave we have that because $a < 0$ and $A > 0$, $dv_x/dp > 0$.

Since we require a stagnation point p_0 at which $v_x(p_0) = v_z(p_0) = 0$, from equation (5.140b) and the uniqueness theorem for the initial value problem we have $v_z = 0$, and equation (5.140c) reduces to

$$\frac{db_z}{dp} = e'_p \frac{b_z}{\eta},$$

which is the transverse case studied above.

The system of equations (5.134) has been integrated numerically (Muscato, 1988b). For the sake of definiteness we choose the field linearly polarized,

$$b_y = 0.$$

From the invariant J_z we then have, assuming $b_z \neq 0$,

$$v_y = 0.$$

It is convenient to write the system (5.134) in dimensionless form, by introducing the variables

$$\bar{p} = p/p_0, \quad \bar{b}_x = b_x/J_1, \quad \bar{b}_z = b_z/J_1, \quad \bar{\eta} = \eta/p_0,$$

$$\bar{A}=\bar{E}a^2-L\bar{B}^2,\quad \bar{B}=(1+\bar{b}^0a)/\Gamma,\quad \bar{E}=\eta/p_0+L|\bar{b}|^2,$$
$$|\bar{b}|^2=|b|^2/J_1^2,\quad L=J_1^2/p_0,$$

where p_0 is the initial pressure.

Therefore the system (5.134) can be written (omitting the bar)

$$\frac{dv_x}{dp}=\frac{(e'_p-1)}{A\Gamma}a[a^2(\lambda v_x-1)/G+LB/\Gamma\eta], \tag{5.141a}$$

$$\frac{dv_z}{dp}=\frac{(e'_p-1)}{A\Gamma}a[a^2\lambda v_z/G+(b_z-b^0v_z)BL/\eta], \tag{5.141b}$$

$$\frac{db_z}{dp}=-\frac{B\Gamma v_z}{a}\left[\frac{(e'_p-1)a^2}{AG}+\frac{1}{\eta}\right]+\frac{b_z}{\eta A}[e'_pEa^2-LB^2]. \tag{5.141c}$$

Now we investigate the signs of the derivatives dv_x/dp, dv_z/dp, db_z/dp. At the stagnation point $(v_x=v_y=v_z=0)$ we have

$$\left.\frac{dv_x}{dp}\right|_0=\frac{a(e'_p-1)}{A}\frac{(\eta+L)(v_A^2-v_\Sigma^2)+L(1+b_z^2)}{G\eta},$$

$$\left.\frac{dv_z}{dp}\right|_0=\frac{(e'_p-1)LaB}{\eta A}b_z,$$

$$\left.\frac{db_z}{dp}\right|_0=\frac{b_z}{\eta A}\frac{L}{v_A^2}(e'_pv_\Sigma^2-v_A^2).$$

For fast magnetoacoustic waves $a<0, A>0, B>0$ and for an initial datum $b_z>0$,

$$(dv_z/dp)_0<0,\quad (db_z/dp)_0>0.$$

The sign of $(dv_x/dp)_0$ depends on the initial datum. For slow magnetoacoustic waves $(dv_x/dp)_0>0$.

From equations (5.141) we see that for $v_z>0$, $b_z>0$ one has $db_z/dp>0$, hence b_z is a monotonically increasing function of p for the fast wave.

When starting to integrate system (5.141), we calculate the four roots $\lambda_1, \lambda_2, \lambda_3, \lambda_4$ and we choose one to be followed during the integration. A suitable algorithm ensures that we are always following the same root; the accuracy of the numerical code has been tested against the exact solutions previously found.

The agreement has been found to be extremely satisfactory. The results of the calculations are plotted in Figs. 5.8–5.11 for various values of the parameter L.

Fig. 5.8a. Progressive fast magnetoacoustic simple waves, v_x as a function of the parameter $z = m/k_B T$ in the ultrarelativistic region, corresponding to an initial propagation speed (at the stagnation point) $\lambda_0 = 0.57$, for the values $L = 1$ and 100 of the parameter $L = J_1^2/p_0$. The two curves ($L = 1$ and $L = 100$) graphically coincide.

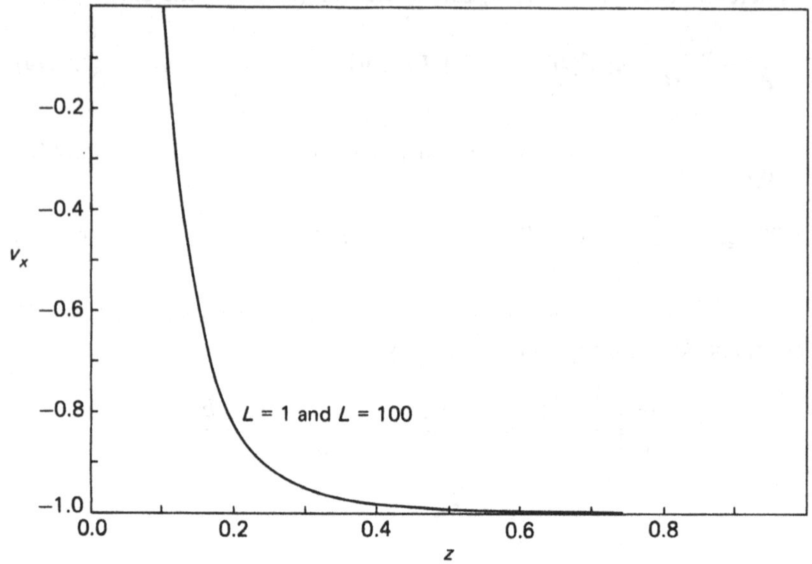

Fig. 5.8b. Same as in Fig. 5.8a but with v_z as a function of z for $L = 1$ and $L = 100$.

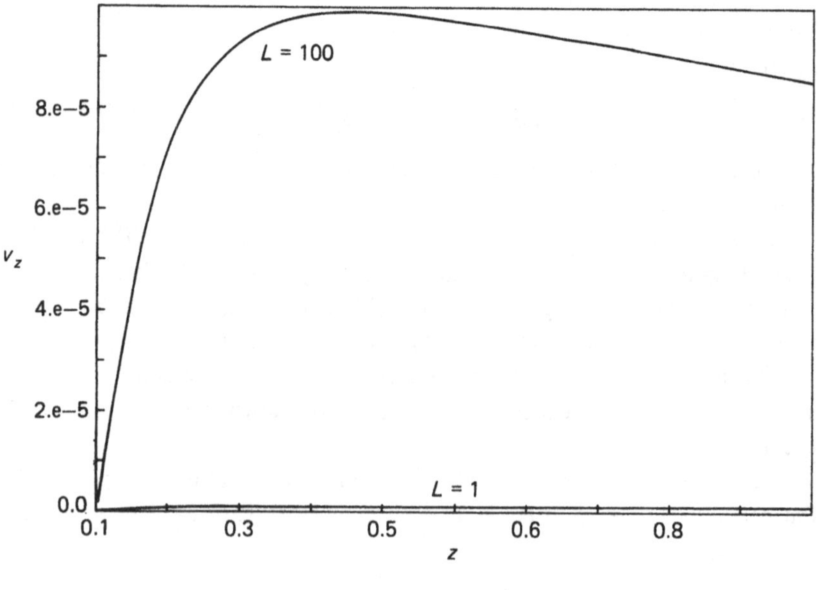

Fig. 5.8c. Same as in Fig. 5.8a but with b_z as a function of the parameter z for $L = 1$ and $L = 100$ (the two curves graphically coincide).

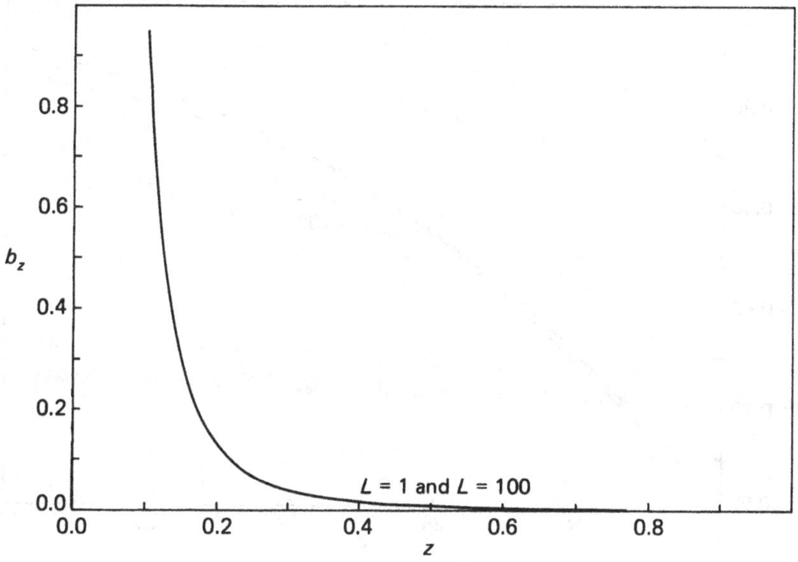

Fig. 5.9a. Progressive fast magnetoacoustic simple waves, v_x as a function of the parameter z in the intermediate regime for the values $L = 1$ and $L = 100$. The initial propagation speed is $\lambda_0 = 0.94$.

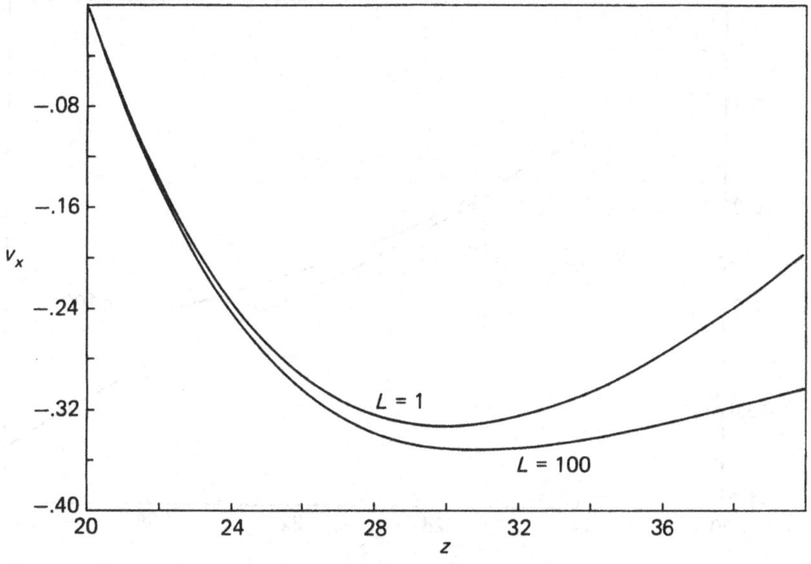

Fig. 5.9b. Same as in Fig. 5.9a but v_z as a function of the parameter z for $L = 1$ and $L = 100$.

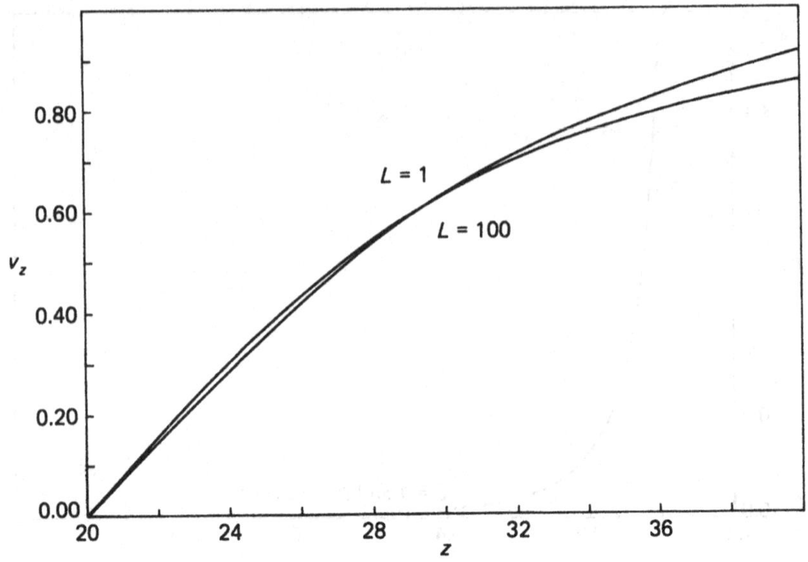

Fig. 5.9c. Same as in Fig. 5.9a but b_z as a function of the parameter z for $L = 1$ and $L = 100$.

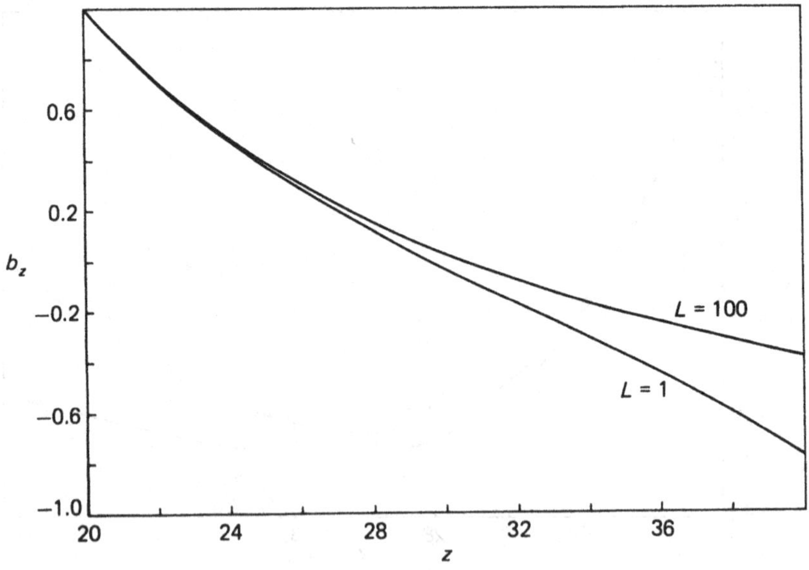

Fig. 5.10a. Progressive slow magnetoacoustic simple waves, v_x as a function of the parameter z in the intermediate relativistic regime for the values $L=1$ and $L=100$ (the two curves graphically coincide). The initial propagation speed is $\lambda_0=0.18$.

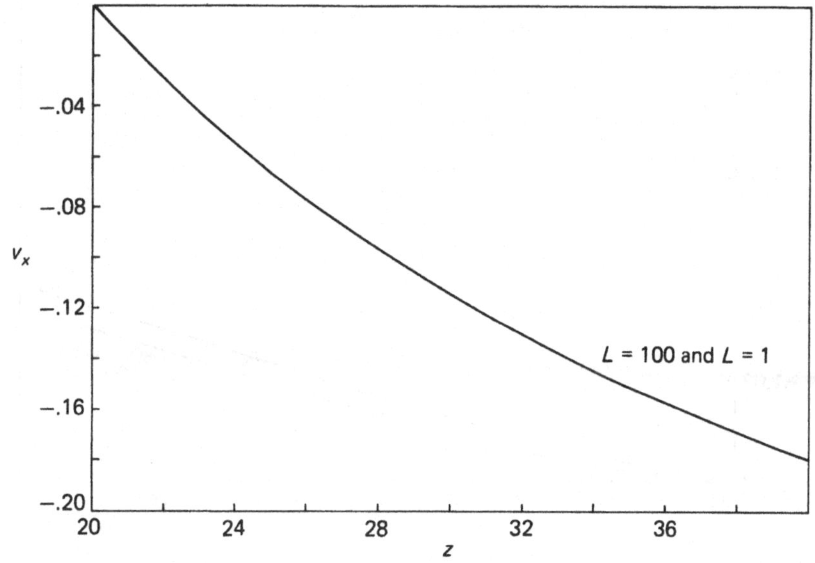

Fig. 5.10b. Same as in Fig. 5.10a but with v_z as a function of the parameter z, for $L=1$ and $L=100$ (the two curves graphically coincide).

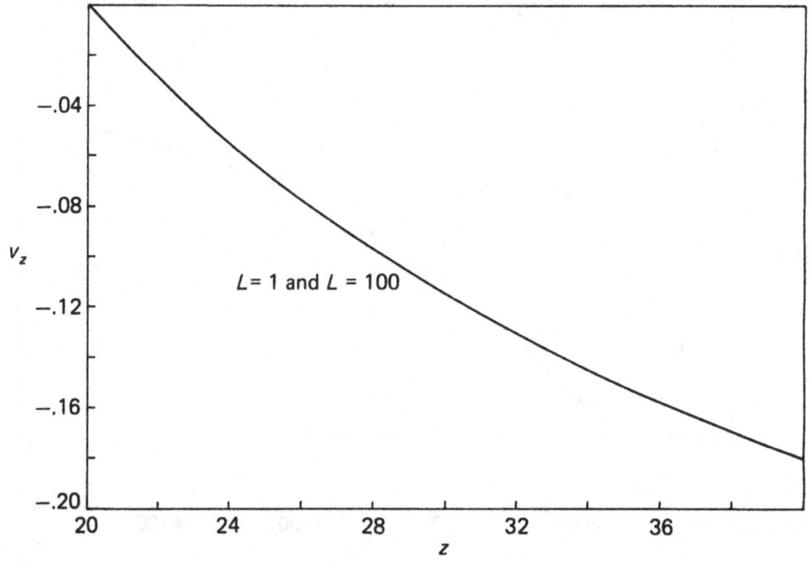

Fig. 5.10c. Same as in Fig. 5.10a but with b_z as a function of the parameter z for $L = 1$ and $L = 100$.

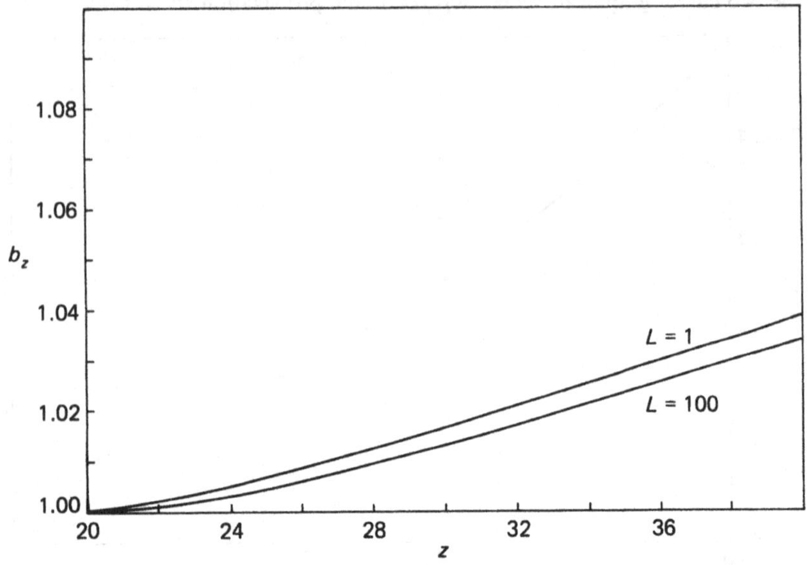

Fig. 5.11a. Progressive fast magnetoacoustic simple wave, v_x as a function of the parameter z in the nonrelativistic regime ($z \gg 1$) for the values $L = 1$ and $L = 100$ (in the Figs. 5.11a–5.11c the two curves graphically coincide). The initial propagation speed is $\lambda_0 = 0.99$.

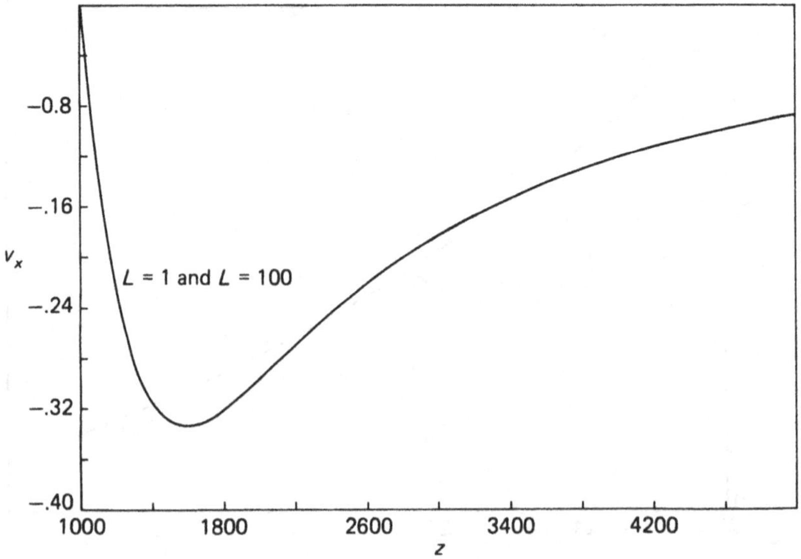

Fig. 5.11b. Same as in Fig. 5.11a but with v_z as a function of the parameter z for $L=1$ and $L=100$.

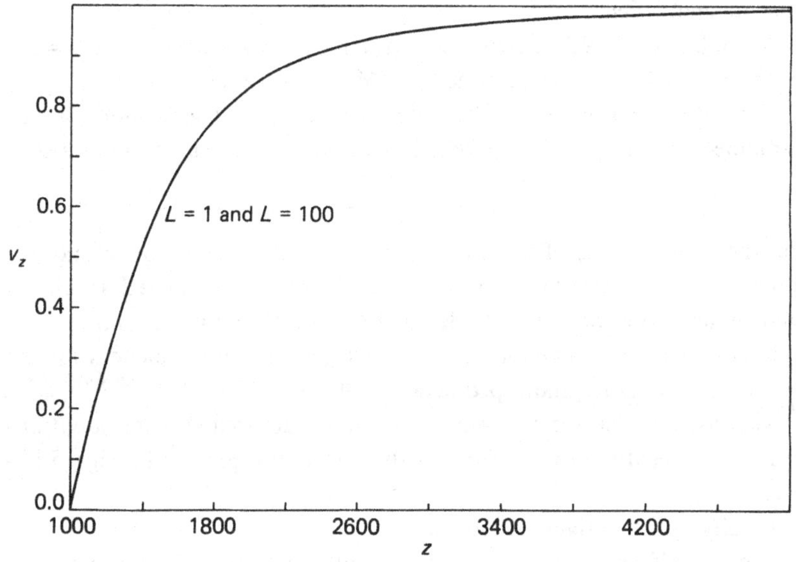

Fig. 5.11c. Same as in Fig. 5.11a but with b_z as a function of the parameter z for $L=1$ and $L=100$.

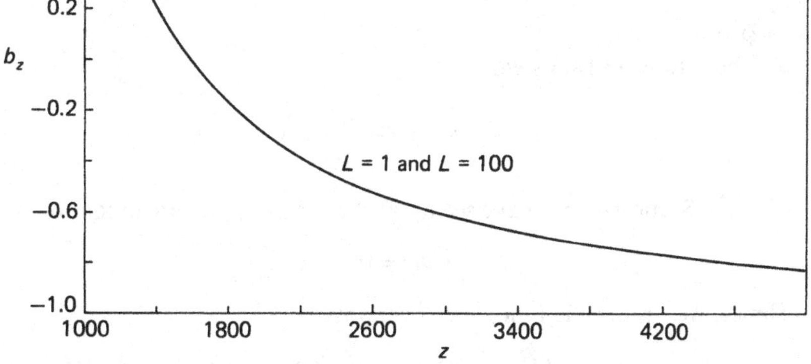

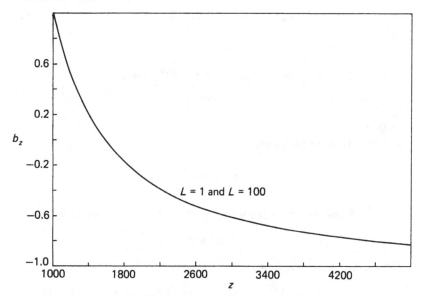

Notice that the velocity v_x is not a monotone function of z, at striking variance with the nonrelativistic results and with the relativistic case with zero magnetic field.

A similar result is found for the ultrarelativistic state equation ($e = 3p$) but in this case for the component v_z (Muscato, 1987).

This phenomenon is peculiar of the relativistic regime and can be explained as follows. In a relativistic framework one has the limitation

$$v_x^2 + v_z^2 \leq 1$$

for the speeds v_x, v_z. Therefore, v_x and v_z cannot both be monotone functions of p, otherwise the above inequality would be violated. It follows that at least one of the variables v_x or v_z must show a nonmonotone behavior. This phenomenon could have important consequences for the relativistic MFD Riemann problem.

The critical time for breaking can also be computed from equations (5.10)–(5.10′) (Muscato, 1988a) and the results are plotted in Figs. 5.12–5.13.

Finally, we investigate Alfvén simple waves.

Because Alfvén waves correspond to multiple roots of the characteristic equation, the methods employed in the previous sections are not directly applicable.

In this case it is convenient to resort to a general method due to Boillat (1982a). We start with the conservation equations

$$\partial_\alpha f^{\alpha A} = 0, \quad A = 0, 1, 2, \ldots, 8, \tag{5.142}$$

with

$$f^\alpha = (T^{\alpha\beta}, \rho u^\alpha, u^\alpha b^\beta - u^\beta b^\alpha)^{\mathrm{T}}.$$

We look for one dimensional solutions of the kind

$$f^\alpha = f^\alpha(\varphi),$$

with $\varphi = x - \lambda(\varphi)t$.

The equations (5.142) yield

$$\frac{\mathrm{d}}{\mathrm{d}\varphi}(\varphi_\alpha f^{\alpha A}) + f^{0A}\frac{\mathrm{d}\lambda}{\mathrm{d}\varphi} = 0.$$

Now, in Section 4.4 we have seen that the Alfvén waves are exceptional,

$$\mathrm{d}\lambda/\mathrm{d}\varphi = 0.$$

Hence, we obtain the following invariants:

$$f^{1A} - \lambda f^{0A} = \text{const.}, \quad A = 0, 1, 2, \ldots, 8. \tag{5.143}$$

Fig. 5.12. Barotropic RMFD with $\gamma = 4/3$. Plot of (T) versus (v). The lower curve corresponds to $W = 0$, the intermediate one to $W = 10$, and the upper one to $W = 100$.

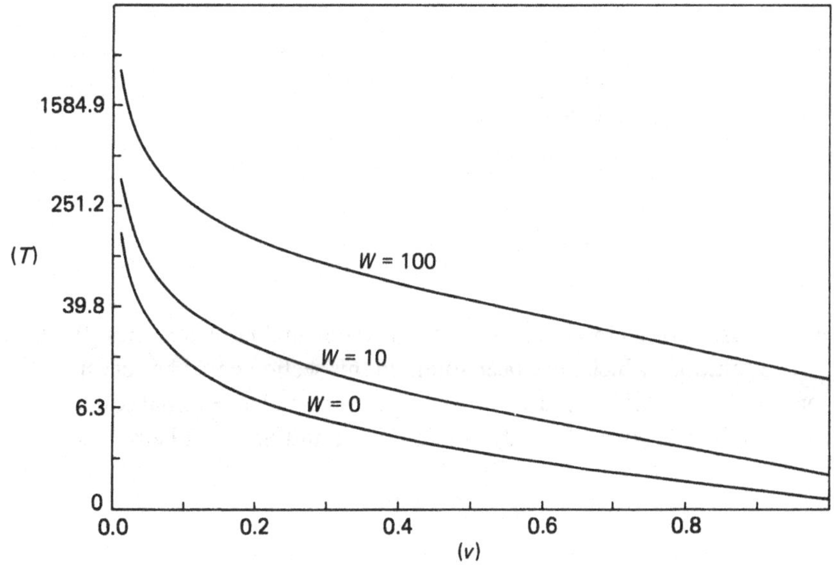

Fig. 5.13. Polytropic RMFD and nonrelativistic MFD with $\gamma = 5/3$, $\bar{c} = 0.4$. Plot of (T) versus (v). The lower curve corresponds to $W = 1$ and the intermediate and upper ones to $W = 10$, 100, respectively.

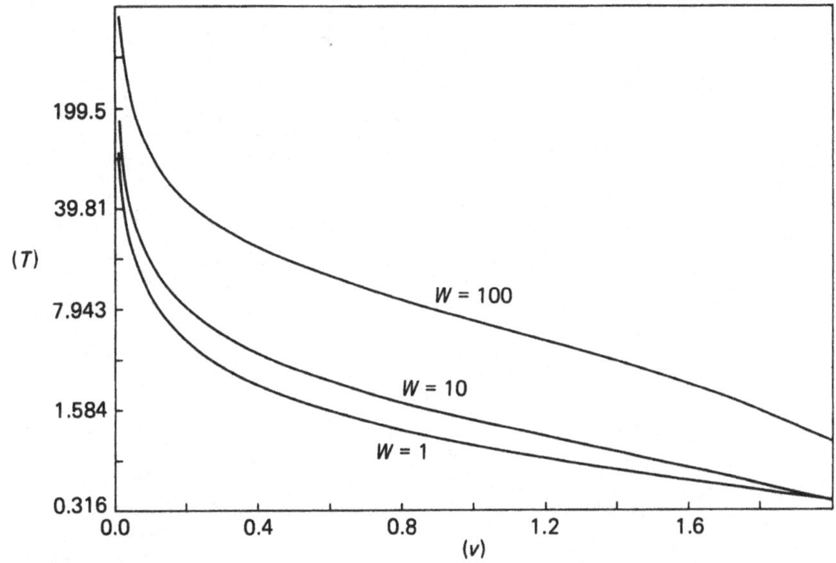

From these we obtain, besides the already known invariants J_1 and S,

$$\bar{J}_2 = |b|^2,$$

$$\bar{J}_3 = \lambda = v_x - \frac{J_1}{\pm\Gamma^2 E^{1/2} - b^0\Gamma},$$

$$\bar{J}_4 = b_y - B\Gamma v_y/a,$$

$$\bar{J}_5 = b_z - B\Gamma v_z/a,$$

$$\bar{J}_6 = p,$$

$$\bar{J}_7 = b_x^0 - \lambda b^0,$$

$$\bar{J}_8 = \Gamma(v_x - \lambda),$$

which correspond closely to the nonrelativistic analogs (Cabannes, 1970).

The solutions which have been found in this section could be very useful for checking numerical codes for the full set of partial differential equations of relativistic magneto-fluid dynamics (Sloan and Smarr, 1986).

6

Relativistic geometrical optics

6.0. Introduction

In the previous chapters we have studied waves which can be represented either as exact solutions of the field equations (as in the case of simple waves) or as propagating surfaces of discontinuities, for which exact transport laws can be obtained. These classes of waves, although very important for testing the mathematical structure of the theory, have a limited range of applicability. In particular, simple waves are restricted to one dimensional propagation into a constant state whereas propagating surfaces can model only impulsive waves. This leaves out the vast class of harmonic waves propagating into an arbitrary nonuniform state, which comprises most of the interesting applications. In order to treat the latter case it is necessary, in general, to resort to perturbation methods. Most of the perturbation methods used for this purpose are based on the so-called geometrical optics approximation, or variants thereof such as the high-frequency expansion or the method of multiple scales.

The idea underlying these methods is that there are at least two widely different length (or time) scales in the problem, the length L characteristic of the variation of the background state into which the wave is propagating, and the mean wavelength λ of the wavetrain, with the ordering $\lambda \ll L$. One then introduces a parameter $\varepsilon = \lambda/L$ into the problem (usually through the appearance of one or several rapidly varying phase functions and also through other stretched variables, depending on ε) and seeks asymptotic solutions of the field equations as appropriate power series in ε. One finds that one can define a class of curves, the *rays* along which the various terms of the asymptotic expansion propagate according to transport laws.

It is apparent that the asymptotic expansions which are used are valid only in a restricted class of coordinates (in order to convince oneself of this it suffices to think of coordinate transformations depending on the parameter ε which could destroy the ordering of the asymptotic expansion). However, this is not a hindrance to using these methods for relativistic problems because *the existence of a restricted class of coordinates enjoying a definite set of properties is in itself an invariant statement.* However, in concrete examples it may be very difficult to prove the existence of such a class of preferred coordinates (de Arajuro, 1986).

In this chapter we shall introduce the basic concepts underlying the perturbation methods referred to as the high-frequency expansion and the two-timing methods, within a relativistic framework. First we shall expound the high-frequency expansion method by applying it to Maxwell's equations in vacuo and in a linear isotropic medium, in general relativity (notice that these equations are linear). Then we shall introduce the two-timing method both in the linear and in the weakly nonlinear case by applying it to Maxwell's equations in the presence of a cold plasma. The high-frequency expansion method in a nonlinear framework (also called the method of asymptotic waves) will be the subject of the next chapter.

The theory developed in this chapter is motivated by applications to laboratory plasma physics and astrophysics. The interaction of an electromagnetic wave with a plasma (a topic treated in Sections 6.3 and 6.4 in the cold plasma approximation and under a simplified hypothesis) is definitely a relativistic problem and therefore should be treated in a covariant framework (although for some specific problems it might be more convenient to use a noncovariant formalism; Shukla et al., 1986). The covariant theory of the interaction of an electromagnetic wave with a plasma could be relevant for several laboratory plasma problems (for example, the stimulated emission by intense relativistic electron beams; Miller, 1985) and for diverse astrophysical problems. One of the most interesting applications to fundamental astrophysical theory is in the derivation of the radiative transfer equation. When considering the transfer of radiation in cosmology, as in the case of the microwave background radiation or the light emitted from quasars, or the radiation emitted from nearby a neutron star or black hole, one needs to consider the relativistic radiative transfer equation (Ellis, 1971; Weinberg, 1972; Thorne, 1981). In its turn the relativistic radiative transfer equation can be obtained either from relativistic kinetic theory (starting from the relativistic Boltzmann equation for photons; Ehlers, 1971) or from relativistic geometrical optics (Ellis, 1971; Anile and Breuer, 1974; Anile, 1976) or, precisely, from the propagation laws for the amplitude and polarization of the wave.

The latter approach is more satisfactory when dealing with polarized radiation or with radiation propagating through a dispersive medium. In fact, in the kinetic approach, in these cases one has to postulate the propagation laws for the photons (Bicak and Hadrava, 1975) or resort to geometrical optics itself. Instead, as will be seen in Sections 6.2 and 6.3, relativistic geometrical optics is capable of deducing the propagation laws for the amplitude and polarization directly from Maxwell's equations.

The plan of the chapter is the following.

In Section 6.1 we investigate high-frequency waves for Maxwell's equations in vacuo in an arbitrary space-time, following an earlier treatment by Ehlers (1967). In this way we prove that, to the main order in ε, the electromagnetic field satisfies the laws of geometrical optics in general relativity, that is, the area-intensity law and the parallel propagation law for the polarization (Synge, 1960).

In Section 6.2, following earlier treatments by Ehlers (1967) and Anile and Moschetti (1979), we investigate high-frequency waves for Maxwell's equations in a linear isotropic refractive medium. Proceeding as in the previous section, we obtain the corresponding laws of geometrical optics, that is, the modified area-intensity law (taking into account the refractive index) and the modified propagation law (differing from parallel propagation by a term representing the interaction of the polarization with the vorticity of the medium).

In Section 6.3 we deal with the two-timing method. This is necessary if one wants to incorporate dispersive effects, as in the case of waves in plasmas. We discuss linear locally plane electromagnetic waves propagating in a cold relativistic plasma and base our treatment on that of Anile and Pantano (1977, 1979). Further extensions of the theory to cases where the plasma is magnetized can be found in the articles by Breuer and Ehlers (1980, 1981). By applying the two-timing method we are able to obtain the laws of geometrical optics in general relativistic dispersive media, which had been postulated by Synge (1960), without deriving them from Maxwell's equations. Moreover, we obtain also a propagation law for the polarization which differs from the parallel propagation law by a term representing the interaction with the plasma vorticity. Such a term is analogous to that occurring in the case of propagation in a refractive medium.

In Section 6.4 we consider the same problem but modify the perturbation method (by introducing several phase functions) in order to treat the weakly nonlinear modulation of a locally plane harmonic wave. The calculations are rather lengthy and the final results are the following: (i) the wave complex amplitude satisfies a propagation law along the rays which is of the form of a generalized nonlinear Schrödinger equation; (ii) the polarization vector of the wave satisfies the same transport law as in the linear case. We remark that in all cases treated in this chapter the only way a background gravitational field can affect propagating radiation is via its effect on the rays (they can be bent or focussed by space-time curvature). This is obviously consistent with the usual interpretation of the equivalence principle which in the approximation we are dealing with (geometrical optics) amounts to stating that a background gravitational field can have no local effect.

6.1. Geometrical optics in general relativity

Maxwell's equations in vacuo are equations (4.81)–(4.82) in the absence of charges, $J^\alpha = 0$, and they are written

$$(\nabla_\alpha F_{\beta\gamma} + \nabla_\beta F_{\gamma\alpha} + \nabla_\gamma F_{\alpha\beta} = 0), \tag{6.1}$$

$$\nabla_\alpha F^{\alpha\beta} = 0, \tag{6.2}$$

where $F_{\alpha\nu}$ is the electromagnetic field tensor.

A high-frequency wave is defined by the following formal asymptotic expansion

$$F_{\alpha\beta} = \mathrm{Re}\left\{ e^{i\omega\varphi} \sum_{n=0}^{\infty} \omega^{-n} \underset{(n)}{F}_{\alpha\beta} \right\} \tag{6.3}$$

where $\varphi(x)$ is a real function to be determined, called the phase, and ω is a real parameter, $\omega > 0$ (related to the "frequency" of the wave).

Obviously this expansion is meaningful only in a selected class of coordinate systems.

We substitute the expansion (6.3) into equations (6.1)–(6.2) and proceed by equating to zero termwise the coefficients of the resulting formal series in $1/\omega$.

Then, to the zeroth order, we obtain, writing $l_\alpha = \nabla_\alpha \varphi$,

$$l_\alpha \underset{(0)}{F}_{\beta\gamma} + l_\beta \underset{(0)}{F}_{\gamma\alpha} + l_\gamma \underset{(0)}{F}_{\alpha\beta} = 0, \tag{6.4a}$$

$$l_\alpha \underset{(0)}{F}{}^{\alpha\beta} = 0. \tag{6.4b}$$

To the nth order we have

$$i(l_\alpha \underset{(n+1)}{F}_{\beta\gamma} + l_\beta \underset{(n+1)}{F}_{\gamma\alpha} + l_\gamma \underset{(n+1)}{F}_{\alpha\beta}) + \nabla_\alpha \underset{(n)}{F}_{\beta\gamma} + \nabla_\beta \underset{(n)}{F}_{\gamma\alpha} + \nabla_\gamma \underset{(n)}{F}_{\alpha\beta} = 0, \tag{6.5a}$$

$$il_\alpha \underset{(n+1)}{F}{}^{\alpha\beta} + \nabla_\alpha \underset{(n)}{F}{}^{\alpha\beta} = 0. \tag{6.5b}$$

From (6.4a)–(6.4b), because $\underset{(0)}{F}{}^{\alpha\beta} \neq 0$, we obtain the compatibility relation

$$l_\alpha l^\alpha = 0. \tag{6.6}$$

This equation shows that the hypersurfaces of constant phase are null hypersurfaces (coinciding with the characteristic hypersurfaces), in analogy with the weak discontinuity case (Section 4.5).

Let n^α be a vector such that $n^\alpha l_\alpha = 1$. Then by contracting (6.4a) with n^α, we see that $\underset{(0)}{F}_{\alpha\beta}$ has the following expression

$$\underset{(0)}{F}_{\alpha\beta} = \psi_\alpha l_\beta - l_\alpha \psi_\beta, \tag{6.7}$$

where ψ_α, which is determined up to the "gauge" transformation $\psi_\alpha \rightarrow \psi_\alpha + \lambda l_\alpha$ ($\lambda \in \mathbb{R}$), is such that

$$\psi_\alpha l^\alpha = 0.$$

This result is also analogous to equation (4.100) of the weak discontinuity case.

The rays are introduced as the bicharacteristic curves, that is, $x^\alpha = x^\alpha(\tau)$ such that

$$\frac{dx^\alpha}{d\tau} = l^\alpha. \tag{6.8}$$

To the first order, equations (6.5a)–(6.5b) read

$$i(l_\alpha \underset{(1)}{F}_{\beta\gamma} + l_\beta \underset{(1)}{F}_{\gamma\alpha} + l_\gamma \underset{(1)}{F}_{\alpha\beta}) + l_\alpha(\nabla_\beta \psi_\gamma - \nabla_\gamma \psi_\beta) + l_\beta(\nabla_\gamma \psi_\alpha - \nabla_\alpha \psi_\gamma) + l_\gamma(\nabla_\alpha \psi_\beta - \nabla_\beta \psi_\alpha) = 0. \tag{6.9a}$$

$$il_\alpha \underset{(1)}{F}^{\alpha\beta} + (\nabla_\alpha \psi^\alpha) l^\beta + \psi^\alpha \nabla_\alpha l^\beta - (\nabla_\alpha l^\alpha)\psi^\beta - l^\alpha \nabla_\alpha \psi^\beta = 0. \tag{6.9b}$$

Let $\underset{(1)}{\psi}_\beta = \underset{(1)}{F}_{\alpha\beta} l^\alpha$, $Z_\beta = l^\alpha(\nabla_\alpha \psi_\beta - \nabla_\beta \psi_\alpha)$.

Then equation (6.9a) by contraction with l^α yields,

$$i\underset{(1)}{\psi}_\beta + Z_\beta = kl_\beta, \quad k \text{ real scalar,} \tag{6.10}$$

whence

$$l^\alpha \nabla_\alpha \psi_\beta + \psi_\beta \nabla_\alpha l^\alpha - \psi^\alpha \nabla_\alpha l_\beta - (\nabla_\alpha \psi^\alpha) l_\beta + Z_\beta - kl_\beta = 0.$$

Now $Z_\beta = l^\alpha \nabla_\alpha \psi_\beta + \psi^\alpha \nabla_\beta l_\alpha$, and therefore

$$2l^\alpha \nabla_\alpha \psi_\beta + \psi_\beta \nabla_\alpha l^\alpha + k' l_\beta = 0, \tag{6.11}$$

with $k' = -k - \nabla_\alpha \psi^\alpha$.

Equation (6.11) provides transport equations for both the amplitude and the polarization of the wave.

In fact, by contracting with the complex conjugate $\bar{\psi}^\beta$, we obtain

$$l^\alpha \nabla_\alpha |\psi|^2 + |\psi|^2 \nabla_\alpha l^\alpha = 0, \tag{6.12}$$

which is the transport equation for the amplitude.

This equation can also be written as a conservation law,

$$\nabla_\alpha(l^\alpha|\psi|^2)=0. \tag{6.13}$$

The transport equation for the polarization can be obtained as follows.

Let u^μ be the four-velocity of an arbitrary observer, $u_\mu u^\mu=-1$. The frequency Ω of the wave relative to this observer is defined by (Synge, 1960)

$$\Omega=l_\alpha u^\alpha.$$

Let us introduce, at least locally, an orthonormal tetrad

$$\{u^\alpha, v^\alpha, \underset{(1)}{e}{}^\alpha, \underset{(2)}{e}{}^\alpha\},$$

where v^α is the wave's propagation unit vector in the observer's rest frame,

$$v^\alpha=\frac{1}{|\Omega|}(l^\alpha+\Omega u^\alpha),$$

and $\underset{(1)}{e}{}^\alpha$, $\underset{(2)}{e}{}^\alpha$ are two spacelike unit vectors, orthogonal to each other and to u^α, v^α.

By exploiting the "gauge" freedom $\psi_\alpha\to\psi_\alpha+\lambda l_\alpha$ we can always make

$$\psi_\alpha u^\alpha=0. \tag{6.14}$$

It follows that ψ_α lies in the two-space spanned by $\underset{(1)}{e}{}^\alpha$, $\underset{(2)}{e}{}^\alpha$, hence we can write

$$\psi^\alpha=|\psi|e^\alpha, \tag{6.15}$$

where e^α is the polarization vector,

$$e^\alpha=\cos\theta\,\underset{(1)}{e}{}^\alpha+\sin\theta\,\underset{(2)}{e}{}^\alpha, \tag{6.16}$$

θ being the polarization angle.

At each point, $\underset{(1)}{e}{}^\alpha$, $\underset{(2)}{e}{}^\alpha$ are defined up to an orthogonal transformation. We exploit this arbitrariness in the definition of $\underset{(1)}{e}{}^\alpha$, $\underset{(2)}{e}{}^\alpha$ by choosing them to satisfy

$$\underset{(2)}{e}{}^\alpha l^\mu\nabla_\mu\underset{(1)}{e}{}_\alpha=0. \tag{6.17}$$

Then from (6.11), (6.12), and (6.17) it follows that

$$l^\alpha\nabla_\alpha\theta=0, \tag{6.18}$$

that is, the polarization angle with respect to the basis $\underset{(1)}{e}^{\alpha}$, $\underset{(2)}{e}^{\alpha}$ is propagated in a parallel manner.

The conservation law (6.13) has a simple geometric and physical interpretation. Let Σ be the null hypersurface given by

$$\varphi(x^{\alpha}) = c = \text{const.}$$

Let $\mathscr{B}$ be a bicharacteristic tube and σ a cross section of such a tube and denote by $d\sigma$ its area element. Then one can state the following theorem.

THEOREM 6.1. *For all cross sections one has*

$$\int_{\sigma} |\psi|^2 \, d\sigma = \text{const.}$$

Proof. We choose coordinates adapted to Σ, $x^0 = \varphi$, x^i are coordinates on Σ. Furthermore, we specialize the coordinates x^i such that

$$l^{\alpha} = (0, 1, 0, 0).$$

From the conditions $l_{\alpha} = (1, 0, 0, 0)$ and $l_{\alpha} l^{\alpha} = 0$ we obtain

$$g_{01} = 1, \quad g_{11} = 0, \quad g_{1A} = 0, \quad A = 2, 3$$

for the metric coefficients. In these coordinates the transport equation (6.13) is written

$$\frac{\partial}{\partial x^1} (\sqrt{|g|} \, |\psi|^2) = 0,$$

where

$$g = \det g_{\mu\nu},$$

whence

$$\sqrt{|g|} \, |\psi|^2 = F(x^2, x^3).$$

A cross section of a tube of rays is given by

$$x^1 = h(x^2, x^3).$$

Now the line element in these coordinates is

$$ds^2 = g_{00}(dx^0)^2 + 2dx^0 dx^1 + 2g_{0A} dx^0 \, dx^A + g_{AB} dx^A dx^B.$$

Therefore,

$$ds^2|_{\Sigma} = g_{AB} dx^A \, dx^B$$

and it is easy to check that

$$|g| = |\det g_{AB}|.$$

It follows that

$$\int_\sigma |\psi|^2 \mathrm{d}\sigma = \int_\sigma |\psi|^2 |g| \,\mathrm{d}x^2\,\mathrm{d}x^3 = \int F(x^2, x^3)\mathrm{d}x^2\,\mathrm{d}x^3 = \text{const.} \qquad \text{Q.E.D.}$$

This result has the following physical meaning. The energy-momentum tensor for the electromagnetic field in vacuo is

$$T^{\alpha\beta} = \frac{1}{4\pi\mu_0}(F^{\alpha\mu}F^{\beta\nu} - \tfrac{1}{4}g^{\alpha\beta}F_{\mu\nu}F^{\mu\nu}). \tag{6.19}$$

By substituting (6.3) into (6.19) we obtain a formal series in $1/\omega$. Now we average out to zero the rapidly oscillating terms containing factors of the kind $e^{-2i\omega\varphi}$, according to the definition

$$\langle T^{\alpha\beta} \rangle = \frac{\omega}{2\pi}\int_0^{2\pi/\omega} T^{\alpha\beta}\,\mathrm{d}\varphi.$$

This averaging is to be interpreted as "ensemble" averaging. Physically it corresponds to averaging over a time scale which is long compared to the period of the wave, but much shorter than the typical time scale of change of the wave's amplitude and of the background gravitational field.

The averaged zeroth order energy-momentum tensor is

$$\langle \underset{(0)}{T^{\alpha\beta}} \rangle = \frac{1}{8\pi\mu_0}|\psi|^2 l^\alpha l^\beta. \tag{6.20}$$

It follows that the averaged energy-density as measured by the observer with four-velocity u^α is

$$\langle E \rangle = \langle \underset{(0)}{T^{\alpha\beta}} \rangle u_\alpha u_\beta = \frac{1}{8\pi\mu_0}|\psi|^2\Omega^2, \tag{6.21}$$

whereas the averaged energy-flux in the wave's propagation direction v is

$$\langle \Phi \rangle = \langle \underset{(0)}{T^{\alpha\beta}} \rangle u_\alpha v_\beta = \frac{1}{8\pi\mu_0}|\psi|^2\Omega^2. \tag{6.22}$$

Therefore, for all observers measuring the same frequency $\Omega = \text{constant}$, Theorem 6.1 expresses that

$$\int_\sigma \langle E \rangle \mathrm{d}\sigma = \int_\sigma \langle \Phi \rangle \mathrm{d}\sigma = \text{const.}$$

for all cross sections of a bicharacteristic tube.

This is the standard area-intensity law of general relativistic geometrical optics (Friedlander, 1975).

The higher order corrections to the geometrical optics field $\underset{(0)}{F}{}^{\alpha\beta}$ as well as the resulting modifications of the area-intensity law have been studied by Anile (1976) using the spinor formalism.

The propagation law (6.13) for the amplitude and (6.18) for the polarization are at the basis of the derivation of the relativistic radiative transfer equation in vacuo (Ellis, 1971; Anile and Breuer, 1974; Anile, 1976). In the next section we shall treat geometrical optics for radiation propagating through a refractive medium.

6.2 Geometrical optics in general relativistic refractive media

Maxwell's equations in current-free refractive media are equations (2.46)–(2.47) with $J^\alpha = 0$, and they read

$$\nabla_\alpha F_{\beta\gamma} + \nabla_\beta F_{\gamma\alpha} + \nabla_\gamma F_{\alpha\beta} = 0, \tag{6.23}$$

$$\nabla_\alpha I^{\alpha\beta} = 0, \tag{6.24}$$

where $F_{\alpha\beta}$, $I_{\alpha\beta}$ are the electromagnetic field tensor and induction tensor, respectively.

As is usual in many applications we shall assume a linear and isotropic refractive medium. The constitutive relations between $F_{\alpha\beta}$ and $I_{\alpha\beta}$ are equations (2.50)—(2.51) and can be written in the form (Pichon, 1965)

$$I_{\alpha\beta} = \tfrac{1}{2}\chi^{\mu\nu}_{\alpha\beta} F_{\mu\nu}, \tag{6.25}$$

with the susceptibility tensor $\chi^{\mu\nu}_{\alpha\beta}$ given by

$$\chi^{\mu\nu}_{\alpha\beta} = \frac{1}{\hat{\mu}}(g^\mu_\alpha g^\nu_\beta - g^\mu_\beta g^\nu_\alpha) - \frac{1-N^2}{\hat{\mu}}(g^\nu_\alpha u^\mu u_\beta - g^\nu_\beta u^\mu u_\alpha - g^\mu_\alpha u^\nu u_\beta + g^\mu_\beta u^\nu u_\alpha), \tag{6.26}$$

where N is the refractive index,

$$N^2 = \hat{\mu}(1 + 4\pi x),$$

u^α is the medium's four-velocity, $\hat{\mu}$ is the magnetic permeability, and x is the electric susceptibility.

We look for high-frequency waves of the form

$$F_{\alpha\beta} = \mathrm{Re}\left\{ e^{i\omega\varphi} \sum_{n=0}^{\infty} \omega^{-n} \underset{(n)}{F}{}_{\alpha\beta} \right\}, \tag{6.27a}$$

$$I_{\alpha\beta} = \mathrm{Re}\left\{ e^{i\omega\varphi} \sum_{n=0}^{\infty} \omega^{-n} \underset{(n)}{I}_{\alpha\beta} \right\}. \tag{6.27b}$$

Then, to the zeroth order, we obtain

$$l_\alpha \underset{(0)}{F}_{\beta\gamma} + l_\beta \underset{(0)}{F}_{\gamma\alpha} + l_\gamma \underset{(0)}{F}_{\alpha\beta} = 0, \tag{6.28a}$$

$$l_\alpha \underset{(0)}{I}^{\alpha\beta} = 0, \tag{6.28b}$$

and to he nth order

$$i(l_\alpha \underset{(n+1)}{F}_{\beta\gamma} + l_\beta \underset{(n+1)}{F}_{\gamma\alpha} + l_\gamma \underset{(n+1)}{F}_{\alpha\beta}) + \nabla_\alpha \underset{(n)}{F}_{\beta\gamma} + \nabla_\beta \underset{(n)}{F}_{\gamma\alpha} + \nabla_\gamma \underset{(n)}{F}_{\alpha\beta} = 0, \tag{6.29a}$$

$$il_\alpha \underset{(n+1)}{I}^{\alpha\beta} + \nabla_\alpha \underset{(n)}{I}^{\alpha\beta} = 0. \tag{6.29b}$$

From (6.28b), by using the constitutive relation (6.26) we obtain

$$l^\mu \underset{(0)}{F}_{\mu\beta} - (1 - N^2)(\Omega u^\nu \underset{(0)}{F}_{\beta\nu} - l^\mu \underset{(0)}{F}_{\mu\nu} u^\nu u_\beta) = 0. \tag{6.30}$$

Let $Y_\beta = \underset{(0)}{F}_{\alpha\beta} l^\alpha$. Then equation (6.30) is rewritten

$$Y_\beta + (N^2 - 1)(\Omega u^\nu \underset{(0)}{F}_{\beta\nu} - Y_\nu u^\nu u_\beta) = 0. \tag{6.30$'$}$$

By contracting with u^β it follows that, since $N \neq 0$,

$$Y_\beta u^\beta = 0. \tag{6.31}$$

Then, from (6.30′),

$$\underset{(0)}{F}_{\nu\beta} k^\nu = 0, \tag{6.32}$$

where

$$k^\nu = \frac{1}{\mu}(l^\nu - (N^2 - 1)\Omega u^\nu). \tag{6.33}$$

Notice that k^ν is tangent to Σ, because $k^\nu l_\nu = 0$. Now let n^α be a vector such that

$$l^\alpha n_\alpha = 1.$$

By contracting equation (6.28a) with n^α we obtain

$$\underset{(0)}{F}_{\alpha\beta} = \psi_\alpha l_\beta - \psi_\beta l_\alpha, \tag{6.34a}$$

with ψ_α determined up to the gauge transformation

$$\psi_\alpha \to \psi_\alpha + \lambda l_\alpha, \quad \lambda \in \mathbb{R}. \tag{6.34b}$$

Now we prove that for a nontrivial solution we must have $l^\alpha l_\alpha \neq 0$. In fact, if $l^\alpha l_\alpha = 0$, we can always choose the gauge transformation such that ψ_α satisfies

$$\psi_\alpha u^\alpha = 0.$$

Then, from the condition (6.31) it follows that, since $\Omega \neq 0$,

$$\psi_\alpha l^\alpha = 0,$$

hence $Y_\beta = 0$. Therefore, from equation (6.30) we have $\psi_\beta = 0$. Having established that $l_\alpha l^\alpha \neq 0$, we can choose ψ_α, by applying the gauge transformation, such that

$$\psi_\alpha l^\alpha = 0, \tag{6.34c}$$

and then equation (6.31) yields

$$\psi_\alpha u^\alpha = 0. \tag{6.34d}$$

Finally, from (6.30′) we obtain the compatibility relation

$$\frac{l_\alpha l^\alpha}{\Omega^2} = N^2 - 1. \tag{6.35}$$

By the way in which it has been obtained it is apparent that equation (6.35) coincides with the characteristic equation for the system of Maxwell's equations (6.23)–(6.26) (Pichon, 1965). It is also easy to see that

$$\underset{(0)}{I}{}_{\alpha\beta} = \psi_\alpha k_\beta - \psi_\beta k_\alpha. \tag{6.36}$$

To the first order, equations (6.29a)–(6.29b) give

$$i(l_\alpha \underset{(1)}{F}{}_{\beta\gamma} + l_\beta \underset{(1)}{F}{}_{\gamma\alpha} + l_\gamma \underset{(1)}{F}{}_{\alpha\beta}) + l_\alpha(\nabla_\beta\psi_\gamma - \nabla_\gamma\psi_\beta)$$

$$+ l_\beta(\nabla_\gamma\psi_\alpha - \nabla_\alpha\psi_\gamma) + l_\gamma(\nabla_\alpha\psi_\beta - \nabla_\beta\psi_\alpha) = 0, \tag{6.37a}$$

$$ik_\alpha \underset{(1)}{F}{}^{\alpha\beta} - i\frac{(N^2-1)}{\hat{\mu}} \underset{(1)}{F}{}^{\alpha\mu} l_\alpha u_\mu u^\beta + \nabla_\alpha(\psi^\alpha k^\beta - \psi^\beta k^\alpha) = 0. \tag{6.37b}$$

Let

$$\underset{(1)}{\psi}{}_\beta = \underset{(1)}{F}{}_{\alpha\beta} k^\alpha,$$

$$Z_\beta = k^\alpha \nabla_\alpha \psi_\beta - k^\alpha \nabla_\beta \psi_\alpha.$$

Then equation (6.37a), contracted with k^α, yields

$$i\underset{(1)}{\psi}{}_\beta + Z_\beta = \tilde{k} l_\beta,$$

with $\tilde{k}$ an arbitrary scalar, which, after substitution into equation (6.37b), gives

$$-2k^\alpha\nabla_\alpha\psi^\beta + \tilde{k}l^\beta + \psi^\alpha(\nabla_\alpha k^\beta - \nabla^\beta k_\alpha) + (\nabla_\alpha\psi^\alpha)k^\beta - \psi^\beta\nabla_\alpha k^\alpha$$
$$-i\frac{(N^2-1)}{\hat{\mu}}\underset{(1)}{F}{}^{\alpha\mu}l_\alpha u_\mu u^\beta = 0. \tag{6.37c}$$

From this equation we derive the amplitude transport law

$$\nabla_\alpha(|\psi|^2 k^\alpha) = 0. \tag{6.38}$$

The rays are the curves tangent to k^α and coincide with the bicharacteristics.

Let

$$\{u^\alpha, v^\alpha, \underset{(1)}{e}{}^\alpha, \underset{(2)}{e}{}^\alpha\}$$

be the orthonormal frame as introduced in the previous section, but with

$$v^\alpha = \frac{l^\alpha + \Omega u^\alpha}{N\Omega}.$$

Then, from equation (6.37c) one obtains the transport law for the polarization,

$$2k^\alpha\nabla_\alpha\theta + \frac{l_\sigma l^\sigma}{\Omega}\omega^{\mu\nu}\underset{(2)}{e}{}_\mu\underset{(1)}{e}{}_\nu = 0, \tag{6.39}$$

where $\omega_{\mu\nu}$ is the vorticity tensor of the vector field u^α, defined by

$$\omega_{\mu\nu} = h^\alpha_\mu h^\beta_\nu(\nabla_\alpha u_\beta - \nabla_\beta u_\alpha),$$

with

$$h^\alpha_\mu = \delta^\alpha_\mu + u_\mu u^\alpha,$$

and θ is defined by

$$\psi^\alpha = |\psi|(\underset{(1)}{e}{}^\alpha\cos\theta + \underset{(2)}{e}{}^\alpha\sin\theta).$$

A physical interpretation of the conservation law (6.38) can be obtained as follows.

Let us assume u^α is hypersurface orthogonal and that $u = \mathrm{d}\tau$, where τ is a scalar function. Let Σ be the hypersurface $\varphi(x) = c$, $c \in \mathbb{R}$, $\sigma_\tau = \{x \in \Sigma : \tau(x) = \text{const.}\}$. Then it is possible to prove the following result.

THEOREM 6.2. *One has*

$$\int_{\sigma_\tau}\frac{1}{\hat{\mu}}N|\psi|^2\,\mathrm{d}\sigma_\tau = \text{const.}$$

Proof. Let (y^0, x^i) be normal coordinates based on Σ (Synge, 1960), which is given by $y^0 = 0$. In these coordinates

$$l^\alpha = (L, 0, 0, 0),$$
$$l_\alpha = (1, 0, 0, 0),$$

where $L = l_\mu l^\mu$ and the metric is given by

$$ds^2 = \frac{1}{L}(dy^0)^2 + g_{ij}\,dx^i\,dx^j.$$

Now, on Σ the coordinates x^i are arbitrary. Let

$$t = \tau(0, x^i).$$

We choose the coordinates x^i on Σ in such a way that they are normal coordinates based on the family of two-surfaces $t = \text{const.}$ On Σ the vector

$$w_\alpha = \partial_\alpha t = u_\alpha - \frac{\Omega}{L} l_\alpha$$

has components $w_\alpha = (0, 1, 0, 0)$ and

$$w^\alpha = \left(0, -1 - \frac{\Omega^2}{L}, 0, 0\right).$$

In these coordinates the metric is

$$ds^2 = \frac{1}{L}(dy^0)^2 + \frac{1}{w}(dy^1)^2 + \tilde{g}_{ij}\,dy^i\,dy^j,$$

with $w = w_\alpha w^\alpha$. The metric on the two-surfaces $t = \text{const.}$ is given by

$$ds^2|_t = \tilde{g}_{ij}\,dy^i\,dy^j.$$

Also,

$$k^\alpha = \frac{1}{\hat{\mu}}(0, \Omega N^2, 0, 0).$$

Let $|\tilde{g}|^{1/2}$ be the surface element induced on σ_τ by the metric $\tilde{g}_{ij}$. Then (6.38) reads,

$$\frac{\partial}{\partial \tau}\left(\frac{1}{\hat{\mu}}|\psi|^2 N |\tilde{g}|^{1/2}\right) = 0,$$

which proves the theorem. Q.E.D.

Now the Minkowski $\underset{(M)}{T}{}^{\alpha\beta}$ energy-momentum tensor for the electromagnetic field in matter is given by equation (2.59).

By proceeding as in the previous section we obtain, at the lowest order,

$$\langle \underset{(M)}{T^{\alpha\beta}} \rangle = \frac{1}{8\pi} |\psi|^2 l^\alpha k^\beta, \tag{6.40}$$

which gives an energy-density

$$\langle E \rangle = \frac{1}{8\pi\hat{\mu}} N^2 \Omega^2 |\psi|^2 \tag{6.41}$$

and an energy flux, in the wave's propagation direction,

$$\langle \Phi \rangle = \frac{1}{8\pi\hat{\mu}} N \Omega^2 |\psi|^2. \tag{6.42}$$

Therefore, for all observers measuring the same frequency $\Omega = \text{const.}$, one has

$$\int_{\sigma_\tau} \frac{E}{N} \mathrm{d}\sigma_\tau = \int_{\sigma_\tau} \Phi \, \mathrm{d}\sigma_\tau = \text{const.}, \tag{6.43}$$

which is the general-relativistic formulation of the area-flux law of geometrical optics in isotropic refractive media (Anile and Moschetti, 1979).

By using this law and equation (6.39) for the polarization one could derive the transfer equation for radiation propagating through refractive media. However, in a plasma, refractive properties are usually associated with dispersion. In the next section we will consider a very simple example for a dispersive medium: a cold electron fluid, which is a model of notable interest for some astrophysical and laboratory plasma.

6.3. Electromagnetic waves in a cold relativistic plasma: linear theory

Maxwell's equations in vacuo in the presence of charges and currents are obtained from equation (4.81)–(4.82) and are written

$$\nabla_\alpha F_{\beta\gamma} + \nabla_\beta F_{\gamma\alpha} + \nabla_\gamma F_{\alpha\beta} = 0, \tag{6.44}$$

$$\nabla_\beta F^{\alpha\beta} = 4\pi\mu_0 J^\alpha, \tag{6.45}$$

with μ_0 the vacuum magnetic permeability.

We make the following assumptions for the medium (Madore, 1974):

(i) The medium consists of two noninteracting components, the ion and electron components;

(ii) the energy-momentum tensor for the electron component is that of a pressureless perfect fluid (dust), i.e.,

$$T_{\alpha\beta} = nmu_\alpha u_\beta, \tag{6.46}$$

where n, u^α are the electron number density and four-velocity of the electrons;

(iii) because of the larger proton mass the ions are considered as a fixed background.

The total current J^α is given by

$$J^\alpha = -4\pi e(nu^\alpha - n_i u_i^\alpha) \tag{6.47}$$

where e is the electron charge (absolute value), n_i, u_i^α are the ion number density and four-velocity, respectively.

Then to Maxwell's equations (6.44)–(6.45) we must add the equation of motion for the electron fluid

$$\nabla_\beta T^{\alpha\beta} = -enF^{\alpha\beta}u_\beta,$$

which is equivalent to the electron number conservation equation

$$\nabla_\alpha(nu^\alpha) = 0 \tag{6.48}$$

and the electron momentum equation

$$u^\mu \nabla_\mu u^\alpha = -\frac{e}{m} F^{\alpha\mu}u_\mu. \tag{6.49}$$

Now we linearize equations (6.44)–(6.45) and (6.48)–(6.49) around an unperturbed state with $F_{\alpha\beta} = 0$. Let $\hat{F}_{\alpha\beta}$, $\hat{n}$, $\hat{u}_\alpha$ denote the perturbations to the electromagnetic field, number density, and four-velocity, respectively. Then the linearized equations give

$$\nabla_\alpha \hat{F}_{\beta\gamma} + \nabla_\beta \hat{F}_{\gamma\alpha} + \nabla_\gamma \hat{F}_{\alpha\beta} = 0, \tag{6.50}$$

$$\nabla_\beta \hat{F}^{\alpha\beta} = -4\pi\mu_0 e(\hat{n}u^\alpha + n\hat{u}^\alpha), \tag{6.51}$$

$$\nabla_\alpha(\hat{n}u^\alpha + n\hat{u}^\alpha) = 0, \tag{6.52}$$

$$\hat{u}^\mu \nabla_\mu u^\alpha + u^\mu \nabla_\mu \hat{u}^\alpha = -\frac{e}{m}\hat{F}^{\alpha\mu}u_\mu, \tag{6.53}$$

$$u_\alpha \hat{u}^\alpha = 0, \tag{6.54}$$

the latter arising from the normalization condition $u_\alpha u^\alpha = -1$.

Now we analyze the system (6.50)–(6.54) by using the two-timing method (Whitham, 1974; Jeffrey and Kawahara, 1982). We assume that, in a selected class of coordinates (x^α), the quantities $\hat{F}^{\alpha\beta}, \hat{n}, \hat{u}^\alpha$ are of the following form:

$$\hat{F}^{\alpha\beta} = \hat{F}^{\alpha\beta}\left(\varepsilon x^\mu, \frac{1}{\varepsilon}\Theta(\varepsilon x^\mu)\right), \quad \hat{n} = \hat{n}\left(\varepsilon x^\mu, \frac{1}{\varepsilon}\Theta(\varepsilon x^\mu)\right), \text{ etc.,}$$

with Θ a function to be determined and $\varepsilon > 0$ a real parameter. The unperturbed background fields $g_{\alpha\beta}$, n, u^α are assumed to vary on the slow scale, that is, they are of the form

$$g_{\alpha\beta} = G_{\alpha\beta}(\varepsilon x^\mu), \quad n = N(\varepsilon x^\mu), \text{ etc.}$$

It is convenient to introduce the auxiliary variables $X^\mu = \varepsilon x^\mu$, $\varphi = \frac{1}{\varepsilon}\Theta(\varepsilon x^\mu)$. For fixed $\varepsilon > 0$ the X^α can be interpreted as "slow" space-time coordinates. The parameter ε measures the ratio of the fast length scale to the slow one. If we write $f = f(X^\alpha, \varphi)$ and put

$$f_{,\alpha} = \frac{\partial f}{\partial X^\alpha}, \quad \dot{f} = \frac{\partial f}{\partial \varphi},$$

then

$$\frac{\partial f}{\partial x^\alpha} = \varepsilon f_{,\alpha} + l_\alpha \dot{f},$$

where

$$l_\alpha = \frac{\partial \varphi}{\partial x^\alpha} = \frac{\partial \Theta}{\partial X^\alpha} = \Theta_{,\alpha}$$

is the normal to the wavefront $\varphi = \text{const}$.

For the connection coefficients $\Gamma^\alpha_{\beta\gamma}(x^\mu, \varepsilon)$ we have

$$\Gamma^\alpha_{\beta\gamma} = \varepsilon \hat{\Gamma}^\alpha_{\beta\gamma},$$

with

$$\hat{\Gamma}^\alpha_{\beta\gamma} = \tfrac{1}{2} G^{\alpha\delta}(G_{\delta\beta,\gamma} + G_{\delta\gamma,\beta} + G_{\beta\gamma,\delta}),$$

whence $\Gamma^\alpha_{\beta\gamma} = 0(\varepsilon)$.

Henceforth we shall use a semicolon in order to indicate the covariant derivative with respect to the slow variables X^α, that is

$$A_{\alpha;\beta} = A_{\alpha,\beta} - \hat{\Gamma}^\mu_{\alpha\beta} A_\mu.$$

In terms of slow and fast variables equations (6.50)–(6.54) can be rewritten

$$l_\alpha \dot{\hat{F}}_{\beta\gamma} + l_\beta \dot{\hat{F}}_{\gamma\alpha} + l_\gamma \dot{\hat{F}}_{\alpha\beta} + \varepsilon(\hat{F}_{\beta\gamma;\alpha} + \hat{F}_{\gamma\alpha;\beta} + \hat{F}_{\alpha\beta;\gamma}) = 0, \tag{6.55}$$

$$l_\beta \dot{\hat{F}}^{\alpha\beta} + \varepsilon \hat{F}^{\alpha\beta}_{;\beta} = -4\pi\mu_0 e(\hat{n}u^\alpha + n\hat{u}^\alpha), \tag{6.56}$$

$$u^\mu l_\mu \dot{\hat{u}}^\alpha + \varepsilon \hat{u}^\mu u^\alpha_{;\mu} + \varepsilon u^\mu \hat{u}^\alpha_{;\mu} = -\frac{e}{m}\hat{F}^{\alpha\mu}u_\mu, \tag{6.57}$$

$$l_\alpha(\dot{\hat{n}}u^\alpha + n\dot{\hat{u}}^\alpha) + \varepsilon(\hat{n}u^\alpha + n\hat{u}^\alpha)_{;\alpha} = 0. \tag{6.58}$$

A *locally plane wave* is defined by the following formal asymptotic expansion (Anile and Pantano, 1977)

$$\hat{n} = \left(\sum_{q=0}^{\infty} \varepsilon^q \underset{(q)}{\hat{n}} \right) \exp(i\varphi),$$

$$\hat{F}^{\alpha\beta} = \left(\sum_{q=0}^{\infty} \varepsilon^q \underset{(q)}{\hat{F}^{\alpha\beta}} \right) \exp(i\varphi),$$

$$\hat{u}^{\alpha} = \left(\sum_{q=0}^{\infty} \varepsilon^q \underset{(q)}{\hat{u}^{\alpha}} \right) \exp(i\varphi).$$

Substituting into equations (6.55)–(6.58) and termwise equating to zero the coefficients of the resulting formal power series in ε, we obtain at the zeroth order

$$l_\alpha \underset{(0)}{\dot{\hat{F}}}_{\beta\gamma} + l_\beta \underset{(0)}{\dot{\hat{F}}}_{\gamma\alpha} + l_\gamma \underset{(0)}{\dot{\hat{F}}}_{\alpha\beta} = 0, \tag{6.59}$$

$$il_\beta \underset{(0)}{\hat{F}^{\alpha\beta}} + 4\pi\mu_0 e \left(\underset{(0)}{\hat{n}} u^\alpha + n \underset{(0)}{\hat{u}^\alpha} \right) = 0, \tag{6.60}$$

$$i\Omega \underset{(0)}{\hat{u}^\alpha} + \frac{e}{m} \underset{(0)}{\hat{F}^{\alpha\mu}} u_\mu = 0, \tag{6.61}$$

$$\Omega \underset{(0)}{\hat{n}} + n l_\alpha \underset{(0)}{\hat{u}^\alpha} = 0, \tag{6.62}$$

where $\Omega = u_\mu l^\mu$ is the local frequency of the wave relative to an observer at rest with the medium.

For the higher orders one gets

$$i\left(l_\alpha \underset{(q+1)}{\dot{\hat{F}}}_{\beta\gamma} + l_\beta \underset{(q+1)}{\dot{\hat{F}}}_{\gamma\alpha} + l_\gamma \underset{(q+1)}{\dot{\hat{F}}}_{\alpha\beta} \right) + \underset{(q)}{\hat{F}}_{\beta\gamma;\alpha} + \underset{(q)}{\hat{F}}_{\gamma\alpha;\beta} + \underset{(q)}{\hat{F}}_{\alpha\beta;\gamma} = 0, \tag{6.63}$$

$$il_\beta \underset{(q+1)}{\hat{F}^{\alpha\beta}} + \underset{(q)}{\hat{F}^{\alpha\beta}_{;\beta}} + 4\pi\mu_0 e(\underset{(q+1)}{\hat{n}} u^\alpha + n \underset{(q+1)}{\hat{u}^\alpha}) = 0, \tag{6.64}$$

$$i\Omega \underset{(q+1)}{\hat{u}^\alpha} + \underset{(q)}{\hat{u}^\mu} u^\alpha_{;\mu} + u^\mu \underset{(q)}{\hat{u}^\alpha_{;\mu}} + \frac{e}{m} \underset{(q+1)}{\hat{F}^{\alpha\mu}} u_\mu = 0, \tag{6.65}$$

$$i\Omega \underset{(q+1)}{\hat{n}} + inl_\alpha \underset{(q+1)}{\hat{u}^\alpha} + (\underset{(q)}{\hat{n}} u^\alpha + n \underset{(q)}{\hat{u}^\alpha})_{;\alpha} = 0. \tag{6.66}$$

By proceeding as in the previous section, from equation (6.59) we have

$$\underset{(0)}{\hat{F}}_{\alpha\beta} = \psi_\alpha l_\beta - \psi_\beta l_\alpha, \tag{6.67}$$

with ψ_α determined up to the transformation

$$\psi_\alpha \rightarrow \psi_\alpha + \lambda l_\alpha.$$

Since $\Omega = l_\alpha u^\alpha \neq 0$ we can always choose λ such that

$$\psi_\alpha u^\alpha = 0. \tag{6.68}$$

Then equations (6.60)–(6.61) are rewritten

$$il_\beta l^\beta \psi^\alpha - il^\alpha(\psi^\beta l_\beta) + 4\pi\mu_0 e(\underset{(0)}{\hat{n}}\, u^\alpha + n\, \underset{(0)}{\hat{u}}{}^\alpha) = 0, \tag{6.69}$$

$$i\, \underset{(0)}{\hat{u}}{}^\alpha + \frac{e}{m}\psi^\alpha = 0. \tag{6.70}$$

From equations (6.62) and (6.69)–(6.70) we obtain

$$\left(\Omega^2 - \frac{4\pi\mu_0 e^2 n}{m}\right) l_\alpha \underset{(0)}{\hat{u}}{}^\alpha = 0. \tag{6.71}$$

Now we assume that

$$\Omega^2 \neq \frac{4\pi\mu_0 e^2 n}{m},$$

since we are not interested in electrostatic waves. Therefore we have

$$l_\alpha \underset{(0)}{\hat{u}}{}^\alpha = 0, \quad l_\alpha \psi^\alpha = 0 \tag{6.72}$$

and from (6.62),

$$\underset{(0)}{\hat{n}} = 0. \tag{6.73}$$

Then, from equation (6.69), because $\psi^\alpha \neq 0$, we have

$$l_\mu l^\mu = -\frac{4\pi\mu_0 e^2 n}{m} = -\Omega_p^2, \tag{6.74}$$

where Ω_p is the plasma frequency, which is the dispersion relation for electromagnetic plasma waves (Stix, 1962).

Now we derive the transport equation. Equations (6.63)–(6.66) for $q=0$ become

$$i(l_\alpha \underset{(1)}{\hat{F}}{}_{\beta\gamma} + l_\beta \underset{(1)}{\hat{F}}{}_{\gamma\alpha} + l_\gamma \underset{(1)}{\hat{F}}{}_{\alpha\beta}) + \underset{(0)}{\hat{F}}{}_{\beta\gamma;\alpha} + \underset{(0)}{\hat{F}}{}_{\gamma\alpha;\beta} + \underset{(0)}{\hat{F}}{}_{\alpha\beta;\gamma} = 0, \tag{6.75}$$

$$il_\beta \underset{(1)}{\hat{F}}{}^{\alpha\beta} + \underset{(0)}{\hat{F}}{}^{\alpha\beta}_{;\beta} + 4\pi\mu_0 e(\underset{(1)}{\hat{n}}\, u^\alpha + n\, \underset{(1)}{\hat{u}}{}^\alpha) = 0, \tag{6.76}$$

$$i\Omega\, \underset{(1)}{\hat{u}}{}^\alpha + \underset{(0)}{\hat{u}}{}^\mu u^\alpha_{;\mu} + u^\mu \underset{(0)}{\hat{u}}{}^\alpha_{;\mu} + \frac{e}{m} \underset{(1)}{\hat{F}}{}^{\alpha\mu} u_\mu = 0, \tag{6.77}$$

$$i\Omega\, \underset{(1)}{\hat{n}} + inl_\alpha \underset{(1)}{\hat{u}}{}^\alpha + n_{;\alpha} \underset{(0)}{\hat{u}}{}^\alpha + n\, \underset{(0)}{\hat{u}}{}^\alpha_{;\alpha} = 0. \tag{6.78}$$

From equation (6.75), by contracting with l^α, we obtain

$$i\underset{(1)}{\hat{F}}_{\alpha\beta} = \psi_{\alpha;\beta} - \psi_{\beta;\alpha} + Z_\alpha l_\beta - Z_\beta l_\alpha, \tag{6.79}$$

with Z_α determined up to the transformation

$$Z_\alpha \rightarrow Z_\alpha + \lambda l_\alpha.$$

Since $l^\alpha l_\alpha \neq 0$, we can choose Z_α in such a way that

$$Z_\alpha l^\alpha = 0. \tag{6.80}$$

Using equation (6.80) in equations (6.76)–(6.77) and also

$$l_\beta \psi^\beta_{\ ;\alpha} = -\psi^\beta l_{\beta;\alpha} = -\psi^\beta l_{\alpha;\beta},$$

gives

$$2l^\beta \psi^\alpha_{\ ;\beta} + \psi^\alpha l^\beta_{\ ;\beta} - \psi^\beta_{\ ;\beta} l^\alpha + Z^\alpha l^\beta l_\beta + 4\pi\mu_0 e \underset{(1)}{\hat{n}} u^\alpha + 4\pi\mu_0 en \underset{(1)}{\hat{u}}{}^\alpha = 0, \tag{6.81}$$

$$\Omega \underset{(1)}{\hat{u}}{}^\alpha + \frac{e}{m}\psi^\mu u^\alpha_{\ ;\mu} + \frac{e}{m}\psi^{\mu;\alpha} u_\mu - \frac{e}{m} Z^\mu u_\mu l^\alpha + \frac{e}{m} Z^\alpha \Omega = 0. \tag{6.82}$$

Using (6.82) in (6.81) gives

$$2l^\beta \psi^\alpha_{\ ;\beta} + \psi^\alpha l^\beta_{\ ;\beta} - \psi^\beta_{\ ;\beta} l^\alpha - \frac{4\pi\mu_0 e^2 n}{m}(\psi^\mu u^\alpha_{\ ;\mu} + \psi^{\mu;\alpha} u_\mu - Z^\mu u_\mu l^\alpha) = 0. \tag{6.83}$$

By contracting equation (6.83) with $\bar{\psi}^\alpha$, we get

$$l^\beta |\psi|^2_{;\beta} + |\psi|^2 l^\beta_{\ ;\beta} = 0, \tag{6.84}$$

which is the transport law for the amplitude.

Proceeding as in the previous section, we introduce the orthonormal frame

$$(u^\alpha, v^\alpha, \underset{(1)}{e}{}^\alpha, \underset{(2)}{e}{}^\alpha),$$

with

$$v^\alpha = \frac{l^\alpha + \Omega u^\alpha}{|\Omega^2 - \Omega_p^2|}$$

and $\underset{(1)}{e}{}^\alpha$, $\underset{(2)}{e}{}^\alpha$ two orthonormal spacelike vectors, orthogonal to both u^α and l^α and such that

$$\underset{(2)}{e}{}^\alpha l^\mu \underset{(1)}{e}{}_{\alpha;\mu} = 0.$$

Then we can write

$$\psi^\alpha = |\psi|(\underset{(1)}{e}{}^\alpha \cos\theta + \underset{(2)}{e}{}^\alpha \sin\theta), \tag{6.85}$$

and by substituting into equation (6.83) and contracting with $\underset{(2)}{e}_\alpha$, we obtain the polarization angle transport equation

$$2l^\beta\theta_{;\beta} + \frac{4\pi\mu_0 e^2 n}{\Omega m}\omega^{\mu\alpha}\underset{(1)}{e}_\mu \underset{(2)}{e}_\alpha = 0, \tag{6.86}$$

where $\omega^{\mu\alpha}$ is the electron fluid vorticity.

An alternative approach which leads to equivalent results is based on the averaged Lagrangian formalism (Dougherty, 1970).

The propagation equation (6.84) for the amplitude can be interpreted in terms of the area-amplitude law, at least in some cases (Moschetti, 1987). The main result which has been obtained is the derivation from Maxwell's equations of the propagation laws (6.84)–(6.86) for the amplitude and polarization of an electromagnetic wave in a cold plasma. These laws form the basis of relativistic geometrical optics in a dispersive medium. They could be used in order to obtain the relativistic transfer equation for polarized radiation in a dispersive medium (the effect of a background magnetic field could easily be taken into account; Anile and Pantano, 1979; Breuer and Ehlers, 1980, 1981). In the next section we shall reconsider the problem we have treated, trying to take into account weakly nonlinear effects.

6.4. Electromagnetic waves in a cold relativistic plasma: weakly nonlinear analysis

We start with equations (6.44)–(6.45) and (6.48)–(6.49), which we rewrite as

$$\nabla_\alpha F_{\beta\gamma} + \nabla_\beta F_{\gamma\alpha} + \nabla_\gamma F_{\alpha\beta} = 0, \tag{6.87}$$

$$\nabla_\beta F^{\alpha\beta} = 4\pi\mu_0 e(n_i u_i^\alpha - n u^\alpha), \tag{6.88}$$

$$\nabla_\alpha(n u^\alpha) = 0, \tag{6.89}$$

$$u^\mu \nabla_\mu u^\alpha = -\frac{e}{m}F^{\alpha\mu}u_\mu. \tag{6.90}$$

In order to study nonlinear effects it is necessary to extend the two-timing method introduced in the latter section and to allow several scales.

Let us introduce the following quantities

$$\hat{X}^\alpha = \varepsilon^2 x^\alpha \quad \text{(very slow variables)},$$

$$X^\alpha = \varepsilon x^\alpha \quad \text{(slow variables)},$$

and the phase functions (to be determined later)

$$\xi^{(\alpha)} = \xi^{(\alpha)}(X^\beta), \quad \varphi = \frac{1}{\varepsilon^2}\Theta(\hat{X}^\alpha),$$

where $\varepsilon > 0$ is a real parameter related to the slow modulation of the waveform.

We shall assume that in a selected class of coordinates (x^α) we have the following asymptotic expansions

$$F_{\alpha\beta} = \sum_{a=1}^{\infty} \varepsilon^a \overset{(a)}{F}_{\alpha\beta}, \tag{6.91a}$$

$$u^\mu = \overset{(0)}{u}{}^\mu + \sum_{a=1}^{\infty} \varepsilon^a \overset{(a)}{u}{}^\mu, \tag{6.91b}$$

$$n = \overset{(0)}{n} + \sum_{a=1}^{\infty} \varepsilon^a \overset{(a)}{n}, \tag{6.91c}$$

with the functional dependence

$$\overset{(0)}{n} = n_i = n_i(\hat{X}^\alpha), \quad \overset{(0)}{u}{}^\mu = u_i^\mu = u_i^\mu(\hat{X}^\alpha), \tag{6.92a}$$

$$\overset{(a)}{F}_{\alpha\beta} = \sum_{p=-\infty}^{+\infty} \overset{(a)}{\underset{(p)}{F}}_{\alpha\beta}(\hat{X}^\alpha, \xi^{(\mu)}) \exp(ip\varphi), \tag{6.92b}$$

$$\overset{(a)}{u}{}^\mu = \sum_{p=-\infty}^{+\infty} \overset{(a)}{\underset{(p)}{u}}{}^\mu(\hat{X}^\alpha, \xi^{(\beta)}) \exp(ip\varphi), \tag{6.92c}$$

$$\overset{(a)}{n} = \sum_{p=-\infty}^{+\infty} \overset{(a)}{\underset{(p)}{n}}(\hat{X}^\alpha, \xi^{(\beta)}) \exp(ip\varphi). \tag{6.92d}$$

Notice that the reality condition implies

$$\overset{(a)}{\underset{(p)}{n}} = \overset{(a)}{\underset{(-p)}{n}}, \quad \text{etc.}$$

We shall also assume for the metric,

$$g_{\alpha\beta} = g_{\alpha\beta}(\hat{X}^\mu)$$

and then, for the Christoffel symbols one has

$$\Gamma^\mu_{\alpha\nu} = \varepsilon^2 \hat{\Gamma}^\mu_{\alpha\nu}.$$

It is convenient to define the covariant derivative $\nabla_{\hat{\alpha}}$ with respect to the very slow variables $\hat{X}^\alpha$, by using the quantities $\hat{\Gamma}^\mu_{\alpha\nu}$.

Equations (6.92b)–(6.92d) are equivalent to assuming a periodic dependence on φ for $\overset{(a)}{F}_{\alpha\beta}$, $\overset{(a)}{u}{}^{\mu}$, $\overset{(a)}{n}$, together with suitable regularity properties.

We define the vector field l_α by

$$l_\alpha = \frac{\partial \varphi}{\partial x^\alpha} = \partial_{\hat{\alpha}} \Theta \tag{6.93}$$

and denote by $\partial_{\hat{\alpha}}$ the derivative with respect to $\hat{X}^\alpha$, and by comma $_{,\alpha}$ the derivative with respect to X^α.

In order to investigate nonlinear effects we need to keep track of the expansion up to the third order in ε. We have

$$\begin{aligned}\nabla_\alpha F_{\beta\gamma} = {} & \varepsilon \sum_p i p l_\alpha \underset{(p)}{\overset{(1)}{F}}{}_{\beta\gamma} \exp(ip\varphi) + \varepsilon^2 \sum_p \left(i p l_\alpha \underset{(p)}{\overset{(2)}{F}}{}_{\beta\gamma} + \xi^{(\mu)}_{,\alpha} \frac{\partial \underset{(p)}{\overset{(1)}{F}}{}_{\beta\gamma}}{\partial \xi^{(\mu)}} \right) \\ & \cdot \exp(ip\varphi) + \varepsilon^3 \sum_p \left(i p l_\alpha \underset{(p)}{\overset{(3)}{F}}{}_{\beta\gamma} + \nabla_{\hat{\alpha}} \underset{(p)}{\overset{(1)}{F}}{}_{\beta\gamma} \right. \\ & \left. + \xi^{(\mu)}_{,\alpha} \frac{\partial \underset{(p)}{\overset{(2)}{F}}{}_{\beta\gamma}}{\partial \xi^{(\mu)}} \right) \exp(ip\varphi) + O(\varepsilon^4),\end{aligned} \tag{6.94}$$

$$\begin{aligned}\nabla_\alpha u^\mu = {} & \varepsilon \sum_p i p l_\alpha \underset{(p)}{\overset{(1)}{u}}{}^{\mu} \exp(ip\varphi) + \varepsilon^2 \nabla_{\hat{\alpha}} \overset{(0)}{u}{}^{\mu} + \varepsilon^2 \sum_p \left(i p l_\alpha \underset{(p)}{\overset{(2)}{u}}{}^{\mu} \right. \\ & \left. + \xi^{(\nu)}_{,\alpha} \frac{\partial \underset{(p)}{\overset{(1)}{u}}{}^{\mu}}{\partial \xi^{(\nu)}} \right) \exp(ip\varphi) + \varepsilon^3 \sum_p \left(i p l_\alpha \underset{(p)}{\overset{(3)}{u}}{}^{\mu} + \nabla_{\hat{\alpha}} \underset{(p)}{\overset{(1)}{u}}{}^{\mu} \right. \\ & \left. + \xi^{(\nu)}_{,\alpha} \frac{\partial \underset{(p)}{\overset{(2)}{u}}{}^{\mu}}{\partial \xi^{(\nu)}} \right) \exp(ip\varphi) + O(\varepsilon^4),\end{aligned} \tag{6.95}$$

$$\begin{aligned}\nabla_\alpha n = {} & \varepsilon \sum_p i p l_\alpha \underset{(p)}{\overset{(1)}{n}} \exp(ip\varphi) + \varepsilon^2 \nabla_{\hat{\alpha}} \overset{(0)}{n} + \varepsilon^2 \sum_p \left(i p l_\alpha \underset{(p)}{\overset{(2)}{n}} + \xi^{(\mu)}_{,\alpha} \frac{\partial \underset{(p)}{\overset{(1)}{n}}}{\partial \xi^{(\mu)}} \right) \\ & \cdot \exp(ip\varphi) + \varepsilon^3 \sum_p \left(i p l_\alpha \underset{(p)}{\overset{(3)}{n}} + \nabla_{\hat{\alpha}} \underset{(p)}{\overset{(1)}{n}} \right. \\ & \left. + \xi^{(\mu)}_{,\alpha} \frac{\partial \underset{(p)}{\overset{(2)}{n}}}{\partial \xi^{(\mu)}} \right) \exp(ip\varphi) + O(\varepsilon^4),\end{aligned} \tag{6.96}$$

By substituting the above expansions into equations (6.87)–(6.90) we obtain, to the first order in ε,

$$p\left(l_\alpha \underset{(p)}{\overset{(1)}{F}}_{\beta\gamma} + l_\beta \underset{(p)}{\overset{(1)}{F}}_{\gamma\alpha} + l_\gamma \underset{(p)}{\overset{(1)}{F}}_{\alpha\beta} \right) = 0, \tag{6.97}$$

$$ipl_\beta \underset{(p)}{\overset{(1)}{F}}{}^{\alpha\beta} = -4\pi\mu_0 e\left(\overset{(0)}{n} \underset{(p)}{\overset{(1)}{u}}{}^{\alpha} + \underset{(p)}{\overset{(1)}{n}} \overset{(0)}{u}{}^{\alpha} \right), \tag{6.98}$$

$$ipl_\mu \overset{(0)}{u}{}^{\mu} \underset{(p)}{\overset{(1)}{u}} = -\frac{e}{m} \overset{(0)}{u}_\mu \underset{(p)}{\overset{(1)}{F}}{}^{\alpha\mu}, \tag{6.99}$$

$$l_\alpha \overset{(0)}{u}{}^{\alpha} \underset{(p)}{\overset{(1)}{n}} + \overset{(0)}{n} l_\alpha \underset{(p)}{\overset{(1)}{u}}{}^{\alpha} = 0. \tag{6.100}$$

The normalization condition

$$u^\alpha u_\alpha = -1$$

up to the third order gives

$$\overset{(0)}{u}_\mu \overset{(0)}{u}{}^{\mu} = -1,$$

$$\overset{(0)}{u}_\mu \overset{(1)}{u}{}^{\mu} = 0,$$

$$2\overset{(0)}{u}_\mu \overset{(2)}{u}{}^{\mu} + \overset{(1)}{u}_\mu \overset{(1)}{u}{}^{\mu} = 0,$$

$$\overset{(0)}{u}_\mu \overset{(3)}{u}{}^{\mu} + \overset{(1)}{u}_\mu \overset{(2)}{u}{}^{\mu} = 0. \tag{6.101}$$

Let $\Omega = \overset{(0)}{u}{}^{\alpha} l_\alpha$ be the local frequency of the wave (as measured by the observer moving with the background four-velocity $\overset{(0)}{u}{}^{\alpha}$). Equations (6.97)–(6.100) can be analyzed as in the previous section. We shall treat the modulation of the lowest harmonic wave, $p = 1$. Hence, we shall put

$$\underset{(p)}{\overset{(1)}{F}}{}^{\mu\nu} = 0, \quad \underset{(p)}{\overset{(1)}{n}} = 0, \quad \underset{(p)}{\overset{(1)}{u}}{}^{\alpha} = 0,$$

for $|p| \neq 1$.

From equation (6.97) we have

$$\underset{(1)}{\overset{(1)}{F}}_{\alpha\beta} = \underset{(1)}{\psi}_\alpha l_\beta - \underset{(1)}{\psi}_\beta l_\alpha \tag{6.102}$$

and, as in the previous section, we can always choose the gauge such that

$$\underset{(1)}{\psi}_\alpha \underset{(0)}{u}^\alpha = 0. \tag{6.103}$$

Then equations (6.98)–(6.99) are rewritten as

$$il_\beta l^\beta \underset{(1)}{\psi}^\alpha - il^\alpha(\underset{(1)}{\psi}^\beta l_\beta) + 4\pi\mu_0 e\left(\underset{(1)}{\overset{(0)(1)}{n\ u}}{}^\alpha + \underset{(1)}{\overset{(1)(0)}{n\ u}}{}^\alpha\right) = 0, \tag{6.104}$$

$$i\underset{(1)}{\overset{(1)}{u}}{}^\alpha + \frac{e}{m}\underset{(1)}{\psi}^\alpha = 0. \tag{6.105}$$

Proceeding as in the previous section we obtain

$$(-\Omega^2 + \Omega_p^2)\underset{(1)}{\overset{(1)}{u}}{}^\alpha l_\alpha = 0. \tag{6.106}$$

We also exclude electrostatic waves and therefore we assume

$$\Omega^2 \neq \Omega_p^2$$

and hence

$$\underset{(1)}{\overset{(1)}{u}}{}^\alpha l_\alpha = 0, \quad \underset{(1)}{\overset{(1)}{n}} = 0, \quad \underset{(1)}{\psi}^\alpha l_\alpha = 0. \tag{6.107}$$

Equation (6.104) then yields

$$(l^\mu l_\mu + \Omega_p^2)\underset{(p)}{\psi}^\alpha = 0 \tag{6.108}$$

and therefore we obtain the dispersion relation

$$l^\mu l_\mu + \Omega_p^2 = 0. \tag{6.109}$$

Note that the above dispersion relation implies

$$\Omega^2 > \Omega_p^2. \tag{6.110}$$

To the second order in ε, equations (6.87)–(6.90) give

$$\begin{aligned} ip\left(l_\alpha \underset{(p)}{\overset{(2)}{F}}{}_{\beta\gamma} + l_\beta \underset{(p)}{\overset{(2)}{F}}{}_{\gamma\alpha} + l_\gamma \underset{(p)}{\overset{(2)}{F}}{}_{\alpha\beta}\right) + \xi^{(\mu)}_{,\alpha}\frac{\partial \underset{(p)}{\overset{(1)}{F}}{}_{\beta\gamma}}{\partial\xi^{(\mu)}} \\ + \xi^{(\mu)}_{,\beta}\frac{\partial \underset{(p)}{\overset{(1)}{F}}{}_{\gamma\alpha}}{\partial\xi^{(\mu)}} + \xi^{(\mu)}_{,\gamma}\frac{\partial \underset{(p)}{\overset{(1)}{F}}{}_{\alpha\beta}}{\partial\xi^{(\mu)}} = 0, \end{aligned} \tag{6.111}$$

$$ipl_\beta \underset{(p)}{\overset{(2)}{F}}{}^{\alpha\beta} + \xi^{(\mu)}_{,\beta}\frac{\partial \underset{(p)}{\overset{(1)}{F}}{}^{\alpha\beta}}{\partial\xi^{(\mu)}} + 4\pi\mu_0 e\left(\overset{(0)}{n}\underset{(p)}{\overset{(2)}{u}}{}^\alpha + \overset{(0)}{u}{}^\alpha \underset{(p)}{\overset{(2)}{n}}\right) = 0, \tag{6.112}$$

$$\overset{(0)}{u}{}^{\alpha}\nabla_{\alpha}\overset{(0)}{n}+\overset{(0)}{n}\nabla_{\alpha}\overset{(0)}{u}{}^{\alpha}+\sum_{p}\left(ip\Omega\underset{(p)}{\overset{(2)}{n}}+ipl_{\alpha}\underset{(p)}{\overset{(2)}{u}}{}^{\alpha}\overset{(0)}{n}+\overset{(0)}{n}\,\xi^{(\mu)}_{,\alpha}\frac{\partial\underset{(p)}{\overset{(1)}{u}}{}^{\alpha}}{\partial\xi^{(\mu)}}\right)\exp(ip\varphi)=0,\tag{6.113}$$

$$\overset{(0)}{u}{}^{\mu}\nabla_{\beta}\overset{(0)}{u}{}^{\alpha}+\overset{(0)}{u}{}^{\mu}\sum_{p}\left(ipl_{\mu}\underset{(p)}{\overset{(2)}{u}}{}^{\alpha}+\xi^{(\nu)}_{,\mu}\frac{\partial\underset{(p)}{\overset{(1)}{u}}{}^{\alpha}}{\partial\xi^{(\nu)}}\right)\exp(ip\varphi)$$

$$+\sum_{p}\left(\sum_{r}irl_{\mu}\underset{(p-r)}{\overset{(1)}{u}}{}^{\mu}\underset{(r)}{\overset{(1)}{u}}{}^{\alpha}\right)\exp(ip\varphi)+\frac{e}{m}\sum_{p}\left(\overset{(0)}{u}_{\mu}\underset{(p)}{\overset{(2)}{F}}{}^{\alpha\mu}\right.$$

$$\left.+\sum_{r}\underset{(p-r)}{\overset{(1)}{F}}{}^{\alpha\mu}\underset{(r)}{\overset{(1)}{u}}_{\mu}\right)\exp(ip\varphi)=0.\tag{6.114}$$

We shall construct the asymptotic solution such that

$$\underset{(0)}{\overset{(2)}{F}}_{\alpha\beta}=0$$

and this choice, on the basis of our ordering, can be interpreted as assuming that the background plasma is unmagnetized. For $p=0$, equation (6.111) is satisfied identically and equation (6.113) yields

$$\overset{(0)}{u}{}^{\alpha}\nabla_{\alpha}\overset{(0)}{n}+\overset{(0)}{n}\nabla_{\alpha}\overset{(0)}{u}{}^{\alpha}=0.\tag{6.115}$$

From equation (6.114) we obtain

$$\overset{(0)}{u}{}^{\mu}\nabla_{\beta}\overset{(0)}{u}{}^{\alpha}=0,\tag{6.116}$$

which shows that the unperturbed neutral plasma moves along geodesic lines, and from equation (6.112) we have

$$\overset{(0)}{n}\underset{(0)}{\overset{(2)}{u}}{}^{\alpha}+\overset{(0)}{u}{}^{\alpha}\underset{(0)}{\overset{(2)}{n}}=0.\tag{6.117}$$

The normalization condition (6.101), for $p=0$, gives

$$\overset{(0)}{u}_{\mu}\underset{(0)}{\overset{(2)}{u}}{}^{\mu}+\frac{e^2}{m^2}\underset{(1)}{\psi}{}^{\mu}\underset{(1)}{\bar{\psi}}_{\mu}=0$$

and by using it in conjunction with equation (6.117) we get

$$\underset{(0)}{\overset{(2)}{n}}=-\overset{(0)}{n}\frac{e^2}{m^2}\underset{(1)}{\psi}{}^{\mu}\underset{(1)}{\bar{\psi}}_{\mu},\tag{6.118a}$$

$$\underset{(0)}{\overset{(2)}{u}}{}^{\alpha}=\frac{e^2}{m^2}\underset{(1)}{\psi}{}^{\mu}\underset{(1)}{\bar{\psi}}_{\mu}\overset{(0)}{u}{}^{\alpha}.\tag{6.118b}$$

For $p = 2$, contracting equation (6.111) with l_α gives

$$\overset{(2)}{\underset{(2)}{F}}_{\alpha\beta} = \overset{(2)}{\underset{(2)}{\psi}}_\alpha l_\beta - \overset{(2)}{\underset{(2)}{\psi}}_\beta l_\alpha \tag{6.119}$$

and $\overset{(2)}{\underset{(2)}{\psi}}_\alpha$ can be chosen such that

$$\overset{(2)}{\underset{(2)}{\psi}}_\alpha u^\alpha = 0. \tag{6.120}$$

From equation (6.114) for $p = 2$, after some manipulations we obtain

$$\overset{(2)}{\underset{(2)}{\psi}}^\mu = -2i\frac{m}{e}\overset{(2)}{\underset{(2)}{u}}^\mu + i\frac{e}{m\Omega} l^\mu \underset{(1)}{\psi}_\alpha \underset{(1)}{\psi}^\alpha. \tag{6.121}$$

Using equation (6.121) in equation (6.112) for $p = 2$ and taking into account the normalization conditions (6.101) gives

$$\overset{(2)}{\underset{(2)}{n}} = -\frac{3}{2}\left(\frac{e}{m}\right)^2 \frac{\overset{(0)}{n}}{1-\left(\dfrac{2\Omega}{\Omega_p}\right)^2} \underset{(1)}{\psi}^\alpha \underset{(1)}{\psi}_\alpha \tag{6.122}$$

and, finally,

$$\overset{(2)}{\underset{(2)}{u}}^\mu = \frac{1}{3}\left(\overset{(0)}{u}^\alpha + \frac{4l^\alpha\Omega}{\Omega_p^2}\right)\overset{(2)}{\underset{(2)}{n}} / \overset{(0)}{n}. \tag{6.123}$$

For $p = 1$, equations (6.111)–(6.114) read

$$i\left(l_\alpha \overset{(2)}{\underset{(1)}{F}}_{\beta\gamma} + l_\beta \overset{(2)}{\underset{(1)}{F}}_{\gamma\alpha} + l_\gamma \overset{(2)}{\underset{(1)}{F}}_{\alpha\beta}\right) + \xi^{(\mu)}_{,\alpha}\frac{\partial \overset{(1)}{\underset{(1)}{F}}_{\beta\gamma}}{\partial \xi^{(\mu)}}$$

$$+ \xi^{(\mu)}_{,\beta}\frac{\partial \overset{(1)}{\underset{(1)}{F}}_{\gamma\alpha}}{\partial \xi^{(\mu)}} + \xi^{(\mu)}_{,\gamma}\frac{\partial \overset{(1)}{\underset{(1)}{F}}_{\alpha\beta}}{\partial \xi^{(\mu)}} = 0, \tag{6.124}$$

$$il_\beta \overset{(2)}{\underset{(1)}{F}}^{\alpha\beta} + \xi^{(\mu)}_{,\beta}\frac{\partial \overset{(1)}{\underset{(1)}{F}}^{\alpha\beta}}{\partial \xi^{(\mu)}} + 4\pi\mu_0 e\left(\overset{(0)}{n}\overset{(2)}{\underset{(1)}{u}}^\alpha + \overset{(0)}{u}^\alpha \overset{(2)}{\underset{(1)}{n}}\right) = 0, \tag{6.125}$$

$$i\Omega \overset{(2)}{\underset{(1)}{n}} + i\overset{(0)}{n} l_\alpha \overset{(2)}{\underset{(1)}{u}}^\alpha + \overset{(0)}{n}\xi^{(\nu)}_{,\alpha}\frac{\partial \overset{(1)}{\underset{(1)}{u}}^\alpha}{\partial \xi^{(\nu)}} = 0, \tag{6.126}$$

$$i\Omega \overset{(2)}{\underset{(1)}{u}}^\alpha + \overset{(0)}{u}^\mu \xi^{(\nu)}_{,\mu}\frac{\partial \overset{(1)}{\underset{(1)}{u}}^\alpha}{\partial \xi^{(\nu)}} + \frac{e}{m}\overset{(0)}{u}_\mu \overset{(2)}{\underset{(1)}{F}}^{\alpha\mu} = 0. \tag{6.127}$$

The general solution of equation (6.124) is

$$\underset{(1)}{\overset{(2)}{F}}_{\beta\gamma} = \underset{(1)}{\overset{(2)}{\psi}}_{\beta} l_\gamma - \underset{(1)}{\overset{(2)}{\psi}}_{\gamma} l_\beta + i\left(\xi^{(\mu)}_{,\beta}\frac{\partial \underset{(1)}{\psi}_\gamma}{\partial \xi^{(\mu)}} - \xi^{(\mu)}_{,\gamma}\frac{\partial \underset{(1)}{\psi}_\beta}{\partial \xi^{(\mu)}}\right) \tag{6.128}$$

and $\underset{(1)}{\overset{(2)}{\psi}}_\alpha$ can be chosen such that

$$\underset{(1)}{\overset{(2)}{\psi}}{}^{\alpha} l_\alpha = 0. \tag{6.129}$$

The normalization condition (6.101), for $p = 1$, gives

$$\overset{(0)}{u}_\mu \underset{(1)}{\overset{(2)}{u}}{}^{\mu} = 0. \tag{6.130}$$

By contracting equation (6.125) with $\overset{(0)}{u}_\alpha$ we obtain

$$i l_\beta \underset{(1)}{\overset{(2)}{F}}{}^{\alpha\beta} \overset{(0)}{u}_\alpha + \xi^{(\mu)}_{,\beta}\frac{\partial \underset{(1)}{\overset{(1)}{F}}{}^{\alpha\beta}}{\partial \xi^{(\mu)}} \overset{(0)}{u}_\alpha - 4\pi\mu_0 e \underset{(1)}{\overset{(2)}{n}} = 0. \tag{6.131}$$

Contracting equation (6.127) with l_α gives

$$i\Omega l_\alpha \underset{(1)}{\overset{(2)}{u}}{}^{\alpha} + \frac{e}{m} \overset{(0)}{u}_\mu \underset{(1)}{\overset{(2)}{F}}{}^{\alpha\mu} l_\alpha = 0,$$

whence

$$\Omega \underset{(1)}{\overset{(2)}{u}}{}^{\alpha} l_\alpha = \frac{e}{m}\xi^{(\mu)}_{,\beta}\frac{\partial \underset{(1)}{\overset{(1)}{F}}{}^{\alpha\beta}}{\partial \xi^{(\mu)}} \overset{(0)}{u}_\alpha - \frac{\Omega_p^2}{\overset{(0)}{n}} \underset{(1)}{\overset{(2)}{n}},$$

which, after substituting into equation (6.126), finally yields

$$\underset{(1)}{\overset{(2)}{n}} = 0. \tag{6.132}$$

Equations (6.125)–(6.127) give

$$\left(l_\beta + \frac{\Omega_p^2}{\Omega}\overset{(0)}{u}_\beta\right)\underset{(1)}{\overset{(2)}{F}}{}^{\alpha\beta} - i\left(l^\beta - \frac{\Omega_p^2}{\Omega}\overset{(0)}{u}{}^\beta\right)\xi^{(\mu)}_{,\beta}\frac{\partial \underset{(1)}{\psi}{}^\alpha}{\partial \xi^{(\mu)}} + i\xi^{(\mu)}_{,\beta} l^\alpha \frac{\partial \underset{(1)}{\psi}{}^\beta}{\partial \xi^{(\mu)}} = 0$$

from which, by taking into account equation (6.128) and contracting with

$\overset{(0)}{u}{}^{\alpha}$, it follows that

$$\overset{(0)}{u}_{\beta}\overset{(2)}{\underset{(1)}{\psi}}{}^{\beta}=\frac{i\Omega}{\Omega_p^2}\xi^{(\mu)}_{,\beta}\frac{\partial\underset{(1)}{\psi}{}^{\beta}}{\partial\xi^{(\mu)}} \tag{6.133}$$

and

$$\xi^{(\mu)}_{,\beta}l^{\beta}\frac{\partial\underset{(1)}{\psi}{}^{\alpha}}{\partial\overset{(\mu)}{\xi}}=0.$$

Consistently we shall construct asymptotic solutions such that

$$\xi^{(\mu)}_{,\beta}l^{\beta}=0. \tag{6.134}$$

The latter equation means that the four phases $\xi^{(\mu)}$ are constant along the curves tangent to l^{μ} (the rays).

To the third order in ε, from the field equations (6.87)–(6.90) we obtain, for $p=1$,

$$\begin{aligned}&i\left(l_{\alpha}\overset{(3)}{\underset{(1)}{F}}{}_{\beta\gamma}+l_{\beta}\overset{(3)}{\underset{(1)}{F}}{}_{\gamma\alpha}+l_{\gamma}\overset{(3)}{\underset{(1)}{F}}{}_{\alpha\beta}\right)+\xi^{(\mu)}_{,\alpha}\frac{\partial\overset{(2)}{\underset{(1)}{F}}{}_{\beta\gamma}}{\partial\xi^{(\mu)}}+\xi^{(\mu)}_{,\beta}\frac{\partial\overset{(2)}{\underset{(1)}{F}}{}_{\gamma\alpha}}{\partial\xi^{(\mu)}}\\&+\xi^{(\mu)}_{,\gamma}\frac{\partial\overset{(2)}{\underset{(1)}{F}}{}_{\alpha\beta}}{\partial\xi^{(\mu)}}+\nabla_{\hat\alpha}\overset{(1)}{\underset{(1)}{F}}{}_{\beta\gamma}+\nabla_{\hat\beta}\overset{(1)}{\underset{(1)}{F}}{}_{\gamma\alpha}+\nabla_{\hat\gamma}\overset{(1)}{\underset{(1)}{F}}{}_{\alpha\beta}=0,\end{aligned} \tag{6.135}$$

$$il_{\beta}\overset{(3)}{\underset{(1)}{F}}{}^{\alpha\beta}+4\pi\mu_0e\left(\overset{(0)}{n}\overset{(3)}{\underset{(1)}{u}}+\overset{(0)}{u}{}^{\alpha}\overset{(3)}{\underset{(1)}{n}}\right)+A^{\alpha}=0 \tag{6.136a}$$

with

$$A^{\alpha}=\xi^{(\mu)}_{,\beta}\frac{\partial\overset{(2)}{\underset{(1)}{F}}{}^{\alpha\beta}}{\partial\xi^{(\mu)}}+\nabla_{\hat\beta}\overset{(1)}{\underset{(1)}{F}}{}^{\alpha\beta}+4\pi\mu_0e\left(\overset{(1)}{\underset{(1)}{u}}{}^{\alpha}\overset{(2)}{\underset{(0)}{n}}+\overset{(1)}{\underset{(-1)}{u}}{}^{\alpha}\overset{(2)}{\underset{(2)}{n}}\right), \tag{6.136b}$$

$$i\Omega\overset{(3)}{\underset{(1)}{u}}{}^{\alpha}+\frac{e}{m}\overset{(0)}{u}_{\mu}\overset{(3)}{\underset{(1)}{F}}{}^{\alpha\mu}+A'^{\alpha}=0 \tag{6.137a}$$

with

$$\begin{aligned}A'^{\alpha}=&\overset{(0)}{u}{}^{\mu}\xi^{(\nu)}_{,\mu}\frac{\partial\overset{(2)}{\underset{(1)}{u}}{}^{\alpha}}{\partial\xi^{(\nu)}}+\overset{(0)}{u}{}^{\mu}\nabla_{\hat\mu}\overset{(1)}{\underset{(1)}{u}}{}^{\alpha}+\overset{(1)}{\underset{(1)}{u}}{}^{\mu}\nabla_{\hat\mu}\overset{(0)}{u}{}^{\alpha}+il_{\mu}\overset{(2)}{\underset{(0)}{u}}{}^{\mu}\overset{(1)}{\underset{(1)}{u}}{}^{\alpha}\\&-il_{\mu}\overset{(2)}{\underset{(2)}{u}}{}^{\mu}\overset{(1)}{\underset{(-1)}{u}}{}^{\alpha}+\frac{e}{m}\left(\overset{(1)}{\underset{(-1)}{u}}{}_{\mu}\overset{(2)}{\underset{(2)}{F}}{}^{\alpha\mu}+\overset{(2)}{\underset{(0)}{u}}{}_{\mu}\overset{(1)}{\underset{(1)}{F}}{}^{\alpha\mu}+\overset{(2)}{\underset{(2)}{u}}{}_{\mu}\overset{(1)}{\underset{(-1)}{F}}{}^{\alpha\mu}\right),\end{aligned} \tag{6.137b}$$

$$i\left(\Omega \overset{(3)}{\underset{(1)}{n}} + \overset{(0)}{n}\, l_\alpha \overset{(3)}{\underset{(1)}{u}}{}^\alpha\right) + C = 0 \tag{6.138a}$$

with

$$C = \overset{(1)}{\underset{(1)}{u}}{}^\alpha \nabla_{\hat\alpha} \overset{(0)}{n} + \overset{(0)}{n}\, \nabla_{\hat\alpha} \overset{(1)}{\underset{(1)}{u}}{}^\alpha + \overset{(0)}{u}{}^\alpha \xi^{(\nu)}_{,\alpha} \frac{\partial \overset{(2)}{\underset{(1)}{n}}}{\partial \xi^{(\nu)}} + \overset{(0)}{n}\, \xi^{(\nu)}_{,\alpha} \frac{\partial \overset{(2)}{\underset{(1)}{u}}{}^\alpha}{\partial \xi^{(\nu)}}. \tag{6.138b}$$

The normalization conditions (6.101) give

$$\overset{(3)}{\underset{(1)}{u}}{}^\alpha \overset{(0)}{u}_\alpha = 0 \tag{6.139}$$

because

$$\overset{(1)}{\underset{(1)}{u}}{}^\alpha \overset{(2)}{\underset{(0)}{u}}_\alpha = \overset{(1)}{\underset{(1)}{u}}{}^\alpha \overset{(2)}{\underset{(2)}{u}}_\alpha = 0.$$

By substituting $\overset{(3)}{\underset{(1)}{u}}{}^\alpha$ and $\overset{(3)}{\underset{(1)}{n}}$ from equations (6.137)–(6.138) into equation (6.136) we obtain

$$-\Omega\left(l_\beta + \frac{\Omega_p^2}{\Omega}\overset{(0)}{u}_\beta\right)\overset{(3)}{\underset{(1)}{F}}{}^{\alpha\beta} + \left(\frac{\Omega_p^2}{\Omega}\overset{(0)}{u}_\mu \overset{(3)}{\underset{(1)}{F}}{}^{\nu\mu} l_\nu + 4\pi\mu_0 e \overset{(0)}{n} A'^\nu l_\nu\right)\overset{(0)}{u}{}^\alpha$$

$$- 4\pi\mu_0 e \overset{(0)}{u}{}^\alpha \Omega C + i\Omega A^\alpha - 4\pi\mu_0 e \overset{(0)}{n} A'^\alpha = 0. \tag{6.140}$$

Equation (6.135) is written explicitly as

$$l_\alpha\left(i \overset{(3)}{\underset{(1)}{F}}_{\beta\gamma} + \xi^{(\mu)}_{,\beta}\frac{\partial \overset{(2)}{\underset{(1)}{\psi}}_\gamma}{\partial \xi^{(\mu)}} - \xi^{(\mu)}_{,\gamma}\frac{\partial \overset{(2)}{\underset{(1)}{\psi}}_\beta}{\partial \xi^{(\mu)}} + \nabla_{\hat\beta} \underset{(1)}{\psi}_\gamma - \nabla_{\hat\alpha} \underset{(1)}{\psi}_\beta\right)$$

$$+ \text{cyclic permutation} = 0 \tag{6.141}$$

and gives

$$\overset{(3)}{\underset{(1)}{F}}_{\alpha\beta} = \overset{(3)}{\underset{(1)}{\psi}}_\alpha l_\beta - \overset{(3)}{\underset{(1)}{\psi}}_\beta l_\alpha - i(\nabla_{\hat\beta} \underset{(1)}{\psi}_\alpha - \nabla_{\hat\alpha} \underset{(1)}{\psi}_\beta) - i\left(\xi^{(\mu)}_{,\beta}\frac{\partial \overset{(2)}{\underset{(1)}{\psi}}_\alpha}{\partial \xi^{(\mu)}} - \xi^{(\mu)}_{,\alpha}\frac{\partial \overset{(2)}{\underset{(1)}{\psi}}_\beta}{\partial \xi^{(\mu)}}\right), \tag{6.142}$$

with $\overset{(3)}{\underset{(1)}{\psi}}_\alpha$ determined up to the transformation

$$\overset{(3)}{\underset{(1)}{\psi}}_\alpha \rightarrow \overset{(3)}{\underset{(1)}{\psi}}_\alpha + \lambda l_\alpha.$$

We can choose $\underset{(1)}{\overset{(3)}{\psi}}{}_\alpha$ such that

$$\underset{(1)}{\overset{(3)}{\psi}}{}_\alpha l^\alpha = 0. \tag{6.143}$$

Then equation (6.140) can be written

$$\begin{aligned}
&i\Omega\left(l^\beta+\frac{\Omega_p^2}{\Omega}\overset{(0)}{u}{}^\beta\right)\left(\nabla_{\hat\beta}\underset{(1)}{\psi}{}_\alpha-\nabla_{\hat\alpha}\underset{(1)}{\psi}{}_\beta+\xi^{(\mu)}_{,\beta}\frac{\partial\underset{(1)}{\overset{(2)}{\psi}}{}_\alpha}{\partial\xi^{(\mu)}}-\xi^{(\mu)}_{,\alpha}\frac{\partial\underset{(1)}{\overset{(2)}{\psi}}{}_\beta}{\partial\xi^{(\mu)}}\right)\\
&+\overset{(0)}{u}{}_\alpha\left(i\frac{\Omega_p^2}{\Omega}\overset{(0)}{u}{}^\mu l^\nu\nabla_{\hat\nu}\underset{(1)}{\psi}{}_\mu-i\frac{\Omega_p^2}{\Omega}l^\mu\overset{(0)}{u}{}^\nu\nabla_{\hat\nu}\underset{(1)}{\psi}{}_\mu+4\pi\mu_0 e\overset{(0)}{n}A'^\nu l_\nu\right)\\
&+i\Omega A_\alpha-4\pi\mu_0 e\overset{(0)}{n}A'_\alpha-4\pi\mu_0 e\overset{(0)}{u}{}_\alpha\Omega C\\
&+\Omega_p^2\left(l_\alpha+\frac{\Omega_p^2}{\Omega}\overset{(0)}{u}{}_\alpha\right)\overset{(0)}{u}{}_\mu\underset{(1)}{\overset{(3)}{\psi}}{}^\mu=0.
\end{aligned} \tag{6.144}$$

After lengthy and tedious calculations equation (6.144) simplifies considerably and yields

$$\begin{aligned}
&2l^\beta\nabla_{\hat\beta}\underset{(1)}{\psi}{}_\alpha+\underset{(1)}{\psi}{}_\alpha\nabla_{\hat\beta}l^\beta+\frac{\Omega_p^2}{\Omega}\underset{(1)}{\psi}{}^\beta(\nabla_{\hat\alpha}\overset{(0)}{u}{}_\beta-\nabla_{\hat\beta}\overset{(0)}{u}{}_\alpha)-i\Omega_p^2\frac{e^2}{m^2}\\
&\cdot\left(\underset{(1)}{\psi}{}_\alpha\underset{(1)}{\psi}{}^\mu\underset{(1)}{\bar\psi}{}_\mu-\frac{3}{2\left(1-\dfrac{\Omega^2}{\Omega_p^2}\right)}\underset{(1)}{\bar\psi}{}_\alpha\underset{(1)}{\psi}{}^\mu\underset{(1)}{\psi}{}_\mu\right)-i\xi^{(\mu)}_{,\beta}\xi^{(\tau),\beta}\frac{\partial^2\underset{(1)}{\psi}{}_\alpha}{\partial\xi^{(\mu)}\partial\xi^{(\tau)}}\\
&+\text{terms parallel to } l_\alpha \text{ and } \overset{(0)}{u}{}_\alpha=0.
\end{aligned} \tag{6.145}$$

We can write $\underset{(1)}{\psi}{}_\alpha$ in the form

$$\underset{(1)}{\psi}{}_\alpha=\psi e_\alpha, \tag{6.146}$$

with

$$\psi=\psi(\hat X^\mu,\xi^{(\mu)})$$

a complex function and e_α a real unit vector orthogonal to l^α and to $\overset{(0)}{u}{}^\alpha$. Furthermore, without loss of generality we can assume that e_α is independent of $\xi^{(\mu)}$ (ψ absorbing all the dependence on these variables). By contracting equation (6.145) with e^α gives

$$2l^\beta\nabla_{\hat\beta}\psi+\psi\nabla_{\hat\beta}l^\beta-i\xi^{(\mu)}_{,\beta}\xi^{(\tau),\beta}\frac{\partial^2\psi}{\partial\xi^{(\mu)}\partial\xi^{(\tau)}}-i\Omega_p^2\frac{e^2}{m^2}\frac{8\Omega^2+\Omega_p^2}{8\Omega^2-2\Omega_p^2}\psi|\psi|^2=0, \tag{6.147}$$

which is a generalized nonlinear Schrodinger equation (Asano, 1974).

We can introduce the orthonormal frame

$$\left(\overset{(0)}{u}{}^{\alpha}, v^{\alpha}, \underset{(1)}{e}{}^{\alpha}, \underset{(2)}{e}{}^{\alpha}\right)$$

and $\underset{(1)}{e}{}^{\alpha}$, $\underset{(2)}{e}{}^{\alpha}$ are two orthonormal spacelike vectors orthogonal both to l^{α} and to $\overset{(0)}{u}{}^{\alpha}$ and such that

$$\underset{(2)}{e}{}^{\alpha} l^{\mu} \nabla_{\mu} \underset{(1)}{e}{}_{\alpha} = 0.$$

Then we can write

$$e^{\alpha} = \underset{(1)}{e}{}^{\alpha} \cos\theta + \underset{(2)}{e}{}^{\alpha} \sin\theta,$$

with θ the polarization angle. It is then easy to see that one obtains the same equation for the polarization angle as equation (6.86).

The theory developed in this section could be easily extended to cover the case of a magnetized background plasma (Anile and Carbonaro, 1988).

7
Relativistic asymptotic waves

7.0. Introduction

One of the most useful perturbation methods for dealing with nonlinear waves is that of asymptotic and approximate waves, which is a fruitful extension of the high-frequency method of the linear theory. In its full generality, for arbitrary quasi-linear systems, the method has been developed by Choquet–Bruhat in a series of articles (Choquet–Bruhat, 1969a, 1969b, 1973) where applications to relativistic fluid dynamics, to Einstein's equations in vacuo, and to the Einstein–Maxwell system are also presented. Further applications have considered the Einstein equations coupled with a scalar field (Choquet–Bruhat and Taub, 1977), relativistic cosmology (Anile, 1977), relativistic magneto-fluid dynamics (Anile and Greco, 1978), and supergravity theory (Choquet–Bruhat and Greco, 1983).

A different but substantially equivalent approach is that of the averaged Lagrangian, originally due to Whitham (1974). Extensions of the averaged Lagrangian approach to the relativistic framework have been made, among others, by Dougherty (1970; 1974), Dewar (1977), and Achterberg (1983) for relativistic plasmas and by MacCallum and Taub (1973), Taub (1978), and de Arajuro (1986) for gravitational waves in vacuo.

The method of asymptotic waves is potentially relevant for several problems in relativistic astrophysics and plasma physics. In Chapter 5 we studied the nonlinear evolution of a simple wave. In many situations one deals with waves which cannot be considered as simple waves (for instance, a pulse propagating down a density gradient). In these cases, in general, the only way by which the nonlinear evolution can be studied is by means of perturbation methods. The method of asymptotic waves is the most suitable one in order to study the nonlinear evolution of nondispersive (hyperbolic) waves. This method has also been applied in order to derive necessary nonlinear stability conditions for problems in Newtonian fluid dynamics and plasma physics, such as in the case of an imploding spherical gas shell (Chin et al., 1986). A relativistic covariant formulation of the method could be useful for tackling analogous problems in relativistic fluid dynamics and plasma physics.

In this chapter we shall limit ourselves to treat the method of asymptotic and approximate waves. The plan of the chapter is the following. In Section 7.1, following Choquet–Bruhat (1969a) we introduce the concepts of asymptotic and approximate waves. We derive the transport equation for the wave amplitude and discuss the nonlinear distortion of the wave profile and its breaking. In Section 7.2 we treat asymptotic waves in relativistic fluid dynamics. We derive the transport equation and, for plane waves propagating into a constant state, obtain explicit expressions for the nonlinear distortion of the profile and for the critical time for breaking. In Section 7.3 we discuss asymptotic magnetoacoustic waves in relativistic magneto-fluid dynamics. In Section 7.4 we treat asymptotic and approximate waves for Einstein's equations in vacuo. It is found that, in order for approximate waves of order 1 to exist, the background metric cannot be arbitrary, but must satisfy some requirements. These constraints can be interpreted as implying that the background space-time is curved by the presence of gravitational radiation (which is endowed with positive energy density). The method of asymptotic waves thus permits us to introduce the concept of a gravitational wave in a satisfactory way, describing the propagation of a high-frequency gravitational pulse in terms of geometrical optics concepts. This method could be of great interest when calculating the propagation through the universe of gravitational radiation emitted by astrophysical sources (rapidly rotating neutron stars at their birth, asymmetric gravitational collapse, etc.). The content of this section is entirely based on Choquet–Bruhat's (1969b) analysis.

7.1. Asymptotic waves for quasi-linear systems

In the manifold $\mathscr{M}$ in a local chart (x^α) we consider the following system of partial differential equations

$$L^A(\mathbf{U}) \equiv \mathscr{A}_B^{A\lambda}(x, \mathbf{U})\frac{\partial U^B}{\partial x^\lambda} + \mathscr{B}^A(x, \mathbf{U}) = 0 \tag{7.1}$$

for the field of unknowns $\mathbf{U} = {}^{\mathrm{T}}(U_1, \ldots, U_N)$, $A, B = 1, 2, \ldots, N$.

We assume $\mathscr{A}_B^{A\lambda}(x, \mathbf{U})$ to be C^∞ in x and analytic in $\mathbf{U}$ in an open domain $\mathscr{P}$ of the kind $|\mathbf{U}–\mathbf{U}_0| < K$ in $\mathbb{R}^N$, and to have a power series representation in $\mathscr{P}$

$$\mathscr{A}_B^{A\alpha}(x, \mathbf{U}) = \underset{0}{\mathscr{A}}{}_B^{A\alpha} + \mathscr{A}_{BC}^{A\alpha}(U^C - U_0^C) + \tfrac{1}{2}\mathscr{A}_{BCD}^{A\alpha}(U^C - U_0^C)(U^D - U_0^D) + \cdots, \tag{7.2}$$

$$\mathscr{B}^A(x, \mathbf{U}) = \mathscr{B}_0^A + \mathscr{B}_C^A(U^C - U_0^C) + \cdots, \tag{7.3}$$

where we have put

$$\underset{0}{\mathscr{A}}{}^{A\alpha}_{B} = \mathscr{A}^{A\alpha}_{B}(x, \mathbf{U}_0), \quad \mathscr{B}^A_0 = \mathscr{B}^A(x, \mathbf{U}_0),$$

$$\mathscr{A}^{A\alpha}_{BC} = \frac{\partial \mathscr{A}^{A\alpha}_{B}}{\partial U^C}(x, \mathbf{U}_0), \quad \mathscr{B}^A_C = \frac{\partial \mathscr{B}^A}{\partial U^C}(x, \mathbf{U}_0).$$

Now we look for solutions in the form of formal series of the kind

$$\mathbf{U} = \sum_{q=0}^{\infty} \omega^{-q} \underset{(q)}{\mathbf{U}}(x, \xi), \tag{7.4}$$

where $\omega \in \mathbb{R}^+$, $\xi = \omega\varphi(x)$, φ is a differentiable function, and $\underset{(0)}{\mathbf{U}} = \mathbf{U}_0(x)$ is a solution of the system (7.1) independent of ξ. The formal series (7.4) is said to be an *asymptotic wave* for the system (7.1) if, when substituted into these equations, taking into account the developments (7.2)–(7.3), the resulting series

$$\sum_{q=-1}^{\infty} \omega^{-q} \underset{(q)}{\mathbf{F}}(x, \xi)$$

has all the coefficients $\underset{(q)}{\mathbf{F}} = 0$ vanishing, $\forall x, \xi$ (Choquet-Bruhat, 1969a).

The truncated expansion

$$\mathbf{F} = \sum_{q=0}^{\infty} \omega^{-q} \underset{(q)}{\mathbf{F}}(x, \xi)$$

is said to be an *approximate wave of order* $m-1$, if the following condition holds:

$$\| L(\mathbf{U}) \| \leq C\omega^{-(m+1)},$$

where $\| \quad \|$ is a suitable norm for the space of solutions of $L(\mathbf{U}) = 0$.

If $\| \quad \|$ is the supremum norm, in order for $\mathbf{U}$ to be an approximate wave of order $m-1$ it is sufficient that the functions $\underset{(q)}{\mathbf{U}}$ $q = 0, 1, \ldots, m$ belong to $\mathscr{P}$, be bounded together with their derivatives with respect to (x, ξ), and verify $\underset{(q)}{\mathbf{F}} = 0$, for $q < m$. Now we shall construct asymptotic waves. Let

$$\underset{(q)}{\dot{\mathbf{U}}} = \left(\frac{\partial \mathbf{U}}{\partial \xi}\right)_x, \quad \partial_\lambda \mathbf{U} = \left(\frac{\partial \mathbf{U}}{\partial x^\lambda}\right)_\xi.$$

Then

$$\frac{\partial \mathbf{U}}{\partial x^\lambda} = \sum_{q=0}^{\infty} \omega^{-q+1} (\partial_\lambda \varphi) \underset{(q)}{\dot{\mathbf{U}}} + \sum_{q=0}^{\infty} \omega^{-q} \partial_\lambda \mathbf{U}. \tag{7.5}$$

By substituting (7.4) into (7.2)–(7.3) and taking into account (7.5) we obtain for the coefficients $\underset{(q)}{\mathbf{F}}$

$$\underset{(-1)}{F^B} = \underset{0}{\mathscr{A}}{}^{B\lambda}_{A} \underset{(0)}{\dot{U}^A} \partial_\lambda \varphi, \tag{7.6}$$

$$\underset{(0)}{F}^B = \underset{0}{\mathscr{A}}{}^{B\lambda}_A(\underset{(1)}{\dot U}{}^A\partial_\lambda\varphi + \partial_\lambda \underset{(0)}{U}{}^A) + \mathscr{A}^{B\lambda}_{AC}\underset{(1)}{U}{}^C\underset{(0)}{\dot U}{}^A\partial_\lambda\varphi + \underset{0}{\mathscr{B}}{}^B \tag{7.7}$$

$$\begin{aligned}\underset{(q)}{F}^B = \underset{0}{\mathscr{A}}{}^{B\lambda}_A(\underset{(q+1)}{\dot U}{}^A\partial_\lambda\varphi + \partial_\lambda \underset{(q)}{U}{}^A) &+ \mathscr{A}^{B\lambda}_{AC}\{\underset{(1)}{U}{}^C(\underset{(q)}{\dot U}{}^A\partial_\lambda\varphi + \partial_\lambda \underset{(q-1)}{U}{}^A) \\ + \underset{(2)}{U}{}^C(\underset{(q-1)}{\dot U}{}^A\partial_\lambda\varphi + \partial_\lambda \underset{(q-2)}{U}{}^A) &+ \cdots + \underset{(q)}{U}{}^C(\underset{(1)}{\dot U}{}^A\partial_\lambda\varphi + \partial_\lambda \underset{(0)}{U}{}^A) \\ + \underset{(q+1)}{U}{}^C \underset{(0)}{\dot U}{}^A\partial_\lambda\varphi\} &+ \mathscr{B}^B_C \underset{(q)}{U}{}^C.\end{aligned} \tag{7.8}$$

Let us assume $\underset{(0)}{U}$ to be a solution of the system (7.1) independent of ξ. Then

$$\underset{(0)}{\dot U} = 0, \quad \underset{0}{\mathscr{A}}{}^{B\lambda}_A\partial_\lambda \underset{0}{U}{}^A + \underset{0}{\mathscr{B}}{}^B = 0.$$

It follows that, $\underset{(-1)}{F}{}^B = 0$,

$$\underset{(0)}{F}^B = \underset{0}{\mathscr{A}}{}^{B\lambda}_A\underset{(1)}{\dot U}{}^A\partial_\lambda\varphi = 0, \tag{7.9}$$

$$\begin{aligned}\underset{(1)}{F}^B &= \underset{0}{\mathscr{A}}{}^{B\lambda}_A\partial_\lambda\varphi\underset{(2)}{\dot U}{}^A + \underset{0}{\mathscr{A}}{}^{B\lambda}_A\partial_\lambda\underset{(1)}{U}{}^A + \mathscr{A}^{B\lambda}_{AC}\partial_\lambda\varphi\, \underset{1}{U}{}^C_1\underset{1}{\dot U}{}^A_1 + \left(\mathscr{A}^{B\lambda}_{AC}\partial_\lambda\underset{0}{U}{}^A + \mathscr{B}^B_C\right)\underset{1}{U}{}^C_1 \\ &= 0,\end{aligned} \tag{7.10}$$

$$\begin{aligned}\underset{(q)}{F}^B = \underset{0}{\mathscr{A}}{}^{B\lambda}_A\partial_\lambda\varphi\ \underset{(q+1)}{\dot U}{}^A + \underset{0}{\mathscr{A}}{}^{B\lambda}_A\partial_\lambda\underset{(q)}{U}{}^A &+ \mathscr{A}^{B\lambda}_{AC}\partial_\lambda\varphi\underset{(q)}{\dot U}{}^A\underset{(1)}{U}{}^C + \{\mathscr{A}^{B\lambda}_{AC}(\partial_\lambda\varphi\underset{(1)}{\dot U}{}^A \\ + \partial_\lambda\underset{(0)}{U}{}^A) + \mathscr{B}^B_C\}\underset{(q)}{U}{}^C &+ \underset{(q)}{f}{}^B = 0, \quad q > 1,\end{aligned} \tag{7.11}$$

where $\underset{(q)}{f}{}^B$ depends only on $\underset{(p)}{U}{}^A$, $\underset{(p)}{\dot U}{}^A$, $\partial_\lambda\underset{(p)}{U}{}^A$, for $p < q$.

From (7.9) we have, for nontrivial solutions $\underset{(1)}{\dot{\mathbf{U}}} \neq 0$,

$$\bar A(x, \varphi_\lambda) \equiv \det(\underset{0}{\mathscr{A}}{}^{B\lambda}_A\partial_\lambda\varphi) = 0, \tag{7.12}$$

which says that φ satisfies the characteristic equation in the unperturbed state $\underset{(0)}{\mathbf{U}}$.

First of all, we shall consider the case when φ is a simple root of the characteristic equation (7.12).

Let $\mathbf{R} = (R^A)$, $\mathbf{L} = (L_A)$ be the right and left eigenvectors of the matrix $\underset{0}{\mathscr{A}}{}^{B\lambda}_A\partial_\lambda\varphi$, defined up to a factor, corresponding to the zero eigenvalue,

$$\underset{0}{\mathscr{A}}{}^{B\lambda}_A\partial_\lambda\varphi R^A = 0, \quad L_B\underset{0}{\mathscr{A}}{}^{B\lambda}_A\partial_\lambda\varphi = 0.$$

Then, from (7.9) we obtain

$$\underset{(1)}{\dot{\mathbf{U}}} = \Pi_1(x, \xi)\mathbf{R}(x) \tag{7.13}$$

and by integrating with respect to ξ,

$$\underset{(1)}{\mathbf{U}} = \Pi(x, \xi)\mathbf{R} + \mathbf{V}_1(x), \tag{7.14}$$

where $\Pi(x, \xi)$ is a function to be determined and $\mathbf{V}_1(x)$ arises from the integration.

The compatibility condition for (7.10), considered as a system of linear equations in $\underset{(2)}{\dot{\mathbf{U}}}$, is $L_B \underset{(1)}{F}^B = 0$, and it yields

$$L_B \underset{0}{\mathscr{A}}{}^{B\lambda}_{A} \partial_\lambda \underset{(1)}{U}{}^A + L_B \underset{0}{\mathscr{A}}{}^{B\lambda}_{AC} \varphi_\lambda \underset{(1)}{\dot{U}}{}^A \underset{(1)}{U}{}^C + L_B(\mathscr{A}^{B\lambda}_{AC} \partial_\lambda \underset{(0)}{U}{}^A + \mathscr{B}^B_C) U^C_1 = 0,$$

having set $\varphi_\lambda \equiv \partial_\lambda \varphi$, whence the transport equation,

$$K^\lambda \partial_\lambda \Pi + N\Pi\dot{\Pi} + M\Pi + P\dot{\Pi} + Q = 0, \tag{7.15}$$

where

$$K^\lambda = L_B \underset{0}{\mathscr{A}}{}^{B\lambda}_{A} R^A, \tag{7.16}$$

$$N = L_B \underset{0}{\mathscr{A}}{}^{B\lambda}_{AC} R^A R^C, \tag{7.17}$$

$$M = L_B\{\underset{0}{\mathscr{A}}{}^{B\lambda}_{A} \partial_\lambda R^A + \mathscr{A}^{B\lambda}_{AC} \partial_\lambda \underset{(0)}{U}{}^A R^C + \mathscr{B}^B_C R^C\}, \tag{7.18}$$

$$P = L_B \mathscr{A}^{B\lambda}_{AC} \partial_\lambda R^A V^C_1, \tag{7.19}$$

$$Q = L_B\{\underset{0}{A}{}^{B\lambda}_{A} \partial_\lambda V^A_1 + (\mathscr{A}^{B\lambda}_{AC} \partial_\lambda \underset{(0)}{U}{}^A + \mathscr{B}^B_C) V^C_1\}, \tag{7.20}$$

with $P = 0$ and $Q = 0$ if $\mathbf{V}_1 = 0$.

By repeating the proof of Proposition (4.2) we have that

$$K^\alpha = k \frac{\partial \bar{A}}{\partial \varphi_\alpha}, \tag{7.21}$$

with k a normalization factor.

Similarly, by repeating the proof of Proposition (4.2),

$$N = k \frac{\partial \bar{A}}{\partial U^C} R^C. \tag{7.22}$$

Equation (7.15) can be integrated by the method of characteristics. The bicharacteristic system is

$$\mathrm{d}t = \frac{\mathrm{d}x^\lambda}{K^\lambda} = \frac{\mathrm{d}\xi}{N\Pi + P} = \frac{-\mathrm{d}\Pi}{M\Pi + Q}. \tag{7.23}$$

The hypersurface $\varphi = \text{const.}$, containing the point y^α, is generated by the "rays"

$$x^\lambda = x^\lambda(t, y^\alpha), \quad x^\lambda(0, y^\alpha) = y^\alpha. \tag{7.24}$$

Let Σ be the initial manifold defined by

$$s(y^\alpha) = 0, \quad \{\Pi(x, \xi)\}_{\substack{x=y\\ \xi=\eta}} = W_1(y^\alpha, \eta), \tag{7.25}$$

with s and W_1 given functions. Then, following the standard procedure expounded in Section 4.1, the solution $\Pi(x, \xi)$ satisfying the initial conditions

$$x^\alpha(0, y^\beta) = y^\beta, \quad \Pi(0, y^\alpha, \eta) = W_1(y^\alpha, \eta), \quad \xi(0, y^\alpha, \eta) = \eta$$

is given in parametric form as

$$x^\alpha = x^\alpha(t, y^\beta), \tag{7.26}$$

$$\Pi = \Pi(t, y^\beta, \eta), \tag{7.27}$$

$$\xi = \xi(t, y^\beta, \eta), \quad \text{where } s(y^\alpha) = 0. \tag{7.28}$$

In order to obtain $\Pi(x^\alpha, \xi)$ one must invert equations (7.26)–(7.28) in a neighborhood of $t = 0$. This is possible, locally, if the manifold $s(y^\alpha) = 0$ is differentiable and transverse to the rays (i.e., each ray meets $s(y^\alpha) = 0$ only once).

$\Pi(t, y^\alpha, \eta)$ and $\xi(t, y^\alpha, \eta)$ are given by

$$\Pi = \exp\left\{-\int_0^t M(x(\tau, y^\alpha))\,\mathrm{d}\tau\right\}\left\{W_1(y^\alpha, \eta) + \int_0^t \left\{Q(x(\tau, y^\alpha))\exp\left(\int_0^\tau M\,\mathrm{d}\tau\right)\right\}\mathrm{d}\tau\right\}, \tag{7.29}$$

$$\xi = \eta + \int_0^t \{N(x(\tau, y^\alpha)\Pi(\tau, y^\alpha, \eta) + P(x(\tau, y^\alpha))\}\,\mathrm{d}\tau. \tag{7.30}$$

For the choice $\mathbf{V}_1 = 0$ (convenient in many initial-value problems), one has

$$\Pi = W_1(y, \eta)\Phi(t, y), \quad \Phi(t, y) = \exp\left(-\int_0^t M\,\mathrm{d}\tau\right), \tag{7.31}$$

$$\xi = \eta + W_1(y, \eta)\Psi(t, y), \quad \Psi \equiv \int_0^t N\Phi\,\mathrm{d}\tau. \tag{7.32}$$

Also, one has

$$\frac{\partial \xi}{\partial \eta} = 1 + \frac{\partial W_1}{\partial \eta}\Psi$$

and

$$\frac{\partial \xi}{\partial \eta}(0, y, \eta) = 1.$$

Therefore, for t sufficiently small, $\frac{\partial \xi}{\partial \eta}(t, y, \eta) \neq 0$ and one can solve for $\eta = \eta(t, y, \xi)$.

This is possible until a "time" t^* such that

$$\frac{\partial \xi}{\partial \eta}(t^*, y, \eta) = 0.$$

The critical time t_c is defined as the infimum

$$t_c = \inf_{(y,\eta)} t^* : \frac{\partial \xi}{\partial \eta}(t^*, y, \eta) = 0.$$

In the exceptional case, $N = 0$ and $\Psi = 0$, therefore $\xi = \eta$ and no breaking occurs. Also, in this case, $W_1(y, \eta) = W_1(y, \xi)$ and therefore the initial profile is not distorted.

It is now possible to determine the higher order terms. $\underset{(2)}{\dot{\mathbf{U}}}$ satisfies the equation $\underset{(1)}{\mathbf{F}} = 0$, and therefore it is of the form

$$\underset{(2)}{\dot{\mathbf{U}}} = V_2(x, \xi)\mathbf{R} + \mathbf{V}_2(x, \xi), \tag{7.33}$$

where $V_2(x, \xi)$ is an arbitrary function and $\mathbf{V}_2(x, \xi)$ is a solution of the system

$$\underset{0}{\mathscr{A}}{}^{B\lambda}_{A} \partial_\lambda \varphi V_2^A + g_1^B = 0, \tag{7.34}$$

where

$$g_1^B = \underset{0}{\mathscr{A}}{}^{B\lambda}_{A} \partial_\lambda \underset{(1)}{U}{}^A + \mathscr{A}^{B\lambda}_{AC} \varphi_\lambda U_1^C \dot{U}_1^A + (\mathscr{A}^{B\lambda}_{AC} \partial_\lambda \underset{(0)}{U}{}^A + \mathscr{B}^B_C) \underset{(1)}{U}{}^C. \tag{7.35}$$

By integration of (7.33) we obtain

$$\underset{(2)}{\mathbf{U}} = \Pi_2(x, \xi)\mathbf{R} + \mathbf{\Pi}_2(x, \xi), \tag{7.36}$$

with $\dot{\mathbf{\Pi}}_2$ a solution of the linear system.

$$\underset{0}{\mathscr{A}}{}^{B\lambda}_{A} \varphi_\lambda \dot{\mathbf{\Pi}}_2 + \mathbf{g}_1^B = 0. \tag{7.37}$$

In order to detemine $\Pi_2(x, \xi)$ we consider the equation

$$\begin{aligned} \underset{(2)}{F}{}^B = {} & \underset{0}{\mathscr{A}}{}^{B\lambda}_{A} \varphi_\lambda \underset{(3)}{\dot{U}}{}^A + \underset{0}{\mathscr{A}}{}^{B\lambda}_{A} \partial_\lambda \underset{(2)}{U}{}^A + \mathscr{A}^{B\lambda}_{AC} \varphi_\lambda \underset{(1)}{U}{}^C \underset{(2)}{\dot{U}}{}^A \\ & + \{\mathscr{A}^{B\lambda}_{AC}(\varphi_\lambda \underset{(1)}{\dot{U}}{}^A + \partial_\lambda \underset{(0)}{U}{}^A) + \mathscr{B}^B_C\} \underset{(2)}{U}{}^C + \underset{(2)}{f}{}^B = 0. \end{aligned} \tag{7.38}$$

The compatibility condition $L_B \underset{(2)}{F}^B = 0$ then yields

$$K^\lambda \partial_\lambda \Pi_2 + N(\Pi \dot{\Pi}_2 + \dot{\Pi} \Pi_2) + M\Pi_2 + P\dot{\Pi}_2 + Q_2 = 0, \qquad (7.39)$$

with Q_2 a known function of x, $\underset{(1)}{\mathbf{U}}$, and $\mathbf{\Pi}_2$ (x, ξ).

The linear equation (7.39) can be solved by the method of characteristics.

Notice that the projection of the bicharacteristics on $\mathcal{M}$ $x^\alpha = x^\alpha(t, y)$, coincide with those of equation (7.15).

This procedure can then be extended to all higher orders.

In this way we have constructed an asymptotic wave for the system (7.1).

An approximate wave of order 1 is obtained as

$$\mathbf{U} = \mathbf{U}_0 + \frac{1}{\omega}\mathbf{U}_1(x, \xi) + \frac{1}{\omega^2}\mathbf{U}_2(x, \xi) \qquad (7.40)$$

in a neighborhood of $s(y^\alpha) = 0$, with $\mathbf{U}_1 = \Pi(x, \xi)\mathbf{R}$ given by the previous formulas, with $W_1(y, \eta)$ bounded together with its first derivatives with respect to y^α, ξ.

Furthermore, $\mathbf{U}_2$ must be such that (7.33)–(7.34) hold, and $\mathbf{\Pi}_2$ must be bounded together with its derivatives with respect to x^α, ξ. For this it is sufficient that $\mathbf{U}_1$ be bounded together with its derivatives with respect to x^α, ξ, as can be seen from the expression (7.35). Similarly, as in Section 4.2, it is possible to give an expressive formula for M when the system (7.1) is hyperbolic. Let us assume propagation into a constant state, $\mathbf{U}_0 = \text{const}$. Then M reduces to

$$M = L_B \underset{0}{\mathscr{A}}{}^{B\lambda}_{A} \frac{\partial R^A}{\partial \varphi_\mu} \varphi_{\mu\lambda} + L_B \mathscr{B}^B_C R^C, \qquad (7.41)$$

because $\mathbf{R} = \mathbf{R}(\varphi_\mu)$.

From the proof of the Proposition (4.4) it is easy to see that

$$M = \frac{k}{2}\frac{\partial^2 \bar{A}}{\partial \varphi_\alpha \partial \varphi_\beta} + L_B \mathscr{B}^B_C R^C. \qquad (7.42)$$

Notice that in the linear ($N = 0$) and homogeneous case ($\mathscr{B} = 0$), for propagation into a constant state, the transport equation (7.15) reduces (for the natural choice $\mathbf{V}_1 = 0$) to

$$\partial_\lambda(\bar{A}^\lambda \Pi^2) = 0, \qquad (7.43)$$

with $\bar{A}^\lambda = \dfrac{\partial \bar{A}}{\partial \varphi_\lambda}$.

Equation (7.43) represents the standard area-intensity conservation law of geometrical optics and linear geometrical acoustics.

Now we consider the case of multiple roots.

Let $\varphi(x^\mu)=0$ be a multiple root of the characteristic equation (7.12), with multiplicity r. Then, as shown in Section 4.2,

$$\bar{A}(\varphi_\alpha)=\{\hat{A}(\varphi_\alpha)\}^r\tilde{A}(\varphi_\alpha), \tag{7.44}$$

where $\tilde{A}(\varphi_\alpha)$ is not divisible by $\hat{A}(\varphi_\alpha)$, and $\hat{A}(\varphi_\alpha)$ is of first degree in φ_α.

Let $\mathbf{R}_i$, ${}^T\mathbf{L}_i$ be the corresponding right and left eigenvectors, $i=1,2,\ldots,r$.

Then from (7.9) we have

$$\mathbf{U}_{(1)}=\Pi_i(x,\xi)\mathbf{R}_i \tag{7.45}$$

[where we have put equal to zero $\mathbf{V}_{(1)}(x)$, for the sake of simplicity].

Proceeding as in Section 4.2, we obtain

$$L_{jA}\underset{0}{\mathscr{A}}{}^{A\alpha}_{B}R^B_i=k_{ij}\frac{\partial\hat{A}}{\partial\varphi_\alpha}, \tag{7.46}$$

with k_{ij} a nonsingular matrix.

The compatibility conditions $L_{jB}\underset{(1)}{F}{}^B=0$ then yield the following coupled transport system,

$$k_{ij}\hat{A}^\lambda\partial_\lambda\Pi_j+N_{i,jj}\Pi_j\dot{\Pi}_j+M_{i,j}\Pi_j=0, \tag{7.47}$$

where

$$\hat{A}^\lambda=\frac{\partial\hat{A}}{\partial\varphi_\lambda},$$

$$N_{i,jj}=\mathscr{A}^{B\lambda}_{AC}\varphi_\lambda L_{iB}R^A_jR^C_j, \tag{7.48}$$

$$M_{i,j}=\{\mathscr{A}^{B\lambda}_{AC}\partial_\lambda R^A_{,j}+(\mathscr{A}^{B\lambda}_{AC}\partial_\lambda \underset{0}{U}{}^A+\underset{0}{\mathscr{B}}_c)R^A_j\}L^c_{B,i}. \tag{7.49}$$

Also, proceeding as in Section 4.2, one has

$$N_{i,jj}=k_{ij}\frac{\partial\hat{A}}{\partial U^c}R^c_j, \tag{7.50}$$

and the exceptionality condition then reads

$$\frac{\partial\hat{A}}{\partial U^C}R^C_j=0\quad\forall j=1,2,\ldots,r.$$

7.2. Asymptotic waves in relativistic fluid dynamics

The characteristic equation, the propagation speeds, and the right and left eigenvectors have been studied in Section 2.3 whereas the propagation of weak discontinuities has been considered in Section 4.3.

We treat the case of acoustic waves, corresponding to the roots of

$$\hat{\mathscr{A}} \equiv (u^\alpha \varphi_\alpha)^2 - p'_e h^{\alpha\beta} \varphi_\alpha \varphi_\beta = 0. \tag{7.51}$$

Because we have chosen $\mathbf{V}_1 = 0$, the transport equation reads

$$\hat{\mathscr{A}}^\alpha \partial_\alpha \Pi + \hat{N} \Pi \dot{\Pi} + \hat{M} \Pi = 0, \tag{7.52}$$

where

$$\hat{\mathscr{A}}^\alpha \equiv \frac{1}{2} \frac{\partial \hat{\mathscr{A}}}{\partial \varphi_\alpha} = (a u^\alpha - p'_e h^{\alpha\mu} \varphi_\mu)_0, \tag{7.53}$$

$$\hat{N} = -\frac{a^3}{2} \left\{ 2(1 - p'_e) + \frac{p''_e}{p'_e}(e + p) \right\}_0, \tag{7.54}$$

$$\begin{aligned} \hat{M} = \frac{a}{2} \nabla_\alpha \underset{(0)}{u}{}^\alpha - \frac{(e+p)_0^{-1}}{2} \Big\{ & a \underset{(0)}{u}{}^\alpha (1 - p'_e)_0 - 2\hat{\mathscr{A}}^\alpha \\ & + (p''_e)_0 \frac{(e+p)_0}{(p'_e)_0} (a \underset{(0)}{u}{}^\alpha - 2\hat{\mathscr{A}}^\alpha) \Big\} \partial_\alpha e_{(0)} \\ & - \frac{(e+p)_0^{-1}}{2(p'_e)_0} \{ (p'_s)_0 (a \underset{(0)}{u}{}^\alpha (1 - p'_e)_0 - \hat{\mathscr{A}}^\alpha) \\ & + (e+p)_0 (p''_{es})_0 (-2\hat{\mathscr{A}}^\alpha + a \underset{(0)}{u}{}^\alpha) \} (\partial_\alpha S)_0 + \tfrac{1}{2} \Box \varphi, \end{aligned} \tag{7.55}$$

where

$$\Box \varphi \equiv (- \underset{(0)}{p'_e} h^{\alpha\lambda} + \underset{(0)}{u}{}^\alpha \underset{(0)}{u}{}^\lambda) \nabla_\alpha \varphi_\lambda. \tag{7.56}$$

Let us consider the case of special relativity and propagation into a constant state. In Minkowski coordinates (x^α) we look for plane waves, that is, $\varphi(x^\alpha)$ is of the form

$$\varphi(x) = l_\alpha x^\alpha, \quad l_\alpha = \text{const.} \tag{7.57}$$

We can always choose $\underset{(0)}{u}{}^\alpha = (1, 0, 0, 0)$, and therefore the characteristic equation (7.31) yields

$$(l_0)^2 - p'_e \{ (l_1)^2 + (l_2)^2 + (l_2)^3 \} = 0. \tag{7.58}$$

We can always take $l_2 = l_3 = 0$, $l_0 > 0$ and then, for a rightward progressive wave,

$$\varphi(x) = |l_1| (c_s x^0 - x^1), \tag{7.59}$$

where $c_s = (p'_e)^{1/2}$.

The acoustic ray tangent vector is

$$\hat{\mathscr{A}}^0 = c_s |l_1|, \quad \hat{\mathscr{A}}^1 = c_s^2 |l_1|, \quad \hat{\mathscr{A}}^2 = \hat{\mathscr{A}}^3 = 0. \tag{7.60}$$

Also,

$$\hat{N} = -\frac{|l_1^3|}{2}c_s^3\left[2(1-c_s^2)+\frac{(e+p)_0}{c_s^2}(p_e'')_0\right], \tag{7.61}$$

where $\hat{M}=0$.

Hence the transport equation is

$$\frac{\partial \Pi}{\partial x^0}+c_s\frac{\partial \Pi}{\partial x^1}-\frac{c_s^2|l_1|^2}{2}\left[2(1-c_s^2)+\frac{(e+p)_0}{c_s^2}(p_e'')_0\right]\Pi\frac{\partial \Pi}{\partial \xi}=0. \tag{7.62}$$

The rays are given by

$$x^1=c_st+y^1,\quad x^0=t+y^0,\quad x^2=y^2,\quad x^3=y^3. \tag{7.63}$$

We take as the initial manifold $y^0=0$ and as the initial data

$$\Pi(0,y^\alpha,\eta)=W_1(y^1,\eta).$$

It follows that

$$\xi=\eta-W_1(y^1,\eta)\frac{c_s^2|l_1|^2}{2}\left[2(1-c_s^2)+\frac{(e+p)_0}{c_s^2}(p_e'')_0\right]t. \tag{7.64}$$

An interesting case arises when the original perturbation has a sinusoidal profile,

$$W_1(y^1,\eta)=W_0\sin\eta. \tag{7.65}$$

Then equation (7.64) is rewritten

$$\xi=\eta-E\sin\eta, \tag{7.66}$$

where

$$E=\frac{W_0c_s^2|l_1|^2}{2}\left[2(1-c_s^2)+\frac{(e+p)}{c_s^2}(p_e'')_0\right]t. \tag{7.67}$$

Equation (7.66) can be inverted for $|t|<1$, and its solution is

$$\eta=\xi+\sum_{q=1}^{\infty}\frac{2J_q(qE)}{q}\sin q\xi, \tag{7.68}$$

where J_q is the Bessel function of order q.

Therefore, we see that an initial sinusoidal profile is subsequently distorted by the creation of the higher order harmonics,

$$\Pi=\frac{W_0}{E}\left\{\sum_{q=1}^{\infty}2\frac{J_0(qE)}{q}\sin q\xi\right\}. \tag{7.69}$$

Equation (7.66) ceases to be invertible when

$$\frac{\partial \xi}{\partial \eta} = 1 - E\cos\eta = 0$$

and therefore the critical time t_c is found to be

$$t_c = \frac{2}{|l|^2 c_s^2 W_0 \left[2(1 - c_s^2) + \frac{(e+p)}{c_s^2}(p_e'') \right]_0}. \tag{7.70}$$

Now, from the definition of Π and the right eigenvectors, we have, since

$$u^\mu = \Gamma(1, v, 0, 0),$$
$$\Gamma = 1 + O(1/\omega^2),$$
$$v = \frac{\Pi}{\omega}|l_1|c_s^2.$$

For the chosen initial sinusoidal profile we can write

$$v = v_0 \sin\eta, \quad v_0 > 0,$$

hence

$$v_0 = \frac{W_0}{\omega}|l_1|c_s^2$$

and therefore

$$t_c = \frac{2}{\omega|l_1|v_0 \left[2(1 - c_s^2) + \frac{e+p}{c_s^2} p_e'' \right]_0}. \tag{7.70'}$$

It is easy to check that equation (7.70′) is the $v_0 \to 0$ limit of the exact breaking time t_B for the corresponding simple wave with the same initial profile [for a barotropic fluid, equation (5.95)].

7.3. Asymptotic waves in relativistic magneto-fluid dynamics

We treat the case of magnetoacoustic waves, corresponding to the roots of equation (4.67)

$$N_4 = \eta(e_p' - 1)a^4 - (\eta + e_p'|b|^2)a^2 G + B^2 G = 0. \tag{7.71}$$

The transport equation reads,

$$4N^\alpha \partial_\alpha \Pi + \delta N_4 \Pi \dot{\Pi} + M\Pi = 0, \tag{7.72}$$

where

$$N^\alpha = \frac{1}{4}\frac{\partial N_4}{\partial \varphi_\alpha},$$

and δN_4 is given by equation (4.74).

The general expression for M is rather complicated and can be found in the article by Anile and Greco (1978).

We consider the case of special relativity and propagation into a constant state. From the right eigenvectors (2.86) we have

$$\delta p = Ea^2 A,$$

hence

$$\frac{1}{\omega_{(1)}} p = \Pi E a^2 A$$

and in terms of p the transport equation (7.70′) is written

$$\frac{\partial N_4}{\partial \varphi_\alpha} \partial_\alpha \underset{(1)}{p} + a^2[a^2 K_1 + GK_2]\frac{1}{\omega_{(1)(1)}} p \, \dot{p} = 0, \tag{7.72′}$$

with K_1 and K_2 given by equations (4.75)–(4.76).

In Minkowski coordinates (x^α) we look for plane waves

$$\varphi(x) = x^\alpha l_\alpha, \quad l_\alpha = \text{const.} \tag{7.73}$$

By choosing $\underset{(0)}{u}^\alpha = (1, 0, 0, 0)$ and taking the unperturbed magnetic field along the x^1-direction, $\underset{(0)}{b}^\alpha = (0, b, 0, 0)$, the characteristic equation reads,

$$\begin{aligned} &e'_p(\eta + |b|^2)(l_0)^4 - |l|^2(\eta + e'_p|b|^2 + |b|^2\cos^2\theta)l_0^2 \\ &+ |b|^2|l|^4\cos^2\theta = 0, \end{aligned} \tag{7.74}$$

where $|l|^2 = (l_1)^2 + (l_2)^2 + (l_3)^2$ and θ is the angle between the magnetic field $\underset{(0)}{b}^\alpha$ and the spatial direction defined by l^α, $|l_1| = |l|\cos\theta$.

It is convenient to introduce the following quantities

$$\Delta = \eta + (e'_p + \cos^2\theta)|b|^2,$$
$$F = \{\Delta^2 - 4e'_p(\eta + |b|^2)|b|^2\cos^2\theta\}^{1/2},$$
$$g_\pm = \frac{1}{2e'_p(\eta + |b|^2)}\{\Delta \pm F\}.$$

Then the solutions to equation (7.74) are

$$(l_0)^2 = |l|^2 g_\pm, \tag{7.75}$$

where the double sign refers to the fast and slow magnetoacoustic waves, respectively. Then

$$\begin{aligned} N^0 &= \pm\tfrac{1}{2}l_0|l|^2F, \\ N^1 &= \tfrac{1}{2}l_1|l|^2[|b|^2(1+\cos^2\theta) - g_\pm(\eta + e'_p|b|^2 + |b|^2)], \\ N^2 &= \tfrac{1}{2}l_2|l|^2[|b|^2\cos^2\theta - g_\pm(\eta + e'_p|b|^2)], \\ N^3 &= \tfrac{1}{2}l_3|l|^2[|b|^2\cos^2\theta - g_\pm(\eta + e'_p|b|^2)]. \end{aligned} \tag{7.76}$$

For propagation orthogonal to the magnetic field we have $l_1 = 0$, which implies $g_- = 0$ (nonexistence of the slow magnetoacoustic wave).

Then, because we have chosen $l_3 = 0$ and $l_2 = -|l| < 0$, the transport equation reads

$$\frac{\partial \underset{(1)}{p}}{\partial x^0} + v_s \frac{\partial \underset{(1)}{p}}{\partial x^1} + \frac{|l|}{2\omega}\beta \underset{(1)(1)}{p\,\dot{p}} = 0, \tag{7.77}$$

with

$$v_s = \left[\frac{\eta + e'_p|b|^2}{e'_p(\eta + |b|^2)}\right]^{1/2},$$

$$\beta = \frac{e'_p(\eta + |b|^2)K_2 + (\eta + e'_p|b|^2)(K_1 - K_2)}{(\eta + e'_p|b|^2)^{1/2}[e'_p(\eta + |b|^2)]^{3/2}}.$$

For propagation parallel to the magnetic field, $\theta = 0$, $l_2 = l_3 = 0$, $l_1 = -|l|$, hence

$$\begin{aligned} \Delta &= \Delta'' = \eta + (e'_p + 1)|b|^2, \\ F &= |\eta - (e'_p - 1)|b|^2| < \eta. \end{aligned}$$

In the case $(e'_p - 1)|b|^2 < \eta$, then

$$g_+ = \frac{1}{e'_p}, \quad g_- = \frac{|b|^2}{E}.$$

The transport equation reads, for the fast magnetoacoustic waves,

$$\frac{\partial \underset{(1)}{p}}{\partial x^0} + \frac{1}{(e'_p)^{1/2}}\frac{\partial \underset{(1)}{p}}{\partial x^1} + \frac{|l|(e'_p)^{-3/2}}{\omega}\alpha \underset{(1)(1)}{p\,\dot{p}} = 0, \tag{7.78}$$

with

$$\alpha = e''_p - \frac{2e'_p(e'_p - 1)}{\eta},$$

which coincides with the purely fluid-dynamical one.

7.4. Asymptotic and approximate waves for Einstein's equations in vacuo

We consider Einstein's equations in vacuo, which in local coordinates (x^α) read (Misner et al., 1973)

$$R_{\alpha\beta} \equiv \frac{\partial \Gamma^\lambda_{\alpha\beta}}{\partial x^\lambda} - \frac{\partial \Gamma^\lambda_{\beta\lambda}}{\partial x^\alpha} + \Gamma^\lambda_{\alpha\beta}\Gamma^\mu_{\lambda\mu} - \Gamma^\mu_{\alpha\lambda}\Gamma^\lambda_{\beta\mu} = 0, \tag{7.79}$$

$\Gamma^\alpha_{\beta\gamma}$ being the Christoffel symbols of the metric $g_{\alpha\beta}$.

Let $\xi = \omega\varphi(x)$. We look for a solution of (7.79) in the form of a finite development (Choquet-Bruhat, 1969a)

$$g_{\lambda\mu}(x, \xi) = \overset{0}{g}_{\lambda\mu}(x) + \frac{1}{\omega}\overset{1}{g}_{\lambda\mu}(x, \xi) + \frac{1}{\omega^2}\overset{2}{g}_{\lambda\mu}(x, \xi), \tag{7.80}$$

where $\overset{0}{g}_{\lambda\mu}$ is a differentiable hyperbolic metric.

We shall impose the condition that $\overset{1}{g}_{\lambda\mu}(x, \xi)$, $\overset{2}{g}_{\lambda\mu}(x, \xi)$ are bounded functions of x^α, ξ in an open subset Ω of space-time.

The inverse matrix $g^{\lambda\mu}$ of the matrix $g_{\lambda\mu}$, defined by

$$g^{\lambda\mu}g_{\lambda\alpha} = \delta^\mu_\alpha,$$

is given as a power series in $1/\omega$.

$$g^{\lambda\mu} = \overset{0}{g}{}^{\lambda\mu}(x) + \frac{1}{\omega}\overset{1}{g}{}^{\lambda\mu}(x, \xi) + \frac{1}{\omega^2}\overset{2}{g}{}^{\lambda\mu}(x, \xi) + \cdots, \tag{7.81}$$

where $\overset{0}{g}{}^{\lambda\mu}$ is the inverse matrix of $\overset{0}{g}_{\alpha\beta}$, $\overset{0}{g}{}^{\lambda\mu}\overset{0}{g}_{\lambda\alpha} = \delta^\mu_\alpha$; also,

$$\overset{1}{g}{}^{\beta\alpha} = -\overset{1}{g}_{\lambda\mu}\overset{0}{g}{}^{\lambda\alpha}\overset{0}{g}{}^{\mu\beta} \tag{7.82}$$

and $\overset{p}{g}{}^{\lambda\mu}$ obeys

$$\overset{0}{g}_{\lambda\alpha}\overset{p}{g}{}^{\lambda\mu} + \overset{1}{g}_{\lambda\alpha}\overset{(p-1)}{g}{}^{\lambda\mu} + \overset{2}{g}_{\lambda\alpha}\overset{(p-2)}{g}{}^{\lambda\mu} = 0, \tag{7.83}$$

so that $\overset{p}{g}{}^{\beta\mu}$ is given recursively by

$$\overset{p}{g}{}^{\beta\mu} = -\overset{1}{g}_{\lambda\alpha}\overset{(p-1)}{g}{}^{\lambda\mu}\overset{0}{g}{}^{\alpha\beta} - \overset{2}{g}_{\lambda\alpha}\overset{(p-2)}{g}{}^{\lambda\mu}\overset{0}{g}{}^{\alpha\beta}.$$

Therefore, the series (7.81) is asymptotic for $\omega \to \infty$ provided $\overset{1}{g}_{\lambda\mu}$ and $\overset{2}{g}_{\lambda\mu}$

are bounded functions of x, ξ. Let us write

$$g'_{\alpha\beta} = \left(\frac{\partial g_{\alpha\beta}}{\partial \xi}\right)_{\xi=\omega\varphi}, \quad \partial_\alpha g_{\lambda\mu} = \left\{\frac{\partial g_{\lambda\mu}}{\partial x^\alpha}\right\}_{\xi=\omega\varphi}.$$

Then

$$\frac{\partial g_{\lambda\mu}}{\partial x^\alpha} = \partial_\alpha \overset{0}{g}_{\lambda\mu} + \varphi_\alpha \overset{1}{g'}_{\lambda\mu} + \frac{1}{\omega}\partial_\alpha \overset{1}{g}_{\lambda\mu} + \frac{1}{\omega}\overset{2}{g'}_{\lambda\mu}\varphi_\alpha + \frac{1}{\omega^2}\partial_\alpha \overset{2}{g}_{\lambda\mu}. \tag{7.84}$$

For the Christoffel symbols of $g_{\alpha\beta}$ one obtains

$$\Gamma^\lambda_{\alpha\beta} = \overset{0}{\Gamma}{}^\lambda_{\alpha\beta} + \frac{1}{\omega}\overset{1}{\Gamma}{}^\lambda_{\alpha\beta} + \frac{1}{\omega^2}\overset{2}{\Gamma}{}^\lambda_{\alpha\beta} + \cdots, \tag{7.85}$$

where

$$\overset{0}{\Gamma}{}^\lambda_{\alpha\beta} = \bar{\Gamma}^\lambda_{\alpha\beta} + \tfrac{1}{2}\overset{0}{g}{}^{\lambda\mu}\left(\overset{1}{g'}_{\beta\mu}\varphi_\alpha + \overset{1}{g'}_{\alpha\mu}\varphi_\beta - \overset{1}{g'}_{\alpha\beta}\varphi_\mu\right), \tag{7.86}$$

with $\bar{\Gamma}^\lambda_{\alpha\beta}$ the Christoffel symbols of the background metric $\overset{0}{g}_{\alpha\beta}$, and

$$\begin{aligned}\overset{1}{\Gamma}{}^\lambda_{\alpha\beta} = {} & \tfrac{1}{2}\overset{0}{g}{}^{\lambda\mu}\left(\overset{2}{g'}_{\beta\mu}\varphi_\alpha + \overset{2}{g'}_{\alpha\mu}\varphi_\beta - \overset{2}{g'}_{\alpha\beta}\varphi_\mu\right) \\ & + \tfrac{1}{2}\overset{0}{g}{}^{\lambda\mu}\left(\partial_\alpha \overset{1}{g}_{\beta\mu} + \partial_\beta \overset{1}{g}_{\alpha\mu} - \partial_\mu \overset{1}{g}_{\alpha\beta}\right) \\ & + \tfrac{1}{2}\overset{1}{g}{}^{\lambda\mu}\left(\partial_\alpha \overset{0}{g}_{\beta\mu} + \partial_\beta \overset{0}{g}_{\alpha\mu} - \partial_\mu \overset{0}{g}_{\alpha\beta}\right) \\ & + \tfrac{1}{2}\overset{1}{g}{}^{\lambda\mu}\left(\overset{1}{g'}_{\beta\mu}\varphi_\alpha + \overset{1}{g'}_{\alpha\mu}\varphi_\beta - \overset{1}{g'}_{\alpha\beta}\varphi_\mu\right).\end{aligned} \tag{7.87}$$

Also, the Ricci tensor $R_{\alpha\beta}$ has the following development

$$R_{\alpha\beta} = \omega \overset{-1}{R}_{\alpha\beta} + \overset{0}{R}_{\alpha\beta} + \frac{1}{\omega}\overset{1}{R}_{\alpha\beta} + \cdots. \tag{7.88}$$

The metric $g_{\lambda\mu}$ will represent an approximate wave of *order zero* if $\overset{-1}{R}_{\alpha\beta} = 0$ with $\overset{0}{R}_{\alpha\beta}$ bounded, and an approximate wave of *order 1* if $\overset{-1}{R}_{\alpha\beta} = 0$, $\overset{0}{R}_{\alpha\beta} = 0$, and $\omega R_{\alpha\beta}$ is bounded $\forall x \in \Omega, \forall \xi$.

Let us consider now the following transformation of coordinates in Ω,

("gauge" transformations)

$$x^\alpha = \tilde{x}^\alpha + \frac{1}{\omega^2}\Psi^\alpha(x,\xi). \tag{7.89}$$

One has to the first order in $1/\omega$

$$\frac{\partial x^\alpha}{\partial \tilde{x}^\beta} = \delta^\alpha_\beta + \frac{1}{\omega}\Psi'^\alpha\varphi_\beta + 0\left(\frac{1}{\omega^2}\right),$$

hence

$$\tilde{g}_{\alpha\beta} = \overset{0}{\tilde{g}}_{\alpha\beta} + \frac{1}{\omega}\overset{1}{\tilde{g}}_{\alpha\beta} + \frac{1}{\omega^2}\tilde{g}_{\alpha\beta} + \cdots,$$

with

$$\overset{0}{\tilde{g}}_{\alpha\beta} = \overset{0}{g}_{\alpha\beta},$$

$$\overset{1}{\tilde{g}}_{\alpha\beta} = \overset{1}{g}_{\alpha\beta} + \frac{1}{\omega}\left(\overset{0}{g}_{\alpha\mu}\Psi'^\mu\varphi_\beta + \overset{0}{g}_{\beta\mu}\Psi'^\mu\varphi_\alpha\right). \tag{7.90}$$

The components of the Ricci tensor in the new coordinates are

$$\tilde{R}_{\alpha\beta} = R_{\alpha\beta} + \frac{1}{\omega}(R_{\alpha\mu}\Psi'^\mu\varphi_\beta + R_{\beta\mu}\Psi'^\mu\varphi_\alpha) + \cdots \tag{7.91}$$

and therefore

$$\overset{-1}{\tilde{R}}_{\alpha\beta} = \overset{-1}{R}_{\alpha\beta}, \tag{7.92}$$

$$\overset{0}{\tilde{R}}_{\alpha\beta} = \overset{0}{R}_{\alpha\beta} + \left(\overset{-1}{R}_{\alpha\mu}\varphi_\beta + \overset{-1}{R}_{\beta\mu}\varphi_\alpha\right)\Psi'^\mu. \tag{7.93}$$

It follows that the equation $\overset{-1}{R}_{\alpha\beta} = 0$, as well as the set of equations $\overset{-1}{R}_{\alpha\beta} = 0$, $\overset{0}{R}_{\alpha\beta} = 0$ are invariant under "gauge" transformations.

One obtains for $\overset{-1}{R}_{\alpha\beta}$,

$$\overset{-1}{R}_{\alpha\beta} = \overset{0}{\Gamma}{}'^{\lambda}_{\alpha\beta}\varphi_\lambda - \overset{0}{\Gamma}{}'^{\lambda}_{\beta\lambda}\varphi_\alpha, \tag{7.94}$$

which yields

$$\overset{-1}{R}_{\alpha\beta} = \tfrac{1}{2}\overset{0}{g}{}^{\lambda\mu}\left\{\left(\overset{1}{g}{}''_{\alpha\mu}\varphi_\beta + \overset{1}{g}{}''_{\beta\mu}\varphi_\alpha\right)\varphi_\lambda - \overset{1}{g}{}''_{\alpha\beta}\varphi_\lambda\varphi_\mu - \overset{1}{g}{}''_{\lambda\mu}\varphi_\beta\varphi_\alpha\right\} = 0. \tag{7.95}$$

Now there are two cases:

(i) $\overset{0}{g}{}^{\lambda\mu}\varphi_\lambda\varphi_\mu \neq 0.$ (7.96)

Then $\overset{1}{g}{}''_{\alpha\beta} = \theta_\alpha\varphi_\beta + \theta_\beta\varphi_\alpha$, where θ_α is an arbitrary covariant vector. Hence $\overset{1}{g}{}''_{\alpha\beta}$ can be annulled by a gauge transformation of the kind (7.89).

It follows that $\overset{1}{g}{}'_{\alpha\beta} = K_{\alpha\beta} =$ constant with respect to ξ, whence $\overset{1}{g}_{\alpha\beta} = K_{\alpha\beta}\xi + \hat{K}_{\alpha\beta}$, $\hat{K}_{\alpha\beta}$ is constant with respect to ξ. Since $\overset{1}{g}_{\alpha\beta}$ must be bounded $\forall\xi$, it follows that $K_{\alpha\beta} = 0$ and therefore $\overset{1}{g}_{\alpha\beta}$ cannot depend on ξ. We conclude that this type of perturbation is not physical.

(ii) $\overset{0}{g}{}^{\lambda\mu}\varphi_\lambda\varphi_\mu = 0$ (gravitational waves), (7.97)

which means that $\varphi(x) =$ constant is a characteristic hypersurface for the wave operator of the metric $\overset{0}{g}_{\alpha\beta}$. Then

$$\overset{-1}{R}_{\alpha\beta} = \tfrac{1}{2}\overset{0}{g}{}^{\lambda\mu}\left\{\overset{1}{g}{}''_{\alpha\mu}\varphi_\beta\varphi_\lambda + \overset{1}{g}{}''_{\beta\mu}\varphi_\alpha\varphi_\lambda - \overset{1}{g}{}''_{\lambda\mu}\varphi_\alpha\varphi_\beta\right\} = 0, \tag{7.98}$$

which is written as

$$\overset{-1}{R}_{\alpha\beta} = -\tfrac{1}{2}(\varphi_\alpha\Phi''_\beta + \varphi_\beta\Phi''_\alpha) = 0, \tag{7.99}$$

with

$$\Phi_\beta = \tfrac{1}{2}\overset{0}{g}{}^{\lambda\mu}\left(\tfrac{1}{2}\overset{1}{g}_{\lambda\mu}\varphi_\beta - \overset{1}{g}_{\beta\mu}\varphi_\lambda\right). \tag{7.100}$$

Equation (7.99) implies $\Phi''_\alpha = 0$. In fact, in a local chart adapted to the hypersurface Σ: $\varphi =$ const., that is, $x^0 = \varphi$ and x^i are local coordinates on Σ, we have $\varphi_\alpha = (1, 0, 0, 0)$ and equation (7.99) yields, $\Phi''_0 = 0$ and $\Phi''_i = 0$.

Because $\overset{1}{g}_{\lambda\mu}$ must be bounded with respect to ξ we see that we must have

$$\Phi_\alpha = \tfrac{1}{2}\overset{0}{g}{}^{\lambda\mu}\left(\tfrac{1}{2}\overset{1}{g}_{\lambda\mu}\varphi_\alpha - \overset{1}{g}_{\alpha\mu}\varphi_\lambda\right) = 0. \tag{7.101}$$

Therefore, for an approximate wave of order zero, equations (7.97) and (7.101) must hold. Furthermore, we must have $\overset{0}{g}_{\lambda\mu}, \overset{1}{g}_{\lambda\mu}$ bounded together with their partial derivatives (up to the second order) with respect to x^α and ξ.

In adapted coordinates the condition (7.97) reads

$$\overset{0}{g}{}^{00}=0, \tag{7.102}$$

and (7.101) yields

$$\begin{aligned}\overset{0}{g}{}^{0j}\overset{1}{g}_{ij}&=0,\\ \overset{0}{g}{}^{ij}\overset{1}{g}_{ij}&=0.\end{aligned} \tag{7.103}$$

By raising indexes with the background metric $\overset{0}{g}{}^{\alpha\beta}$, equations (7.103) can be rewritten in the form

$$\varphi^j\overset{1}{g}_{ij}=0, \quad \text{where} \quad \varphi^\alpha=\overset{0}{g}{}^{\alpha\beta}\varphi_\beta, \tag{7.104}$$

and

$$\overset{1}{g}{}^{00}=0, \tag{7.105}$$

which expresses the statement that the hypersurface $\Sigma: \varphi(x)=\text{const.}$ is also a characteristic hypersurface for the metric $g_{\alpha\beta}=\overset{0}{g}_{\alpha\beta}+\dfrac{1}{\omega}\overset{1}{g}_{\alpha\beta}(x,\xi)$.

Now we look for approximate waves of order 1, that is, verifying the equations $\overset{-1}{R}_{\alpha\beta}=0$ and $\overset{0}{R}_{\alpha\beta}=0$.

One has

$$\overset{0}{R}_{\alpha\beta}=\overset{1}{\Gamma}{}'^{\lambda}_{\alpha\beta}\varphi_\lambda-\overset{1}{\Gamma}{}'^{\lambda}_{\beta\lambda}\varphi_\alpha+\partial_\lambda\overset{0}{\Gamma}{}^{\lambda}_{\alpha\beta}-\partial_\alpha\overset{0}{\Gamma}{}^{\lambda}_{\beta\lambda}+\overset{0}{\Gamma}{}^{\lambda}_{\alpha\beta}\overset{0}{\Gamma}{}^{\mu}_{\lambda\mu}-\overset{0}{\Gamma}{}^{\mu}_{\alpha\lambda}\overset{0}{\Gamma}{}^{\lambda}_{\beta\mu}. \tag{7.106}$$

Now, since $\varphi^\lambda\varphi_\lambda=0$, we have

$$\overset{0}{R}_{\alpha\beta}=\tfrac{1}{2}\overset{0}{g}{}^{\lambda\mu}\left(\overset{2}{g}{}''_{\alpha\mu}\varphi_\beta\varphi_\lambda+\overset{2}{g}{}''_{\beta\mu}\varphi_\alpha\varphi_\lambda-\overset{2}{g}{}''_{\lambda\mu}\varphi_\alpha\varphi_\beta\right)+\chi_{\alpha\beta}=0, \tag{7.107}$$

where

$$\chi_{\alpha\beta}=I_{\alpha\beta}+II_{\alpha\beta}+III_{\alpha\beta}+IV_{\alpha\beta}+V_{\alpha\beta}, \tag{7.108}$$

$$I_{\alpha\beta}=\tfrac{1}{2}\overset{1}{g}{}^{\lambda\mu}\left(\overset{1}{g}{}''_{\alpha\mu}\varphi_\beta\varphi_\lambda+\overset{1}{g}{}''_{\beta\mu}\varphi_\alpha\varphi_\lambda-\overset{1}{g}{}''_{\lambda\mu}\varphi_\alpha\varphi_\beta\right) \tag{7.109}$$

$$\begin{aligned}II_{\alpha\beta}=&-\varphi^\lambda\partial_\lambda\overset{1}{g}{}'_{\alpha\beta}+\tfrac{1}{2}\overset{0}{g}{}^{\lambda\mu}\Big(\partial_\lambda\overset{1}{g}{}'_{\alpha\mu}\varphi_\beta+\partial_\lambda\overset{1}{g}{}'_{\beta\mu}\varphi_\alpha-\partial_\alpha\overset{1}{g}{}'_{\lambda\mu}\varphi_\beta\\ &-\partial_\beta\overset{1}{g}{}'_{\lambda\mu}\varphi_\alpha\Big)+\tfrac{1}{2}\varphi^\lambda\left(\partial_\alpha\overset{1}{g}{}'_{\lambda\beta}+\partial_\beta\overset{1}{g}{}'_{\lambda\alpha}\right)\end{aligned} \tag{7.110}$$

$$III_{\alpha\beta} = \tfrac{1}{2}\overset{1}{g}'^{\lambda\mu}\left\{\left(\overset{1}{g}'_{\alpha\mu}\varphi_\beta + \overset{1}{g}'_{\beta\mu}\varphi_\alpha\right)\varphi_\lambda - \overset{1}{g}'_{\lambda\mu}\varphi_\alpha\varphi_\beta - \overset{1}{g}'_{\alpha\beta}\varphi_\lambda\varphi_\mu\right\}$$

$$+ \tfrac{1}{4}\varphi^\lambda \overset{0}{g}^{\rho\sigma}\overset{1}{g}'_{\rho\sigma}\left(\overset{1}{g}'_{\beta\lambda}\varphi_\alpha + \overset{1}{g}'_{\alpha\lambda}\varphi_\beta\right) - \tfrac{1}{4}\overset{0}{g}^{\lambda\mu}\overset{0}{g}^{\rho\nu}\left(\overset{1}{g}'_{\rho\mu}\varphi_\alpha + \overset{1}{g}'_{\alpha\mu}\varphi_\rho - \overset{1}{g}'_{\alpha\rho}\varphi_\mu\right)$$

$$\times\left(\overset{1}{g}'_{\lambda\nu}\varphi_\beta + \overset{1}{g}'_{\beta\nu}\varphi_\lambda - \overset{1}{g}'_{\beta\lambda}\varphi_\nu\right). \tag{7.111}$$

$$IV_{\alpha\beta} = \{-\varphi^\lambda\bar{\Gamma}^\mu_{\alpha\beta} + \tfrac{1}{2}\overset{0}{g}^{\rho\mu}(\bar{\Gamma}^\lambda_{\alpha\rho}\varphi_\beta + \bar{\Gamma}^\lambda_{\beta\rho}\varphi_\alpha)\}\overset{1}{g}'_{\lambda\mu}$$

$$-\tfrac{1}{2}\overset{0}{g}^{\lambda\mu}\overset{1}{g}'_{\lambda\mu}\bar{\nabla}_\alpha\varphi_\beta + \tfrac{1}{2}\overset{1}{g}'_{\beta\mu}\left(\overset{0}{g}^{\lambda\mu}\bar{\nabla}_\lambda\varphi_\alpha + \varphi^\lambda\bar{\Gamma}^\mu_{\alpha\lambda} - \overset{0}{g}^{\rho\lambda}\bar{\Gamma}^\mu_{\lambda\rho}\varphi_\alpha\right)$$

$$+\tfrac{1}{2}\overset{1}{g}'_{\alpha\mu}\left(\overset{0}{g}^{\lambda\mu}\bar{\nabla}_\lambda\varphi_{\beta\alpha} + \varphi^\lambda\bar{\Gamma}^\mu_{\beta\lambda} - \overset{0}{g}^{\rho\lambda}\bar{\Gamma}^\mu_{\lambda\rho}\varphi_\beta\right)$$

$$-\tfrac{1}{2}\overset{1}{g}'_{\alpha\beta}\bar{\nabla}_\lambda\varphi^\lambda, \tag{7.112}$$

where $\bar{\nabla}$ denotes the Riemannian connection of the metric $\overset{0}{g}_{\alpha\beta}$ and

$$V_{\alpha\beta} = \bar{R}_{\alpha\beta}, \tag{7.113}$$

with $\bar{R}_{\alpha\beta}$ the Ricci tensor of the metric $\overset{0}{g}_{\alpha\beta}$.

In adapted coordinates, $\varphi_\alpha = (1, 0, 0, 0)$, and we have

$$\chi_{ij} = 0, \tag{7.114}$$

which, by using (7.103)–(7.105) yields

$$-\varphi^h\left\{\partial_h\overset{1}{g}'_{ij} - \bar{\Gamma}^k_{ih}\overset{1}{g}'_{kj} - \bar{\Gamma}^k_{jh}\overset{1}{g}'_{ki}\right\} - \tfrac{1}{2}\overset{1}{g}'_j\bar{\nabla}_\lambda\varphi^\lambda + \bar{R}_{ij} = 0. \tag{7.115}$$

Equations (7.115) are ordinary differential equations for $\overset{1}{g}'_{ij}$ along the rays given by

$$\frac{dx^h}{dt} = \varphi^h, \tag{7.116}$$

which are tangent curves to Σ.

The system (7.115) can be written in the form

$$\frac{dU}{dt} = A(x)U + B(x), \tag{7.117}$$

with U the matrix with elements $\overset{1}{g}'_{ij}(x, \xi)$, $A(x)$ a matrix, and $B(x)$ the matrix $\bar{R}_{ij}$, on the rays

$$\frac{dx^\lambda}{dt} = \varphi^\lambda.$$

Let S denote an initial hypersurface transverse to the rays, $x(0, y) = y \in S$. Then the rays are given by the solutions of (7.116) $x = x(t, y)$.

The solution of (7.117) is then

$$U(t, y, \xi) = \Phi(t, y)(\Psi(t, y) + \Theta(y, \xi)), \tag{7.118}$$

where

$$\Phi(t, y) = \exp \int_0^t A(x(\tau, y)d\tau,$$

$$\Psi(t, y) = \int_0^t \Phi^{-1}(\tau, y)B(x(\tau, y))d\tau,$$

with $\Theta(y, \xi)$ an arbitrary function such that

$$U(0, y, \xi) = \Theta(y, \xi).$$

From (7.118), by integrating with respect to ξ, we obtain for the matrix V with components $\overset{1}{g}_{ij}$

$$V(t, y, \xi) = \Phi(t, y)\Psi(t, y)\xi + \Phi(t, y)\chi(y, \xi), \tag{7.119}$$

where $\chi(y, \xi)$ is a primitive of $\Theta(y, \xi)$.

Therefore, in order to have $\overset{1}{g}_{ij}$ bounded $\forall \xi$ we must have $\Psi(t, y) = 0$, that is, $B = 0$, which means

$$\bar{R}_{ij} = 0, \tag{7.120}$$

which in general coordinates is written

$$\bar{R}_{\alpha\beta} = m_\alpha \varphi_\beta + m_\beta \varphi_\alpha, \tag{7.121}$$

with m_α an arbitrary vector.

Now we turn to the remaining equations.

By performing an appropriate "gauge" transformation of the kind (7.90) we can set

$$\overset{1}{g}_{0\alpha} = 0.$$

Then the equation

$$\overset{0}{R}_{0i} = 0$$

gives

$$\tfrac{1}{2}\varphi^h \overset{2}{g}{}''_{ih} + \tfrac{1}{2}\overset{0}{g}{}^{kh}\left(\partial_h \overset{1}{g}{}'_{ik} - \bar{\Gamma}^j_{hk}\overset{1}{g}{}'_{ij} - \bar{\Gamma}^j_{ih}\overset{1}{g}{}'_{kj}\right) + \bar{R}_{0i} = 0, \tag{7.122}$$

for which it is apparent that in order to have $\overset{2}{g}{}'_{ij}$ bounded $\forall \xi$, one must require

$$\bar{R}_{0i} = 0. \tag{7.123}$$

In general coordinates the latter equation is written

$$\bar{R}_{\alpha\beta} = \tau\varphi_\alpha\varphi_\beta, \tag{7.124}$$

with τ an arbitrary scalar function. Furthermore, $\overset{1}{g}_{ij}$ must have primitives with respect to ξ which are bounded $\forall \xi$.

Finally, we consider the last equation

$$-\tfrac{1}{2}\overset{0}{g}{}^{ij}\overset{2}{g}{}''_{ij} - \tfrac{1}{4}\left(\overset{1}{g}{}^{ij}\overset{1}{g}_{ij}\right)'' + \tfrac{1}{4}\overset{1}{g}{}'^{ij}\overset{1}{g}{}'_{ij} - \overset{1}{g}{}'_{ij}\overset{0}{g}{}^{hi}\bar{\Gamma}^j_{ch} + \bar{R}_{00} = 0. \tag{7.125}$$

Now, if $a(x^\mu, \xi)$ is any uniformly bounded function of ξ, we have

$$\lim_{T\to\infty} \frac{1}{T}\int_0^T a'(x^\mu, \xi)\,\mathrm{d}\xi = 0,$$

that is, the average of $a'(x^\mu, \xi)$ with respect to ξ vanishes.

Therefore, by averaging equation (7.125) with respect to ξ and assuming that $\overset{2}{g}_{ij}, \overset{1}{g}_{ij}, \overset{2}{g}{}'_{ij}, \overset{1}{g}{}'_{ij}$ are uniformly bounded, we obtain

$$\lim_{T\to\infty} \frac{1}{T}\int_0^T \left(\tfrac{1}{4}\overset{1}{g}{}'^{ij}\overset{1}{g}{}'_{ij} + \bar{R}_{00}\right)\mathrm{d}\xi = 0, \tag{7.126}$$

whence

$$\tau = -\lim_{T\to\infty} \frac{1}{T}\int_0^T \tfrac{1}{4}\overset{1}{g}{}'^{ij}\overset{1}{g}{}'_{ij}\,\mathrm{d}\xi. \tag{7.127}$$

From $\overset{1}{g}{}'^{\lambda\mu} = -\overset{0}{g}{}^{\lambda\alpha}\overset{0}{g}{}^{\mu\beta}\overset{1}{g}{}'_{\alpha\beta}$ we have

$$\overset{1}{g}{}'^{\lambda\mu}\overset{1}{g}{}'_{\lambda\mu} = -\overset{0}{g}{}^{\lambda\alpha}\overset{0}{g}{}^{\mu\beta}\overset{1}{g}{}'_{\alpha\beta}\overset{1}{g}{}'_{\lambda\mu},$$

hence

$$\overset{1}{g}{}'^{ij}\overset{1}{g}{}'_{ij} = -\overset{0}{g}{}^{ik}\overset{0}{g}{}^{jl}\overset{1}{g}{}'_{ij}\overset{1}{g}{}'_{kl} < 0.$$

It follows that we must have

$$\tau > 0. \tag{7.128}$$

Then τ can be interpreted as energy loss due to gravitational radiation. Also, the quantity

$$\mathrm{E} = -\tfrac{1}{4}\overset{1}{g}{}'^{ij}\overset{1}{g}{}'_{ij}$$

can be interpreted as gravitational radiation energy density. It is easily seen that ε satisfies the conservation law

$$\bar{\nabla}_\lambda(\mathrm{E}\varphi^\lambda) = 0. \tag{7.129}$$

In fact, by contracting equation (7.115) with $\overset{1}{g}{}'^{ij}$ one obtains

$$\varphi^h \overset{1}{g}{}'^{ij}\bar{\nabla}_h \overset{1}{g}{}'_{ij} + \tfrac{1}{2}\overset{1}{g}{}'^{ij}\overset{1}{g}{}'_{ij}\bar{\nabla}_\lambda\varphi^\lambda = 0,$$

which is equivalent to equation (7.129) in adapted coordinates.

There are other approaches which are also fruitful for discussing gravitational radiation, in particular those based on a direct application of the Wentzel, Kramers, and Brillouin method to Einstein's equations (Brill and Hartle, 1964; Isaacson, 1968; Anile, 1976).

One of the great advantages of the method of asymptotic waves applied to discuss gravitational radiation is that it introduces, in the high-frequency limit, the local concept of gravitational energy density.

8
Relativistic shock waves

8.0. Introduction

From the results of the previous chapters we have seen that, under suitable assumptions (the compressibility hypothesis) a relativistic compressive nonlinear acoustic or magnetoacoustic pulse steepens and degenerates into a shock wave. Therefore, shocks are common occurrences in nonlinear wave motion and this chapter is devoted to laying out the basic theory of relativistic shock waves. Relativistic shocks are a very important feature in several models of phenomena occurring in astrophysics, plasma physics, and nuclear physics and in the following we shall briefly touch upon some examples.

Supernovas represent one of the most fierce phenomena occurring in the universe. A star suddenly increases its luminosity by many orders of magnitude such that at its maximum its light can outshine the total light from its parent galaxy. This phenomenon is suggestive of an explosion taking place in the star. Several mechanisms have been proposed in order to explain the source of energy driving the explosion (Carbon detonation, neutrino energy deposition, gravitational collapse and bounce, etc.).

In the case of massive stars, in the range between 8 and 100 solar masses, which are thought to be progenitors of type II supernovas, one of the most viable mechanisms for producing an explosion is gravitational collapse and bounce (Van Riper, 1979). At the end of stellar evolution the star will develop a core composed mainly of nuclei near the iron peak and free electrons, with a mass close to the Chandrasekhar limit of about 1.4 solar masses. Due to electron capture and partial dissociation of nuclei, a dynamical instability sets in and the core starts collapsing until hydrodynamical stability is regained. At this point the core overshoots the equilibrium configuration and rebounds acting as a piston. This causes a shock to form outside the inner core, which propagates outward reaching relativistic speeds. If the shock is sufficiently strong it will be able to expel the bulk of the star and leave behind a hot compact remnant which will become a neutron star.

This is one of the most viable models, but much work remains to be done before it can be considered to be on a firm basis. In particular, the precise conditions under which the shock forms at some point with exactly the necessary strength to expel the bulk of the star but still leave behind a remnant remain to be investigated in detail. We remark that the relativistic shock propagates into a medium with a changing equation of state (having the complexity of nuclear matter). Therefore, a simple analysis of the jump conditions for a polytropic or perfect fluid is not adequate and a deep understanding of this problem calls on the full theoretical description of relativistic shocks in a medium with an arbitrary equation of state. The analysis of the jump conditions under very general assumptions on the state equation will be performed in this chapter.

Another complication which might arise is that a significant magnetic field might be present. In fact, one can show that the relative importance of a magnetic field can grow during the collapse. Let $\mathscr{R}$ denote the ratio of the magnetic energy density to the total matter energy density e. Then, in the case of shear-free collapse, it is possible to prove that (Yodzsis, 1976)

$$\frac{\mathrm{d}\mathscr{R}}{\mathrm{d}t} = \frac{\mathscr{R}(e-3p)}{3e(e+p)}\frac{\mathrm{d}e}{\mathrm{d}t}.$$

Therefore, for a collapse one has $\mathrm{d}e/\mathrm{d}t > 0$ and, assuming that $e > 3p$ (the restriction $e > 3p$ definitely holds in the early stages of the collapse) one obtains that $\mathscr{R}$ increases.

In the area of laboratory plasma physics, magnetoacoustic shock waves with speeds of up to 4×10^8 cm/sec have been achieved (Taussig, 1973) with the Columbia University Plasma Laboratory Electromagnetic High-Energy Shock Tube (Gross, 1971). The shock tube is 3 m long, has an outer diameter of 23 cm, and consists of coaxial concentric cylinders with a width of 5 cm between the tubes. The tube length was chosen in order to provide sufficient time for the shock wave to evolve to a steady state. The shock tube is initially filled with room temperature hydrogen or deuterium at a pressure ranging between 10 m Torr and 200 m Torr. Also, there is a magnetic azimuthal field (bias) in the shock tube test region which is typically 0.7 Teslas. In some experiments the speed of sound in the pre-shock gas is typically 1.3×10^5 cm/sec and the Alfvén speed is 2.7×10^7 cm/sec. The shock thickness, ranging from 3 to 80 cm, is greatly reduced by the transverse magnetic field with respect to the purely gas dynamical field. The mean free path in the pre-shock gas is about 2 cm and 140 cm in the post-shock plasma. Therefore, the shock wave can be considered a collisional one. The interpretation of the experiments has been done in a Newtonian framework (Liberman and Velikovich, 1985) and this might just be adequate for the attained speeds. However, the relativistic evolutionary conditions (which will be treated in Section 8.5) seem to be different from the Newtonian ones and therefore a relativistic approach is in the end necessary for a proper theoretical understanding of these phenomena.

In the field of nuclear physics, high-energy collisions among heavy ions can be modeled by using fluid dynamical concepts. The rationale for using hydrodynamical concepts is twofold. One reason is that the collision of two heavy nuclei excites a very large number of degrees of freedom (compared to those usually excited in few-body nuclear reactions) and therefore a statistical thermodynamical and hydrodynamical approach seems appropriate. The second reason is that, due to the lack of a detailed understanding of the nuclear dynamics, a fluid dynamical approach (which essentially relies on the basic conservation laws of energy-momentum) seems a safe first step.

For a fluid dynamical description to be appropriate the following criteria must be satisfied, at least approximately: (1) many degrees of freedom are excited; (2) there is a short mean free path and mean stopping length (the average distance it takes for a nucleon to dissipate its kinetic energy); (3) the reaction time is sufficiently short in order to ensure local thermodynamical equilibrium and; (4) there is a sufficiently short de Broglie wavelength so that a semiclassical particle description is adequate. Simple estimates of the nuclear parameters show that the above conditions are satisfied, at least for near central collisions of heavy nuclei at a moderate bombarding energy (≥ 200 MeV) (Amsden, Harlow, and Nix, 1977).

When a nondissipative hydrodynamical description applies and relativistic effects are not negligible, nuclear matter is described by the equations of relativistic fluid dynamics and all the details of nuclear interactions are incorporated in the state equation, giving the pressure p as a function of nucleon density n and specific entropy S, $p = p(n, S)$. A general constraint on the state equation can be obtained from the relativistic causality principle, which in this case states that the adiabatic speed of sound must not exceed the speed of light (Osnes and Strottman, 1986; Clare and Strottman, 1986, and references therein).

In the collision of two heavy nuclei the relative speed of the two nuclei is supersonic. Drawing an analogy with conventional gas dynamics, one expects that a relativistic shock wave might be produced (Sobel et al., 1975). Also, some current models under investigation predict that relativistic shocks (or relativistic detonation and deflagration waves) might be related to the phase transition from a nuclear matter to a quark-gluon plasma (Barz et al., 1985; Clare and Strottman, 1986).

Relativistic shock waves have been the subject of early investigations in relativistic fluid dynamics and magneto-fluid dynamics. In relativistic fluid dynamics the pioneering work is that of Taub (1948), where the relativistic form of the jump conditions is established (the relativistic analog of the Rankine-Hugoniot relations). A detailed analysis of the thermodynamic properties of relativistic shock waves, patterned after Landau and Lifshitz's treatment of classical shock waves (Landau and Lifshitz, 1959a), is due to Thorne (1973). In Thorne's article the shock adiabat is named the Taub adiabat, a terminology which we adopt here. The compressibility assumption used in the analysis of the Taub adiabat [the relativistic counterpart of the classical Weyl conditions (Courant and Friedrichs, 1976)] has been discussed by Israel (1960) in the framework of relativistic fluid dynamics.

Explicit solutions of the jump conditions have been obtained for special equations of state. Liang (1977a) and Anile, Miller, and Motta (1983) have treated the case of barotropic and polytropic fluids. The case of the Synge gas has been treated numerically by Fujimara and Kennel (1979), by Lanza, Miller, and Motta (1985), and in an analytical way by Majorana (1987). Shock waves in relativistic magneto-fluid dynamics have been investigated extensively and in a rigorous mathematical way by Lichnerowicz (1967, 1971, 1976). Explicit expressions for transverse magnetoacoustic shocks have been provided by Taussig (1973) for the Synge equation of state. The case of an oblique magnetic field has been treated by Majorana and Anile (1987). Detonation and deflagration waves in relativistic magneto-fluid dynamics have been investigated by Coll (1976) under fairly general thermodynamic assumptions. In relativistic fluid dynamics with combustion, the case of special state equations of interest for nuclear physics and cosmology has been treated, among others, by Steinhardt (1982), Gyulassy et al. (1984), Miller and Pantano, (1984), and Cleymans, Gavai, and Suhonen (1986).

The plan of the chapter is the following. In Section 8.1, first of all, we derive the jump conditions for relativistic fluid dynamics by using the methods introduced in Section 3.1. Then the jump conditions for shock waves are used in order to obtain the Taub adiabat. Weak shock waves are studied in detail and the role of the compressibility assumptions is brought to evidence. Shocks of arbitrary strength are investigated by performing a detailed analysis of the properties of the Taub adiabat. We adopt the mathematically rigorous approach of Lichnerowicz (1976), but we apply it to the simpler case of relativistic fluid dynamics (as opposed to magneto-fluid dynamics). This has the advantages that the calculations are simpler and the interpretation of the results is not hindered by a tedious and cumbersome analysis. In particular, the timelike character of the shock hypersurface under the relativistic causality assumption is established. Also, the relationship between the compressibility assumptions and the entropy growth across a shock wave is discussed. Finally, the existence of shock wave solutions is established.

In Section 8.2 explicit solutions of the jump conditions are found for various state equations. In particular, barotropic and polytropic fluids and the Synge equation of state are treated.

In Section 8.3 detonation and deflagration waves are studied for relativistic fluid dynamics. The analysis is based mainly upon Coll's article (1976). The detonation adiabat is defined and its general properties are discussed. In particular, the Jouguet points are introduced and their significance is investigated. Finally, we treat in detail the case of the "bag model" state equation which may be relevant for nuclear matter theory as well as for cosmology.

In Section 8.4 we treat shock waves in relativistic magneto-fluid dynamics. Owing to the availability of Lichnerowicz's excellent monographs (Lichnerowicz, 1967, 1971) our treatment will be cursory. First of all, we derive the jump conditions and from these the shock adiabat, which we call the *Lichnerowicz adiabat*. The properties of the Lichnerowicz adiabat and the theorems guaranteeing the existence of shock waves are simply stated without proofs. The proofs can be reconstructed by applying the techniques introduced in Section 8.1 or can be found in Lichnerowicz (1967, 1971, 1976). At the end we solve the jump conditions for a Synge gas.

Finally, in Section 8.5 we introduce the concept of evolutionary shock. The Lax conditions (Jeffrey, 1976; Liberman and Velikovich, 1985) are then recalled and applied to relativistic fluid dynamics. For the case of relativistic magneto-fluid dynamics we refer to Lichnerowicz's work (Lichnerowicz, 1971). Also, some comments are made on the relativistic shock tube problem.

8.1. The jump conditions in relativistic fluid dynamics

The field equations of relativistic fluid dynamics in the form of conservation laws have been discussed in Section 2.2 and are

$$\nabla_\mu(\rho u^\mu) = 0, \tag{8.1}$$

$$\nabla_\mu T^{\mu\nu} = 0, \tag{8.2}$$

with

$$T^{\mu\nu} = \rho f u^\mu u^\nu + p g^{\mu\nu}.$$

A shock wave is an oriented hypersurface Σ in space-time $\mathcal{M}$, across which the field variables ρ, u^α, p, f suffer a jump discontinuity, i.e., ρu^μ and $T^{\mu\nu}$ are regularly discontinuous.

Let Ω be a neighborhood of Σ, which is described by the equation

$$\varphi(x^\alpha) = 0, \quad d\varphi|_\Sigma \neq 0.$$

Equation (8.1)–(8.2) hold in Ω in the sense of distributions, that is, for the distributions associated with the regularly discontinuous tensor fields ρu^μ, $T^{\mu\nu}$.

Let $\hat{T}^{\mu\nu}$ be the tensor distribution associated with $T^{\mu\nu}$. Then by equation (3.45),

$$\nabla_\mu \hat{T}^{\mu\nu} = [[T^{\mu\nu}]] l_\mu \delta_\Sigma + (\nabla_\mu T^{\mu\nu}\hat{)},$$

where $l_\alpha = \nabla_\alpha \varphi$, hence equation (8.2) implies

$$[[T^{\mu\nu}]] l_\mu = 0. \tag{8.3}$$

Similarly, from equation (8.1) one obtains

$$[[\rho u^\mu]] l_\mu = 0. \tag{8.4}$$

This latter equation yields the invariant m,

$$m = \rho_+ u^\mu_+ l_\mu = \rho_- u^\mu_- l_\mu. \tag{8.5}$$

The case $m = 0$ corresponds to a slip-discontinuity.

In this case equation (8.3) implies

$$[[p]] = 0.$$

Furthermore, under the condition $\rho_+ \rho_- > 0$, from $m = 0$ it follows that

$$u^\mu_+ l_\mu = u^\mu_- l_\mu = 0$$

and the tangential components of $[[u^\alpha]]$ are undetermined.

The case $m \neq 0$ corresponds to shock waves and will be discussed in the following.

Then equation (8.3) gives

$$m f_- u^\nu_- + p_- l^\nu = m f_+ u^\nu_+ + p_+ l^\nu. \tag{8.6}$$

By contracting with $u_{-\nu}, u_{+\nu}$, respectively, and using equation (8.5), equation (8.6) yields

$$f_+^2 - f_-^2 + (p_- - p_+)(\tau_- + \tau_+) = 0, \tag{8.7}$$

with $\tau = f/\rho$ the dynamical volume.

Equation (8.7) is the Taub adiabat, which is the relativistic generalization of the classical Hugoniot adiabat.

General properties of relativistic shock waves are expressed in the following propositions.

Let Π be the half-plane (τ, p), $\tau \geq 0$, and denote by $Z \equiv (\tau, p)$ a point in the half-plane.

Let $\mathscr{H}(Z, Z_+)$ be the Taub function (the relativistic analog of the classical Hugoniot function) defined on Π by

$$\mathscr{H}(Z, Z_+) = -f_+^2 + f^2 + (p_+ - p)(\tau_+ + \tau). \tag{8.8}$$

Then

$$\mathscr{H}(Z_+, Z_+) = 0,$$
$$\mathscr{H}(Z_-, Z_+) = 0,$$

and the Taub adiabat passing through Z_+ is given by

$$\mathscr{H}(Z, Z_+) = 0.$$

Remark 1. One has

$$\mathrm{d}\mathscr{H} = 2fT\mathrm{d}S + (\tau - \tau_+)\mathrm{d}p - (p - p_+)\mathrm{d}\tau. \tag{8.9}$$

PROPOSITION 8.1. *Under the following assumptions on the state equation* $\tau = \tau(p, S)$:

$$\tau'_S \neq 0, \tag{8.10a}$$

$$\tau'_p < 0 \tag{8.10b}$$

the hypersurface Σ *is timelike.*

Proof. Consider the straight line $\mathscr{L}: m^2(\tau - \tau_+) = (p_+ - p)l_\alpha l^\alpha$, connecting the states Z_-, Z_+. On $\mathscr{L}$ one has

$$\mathrm{d}\tau = -\frac{l_\alpha l^\alpha}{m^2}\mathrm{d}p,$$

and therefore, on $\mathscr{L}$,

$$\tau'_S\mathrm{d}S = \mathrm{d}\tau - \tau'_p\mathrm{d}p = -\left(\tau'_p + \frac{l_\alpha l^\alpha}{m^2}\right)\mathrm{d}p. \tag{8.11}$$

Now, since $\tau'_S \neq 0$, $\tau'_p < 0$, if $\dfrac{l_\alpha l^\alpha}{m^2} \leq 0$, it follows that $\dfrac{\mathrm{d}S}{\mathrm{d}p} \neq 0$ along $\mathscr{L}$.

However, on $\mathscr{L}$, from (8.9), $\dfrac{d\mathscr{H}}{dp} = 2fT\dfrac{dS}{dp}$. Because $\mathscr{H}$ vanishes at both ends of $\mathscr{L}$, $\mathscr{H}(Z_+, Z_+) = \mathscr{H}(Z_-, Z_+) = 0$, there exists $Z^* \in \mathscr{L}$, such that $\dfrac{d\mathscr{H}}{dp}(Z^*) = 0$, which is a contradiction. It follows that one must have $l_\alpha l^\alpha > 0$.

Q.E.D.

Remark 2. In the nonrelativistic limit, the dynamical volume $\tau = \dfrac{f}{\rho}$ reduces to the specific volume $V = 1/\rho$. In this case $\left(\dfrac{\partial V}{\partial S}\right)_p = \left(\dfrac{T}{c_p}\right)\left(\dfrac{\partial V}{\partial T}\right)_p$. Therefore, in general $\left(\dfrac{\partial V}{\partial S}\right)_p \neq 0$ and for materials for which the coefficient of thermal dilation is positive, $\left(\dfrac{\partial V}{\partial S}\right)_p > 0$.

Remark 3. One has

$$\tau'_p = \frac{1}{\rho^2}(-e'_p + 1), \tag{8.12}$$

therefore

$$\tau'_p < 0 \quad \text{iff} \quad e'_p > 1.$$

Remark 4. Let

$$P(l_\alpha) = (u^\mu l_\mu)^2 - p'_e h^{\mu\nu} l_\mu l_\nu. \tag{8.13}$$

Then $P(l_\alpha) = 0$ is the characteristic equation for acoustic waves. It can be immediatly checked that

$$\tau'_p + \frac{l_\mu l^\mu}{m^2} = -\frac{P(l_\alpha) e'_p}{m^2},$$

hence equation (8.11) can be written as

$$\tau'_S \frac{dS}{dp} = \frac{P(l_\alpha) e'_p}{m^2}, \text{ on } \mathscr{L}. \tag{8.14}$$

Henceforth, the inequalities (8.10a)–(8.10b) shall be assumed, except when stated otherwise. Then $l_\alpha l^\alpha > 0$ and it is convenient to normalize by defining

$$n_\alpha = \frac{l_\alpha}{(l_\mu l^\mu)^{1/2}} \tag{8.15}$$

and substituting n_α in place of l_α in the jump conditions (8.3)–(8.4), which gives, after some manipulations,

$$\bar{m}^2 = -\frac{p_+ - p_-}{\tau_+ - \tau_-}, \tag{8.16}$$

where $\bar{m} = \rho_+ u_+^\mu n_\mu = \rho_- u_-^\mu n_\mu$, $\bar{m} = m/l_\alpha l^\alpha$.

First of all, the case of weak shocks will be treated.

The Taub adiabat passing through the point Z_+ is

$$\mathscr{H}(Z, Z_+) = f_+^2 - f^2 + (p - p_+)(\tau_+ + \tau) = 0,$$

and the isentropic (Poisson) adiabat through the same point Z_+ is

$$S(\tau, p) = S(Z_+) = \text{const.}$$

PROPOSITION 8.2. *The Taub and isentropic adiabats through Z_+ are tangent at Z_+ and have a second order contact.*

Proof. Expanding $f_-^2 - f_+^2$ around f_+ along the Taub adiabat gives

$$f_-^2 - f_+^2 = (p_- - p_+)\{2\tau_+ + (\tau_p')_+(p_- - p_+) + \tfrac{1}{2}(\tau_p'')_+(p_- - p_+)^2\} + O((p_- - p_+)(S_- - S_+))$$

where we have kept the terms of order $(p_- - p_+)^3$ and neglected the contributions in $(S_- - S_+)$ of order greater than 1, by anticipating the result. Similarly, one obtains by a direct expansion of $f_-^2 - f_+^2$,

$$f_-^2 - f_+^2 = 2\{f_+ T_+(S_- - S_+) + (p_- - p_+)\tau_+ + \tfrac{1}{2}(\tau_p')_+(p_- - p_+)^2 + \tfrac{1}{6}(\tau_p'')_+(p_- - p_+)^3\} + O((p_- - p_+)(S_- - S_+)),$$

having used the first law of thermodynamics $\mathrm{d}f = T\mathrm{d}S + \frac{1}{\rho}\mathrm{d}p$ and the equality $\frac{\partial f^2}{\partial p} = 2\tau$.

Equating the two expressions yields

$$S_- - S_+ = \left(\frac{1}{12 f T}\frac{\partial^2 \tau}{\partial p^2}\right)_+ (p_- - p_+)^3 + O((p_- - p_+)^4). \quad \text{Q.E.D.} \tag{8.17a}$$

Remark 5. Let $u_s = \Gamma_s c_s$, with $\Gamma_s = (1 - c_s^2)^{-1/2}$ the "proper speed of sound." Then the common tangent to the Taub and isentropic adiabats has a negative slope, given by

$$(p'_\tau)_+ = -\rho_+^2 (u_s^2).$$

This is easily seen by using equation (8.12).

PROPOSITION 8.3. *Under the assumptions* (8.10b)–(8.10c) *and*

$$\tau''_p > 0 \quad \text{(compressibility condition)}, \tag{8.17b}$$

$$S_- - S_+ > 0, \tag{8.17c}$$

the following inequalities hold, for weak shocks,

$$p_- > p_+, \tag{8.17d}$$

$$f_- > f_+, \tag{8.17e}$$

$$\rho_- > \rho_+, \tag{8.17f}$$

$$u_- < u_+, \tag{8.17g}$$

$$u_+ > (u_s)_+, \tag{8.17h}$$

$$u_- < (u_s)_-, \tag{8.17i}$$

where u_+ *and* u_- *are defined by*

$$u_+ = \Gamma(v_+)v_+, \quad u_- = \Gamma(v_-)v_-,$$

$$v_\pm = \left\{ \frac{(u_\pm^\mu l_\mu)^2}{h_\pm^{\alpha\beta} l_\alpha l_\beta} \right\}^{1/2}, \tag{8.17j}$$

with $v_\pm$ *the absolute values of the three-velocities of the fluid with respect to the shock front, and* $\Gamma(v_+)$, $\Gamma(v_-)$ *the respective Lorentz factors.*

Proof. Equation (8.17d) follows immediately from equation (8.17a)–(8.17c).

Then (8.17e) follows from (8.7).

From $\tau'_p < 0$ one has

$$\tau_- < \tau_+,$$

hence (8.17f).

From (8.17j) it is easily seen that

$$u_+^2 = \bar{m}^2/\rho_+^2, \quad u_-^2 = \bar{m}^2/\rho_-^2, \tag{8.17k}$$

whence (8.17g) follows.

Equation (8.17h)–(8.17i) can be proved geometrically, from the fact that, as a consequence of $\tau''_p > 0$, the Taub adiabat is concave.

The slope of the chord connecting Z_+ and Z_- is

$$-\bar{m}^2 = -u_+^2 \rho_+^2 = -u_-^2 \rho_-^2.$$

Now, from Fig. 8.1, one obtains

$$|\text{slope } \mathscr{L}| > |\text{slope } a|,$$
$$|\text{slope } \mathscr{L}| < |\text{slope } b|,$$

which, because $\tau_p' = -1/\rho^2 u_s^2$, imply (8.17h)–(8.17i), respectively.

Q.E.D.

Remark 6. For a fluid for which $\tau_p'' < 0$, the requirement $[[S]] > 0$ and equation (8.17a) yield

$$p_- < p_+,$$

and from (8.7),

$$f_- < f_+.$$

Also, from $\tau_p' < 0$, $\tau_- > \tau_+$, hence

$$\rho_- < \rho_+$$

and the shock is expansive. From (8.17k) this implies $u_+ < u_-$ and a simple geometric argument as in Proposition 8.3 shows that (8.17h)–(8.17i) still hold, that is,

$$u_- < (u_s)_-,$$
$$u_+ > (u_s)_+.$$

Fig. 8.1.
$\mathscr{C}$: Taub adiabat connecting the state $Z_+ \equiv (\tau_+, p_+)$ ahead of the shock to the state $Z_- \equiv (\tau_-, p_-)$ behind the shock,
$\mathscr{L}$: straight line $p - p_+ = -m^2(\tau - \tau_+)$,
a: tangent to the isentropic adiabat at Z_+,
b: tangent to the isentropic adiabat at Z_+.

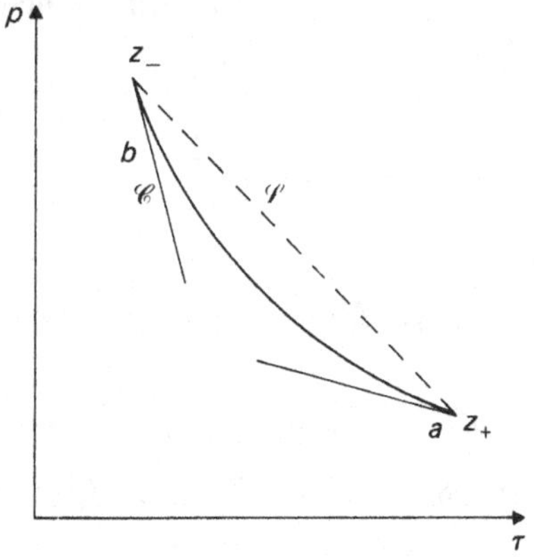

By expanding (8.16) around (τ_+, ρ_+) it is possible to find an expression for the proper Mach number ahead of the shock,

$$\mathcal{M}_+^2 = \frac{u_+^2}{(u_s^2)_+} = 1 + \frac{6f_+T_+\rho_+^2u_+^2}{(p_- - p_+)^2}(S_- - S_+),$$

and similarly

$$\mathcal{M}_-^2 = \frac{u_-^2}{(u_s^2)_-} = 1 - \frac{6f_+T_+\rho_+^2u_+^2}{(p_- - p_+)^2}(S_- - S_+).$$

Now shocks of arbitrary strength will be considered.

PROPOSITION 8.4. *Under the assumptions*

$$\tau_S' \neq 0, \tag{8.10a}$$

$$\tau_p' < 0, \tag{8.10b}$$

$$\tau_p'' > 0, \tag{8.17b}$$

$$S_- - S_+ > 0, \tag{8.17c}$$

and that $\tau(p, S)$ can be inverted giving $S = S(p, \tau)$, one has

$$\tau_- < \tau_+. \tag{8.18a}$$

Furthermore, if

$$\tau_S' > 0, \tag{8.18b}$$

one also has

$$p_- > p_+, \tag{8.18c}$$

$$f_- > f_+, \tag{8.18d}$$

$$\rho_- > \rho_+. \tag{8.18e}$$

Proof. In the case when $p_- \geq p_+$ the following conclusions can be drawn. From

$$\frac{\partial(f^2)}{\partial p} = 2\tau,$$

it follows that

$$f^2(p_-, S_-) - f^2(p_+, S_-) = 2\int_{p_+}^{p_-} \tau(p, S_-)\mathrm{d}p.$$

From $\tau_p' < 0$ one has $\tau_- \leq \tau(p, S_-)$, $p \leq p_-$, hence

$$f^2(p_-, S_-) - f^2(p_+, S_-) \geq 2\tau_-(p_- - p_+).$$

From $f'_s = T > 0$, it follows that $f(p_+, S_-) > f(p_+, S_+) \equiv f_+$.
Then

$$f^2(p_-, S_-) - f^2(p_+, S_+) \equiv f_-^2 - f_+^2 > f_-^2 - f^2(p_+, S_-) \geq 2\tau_-(p_- - p_+),$$

whence

$$f_-^2 - f_+^2 - 2\tau_-(p_- - p_+) > 0.$$

The Taub adiabat (8.7) then gives (8.18a), that is,

$$\tau_- < \tau_+.$$

Now, it can be proved that if $\tau'_S < 0$ then one must have $p_- > p_+$. In fact, let $p_- \leq p_+$.

From

$$f^2(p_+, S_-) - f^2(p_-, S_-) = 2\int_{p_-}^{p_+} \tau(p, S_-)\mathrm{d}p$$

and from the concavity condition (8.17b) one obtains

$$f^2(p_+, S_-) - f_-^2 \leq (p_+ - p_-)(\tau(p_-, S_-) + \tau(p_+, S_-)).$$

Hence, from $\tau'_S < 0$,

$$\begin{aligned} f_+^2 - f_-^2 &< f^2(p_+, S_-) - f_-^2 \leq (p_+ - p_-)(\tau_- + \tau(p_+, S_-)) \\ &< (p_+ - p_-)(\tau_- + \tau_+), \end{aligned}$$

which contradicts the Taub adiabat.

Similarly, if $\tau'_S > 0$, one must have $p_- > p_+$. In fact, let $p_- \leq p_+$. Then, from

$$f^2(p_+, S_+) - f^2(p_-, S_+) = 2\int_{p_-}^{p_+} \tau(p, S_+)\mathrm{d}p$$

and the concavity condition,

$$\begin{aligned} f_+^2 - f_-^2 < f_+^2 - f^2(p_-, S_+) &\leq (p_+ - p_-)(\tau_+ + \tau(p_-, S_+)) \\ &\leq (p_+ - p_-)(\tau_- + \tau_+), \end{aligned}$$

which contradicts the Taub adiabat.

Equations (8.18a)–(8.18e) are easily proved. Q.E.D.

In order to discuss the speeds of propagation of shocks of arbitrary strength, several propositions are needed.

Also, assumptions (8.10a)–(8.10b) and (8.17b) hold, except when stated otherwise.

PROPOSITION 8.5. *Each isentropic adiabat of* Π *is strictly concave.*

Proof. Let $\mathscr{C}$ be an isentropic curve of Π, with equation

$$p = p(\tau, S), \quad S = \text{const.}$$

Then

$$\left(\frac{\mathrm{d}^2 p}{\mathrm{d}\tau^2}\right)_{\mathscr{C}} = -\frac{1}{(\tau'_p)^3}\tau''_p > 0. \qquad \text{Q.E.D.}$$

PROPOSITION 8.6. *Let*

$$\Lambda : p - p_+ = M(\tau - \tau_+)$$

be a straight line of Π. *Then, at each point* Z_s *of* Λ *where S is stationary, one has*

$$\left(\tau'_S \frac{\mathrm{d}^2 S}{\mathrm{d}\tau^2}\right)_{\Lambda}(Z_s) = -\left(\frac{\tau''_p}{\tau'^2_p}\right)(Z_s) < 0.$$

Proof. Along Λ one has

$$\frac{\mathrm{d}^2 p}{\mathrm{d}\tau^2} = 0 = p_{\tau\tau} + 2p_{\tau S}\frac{\mathrm{d}S}{\mathrm{d}\tau} + p_{SS}\left(\frac{\mathrm{d}S}{\mathrm{d}\tau}\right)^2 + p_S\frac{\mathrm{d}^2 S}{\mathrm{d}\tau^2},$$

therefore at Z_s,

$$p_{\tau\tau} + p_S\frac{\mathrm{d}^2 S}{\mathrm{d}\tau^2} = 0.$$

Now, from

$$\mathrm{d}\tau = \tau'_p \mathrm{d}p + \tau'_S \mathrm{d}S,$$

it follows that

$$p_S = -\frac{\tau'_S}{\tau'_p}, \quad p_{\tau\tau} = -\frac{\tau''_p}{(\tau'_p)^3}. \qquad \text{Q.E.D.}$$

PROPOSITION 8.7. *Let* $\mathscr{H} : \mathscr{H}(Z_+, Z) = 0$ *be the Taub adiabat in* Π, *and* $\hat{Z}$ *any point on* $\mathscr{H}, \hat{Z} \neq Z_+$. *Let* Λ *be the straight line joining* Z_+ *and* $\hat{Z}$, *of slope* M,

$$\Lambda : \hat{p} - p_+ = M(\hat{\tau} - \tau_+).$$

Then

$$\left(\tau'_S \frac{\mathrm{d}S}{\mathrm{d}\tau}\right)_{\Lambda}(\hat{Z}) > 0, \quad \left(\tau'_S \frac{\mathrm{d}S}{\mathrm{d}\tau}\right)_{\Lambda}(Z_+) < 0.$$

Proof. Since $\mathscr{H}(Z_+, Z_+) = \mathscr{H}(Z_+, \hat{Z}) = 0$, $\mathscr{H}$ is stationary at a point Z_s of the segment $(Z_+, \hat{Z})$. However, on Λ, from (8.9), $\dfrac{d\mathscr{H}}{d\tau} = 2fT\dfrac{dS}{d\tau}$ and therefore Z_+ is a stationary point also for S. From Proposition (8.6) it follows that Z_s is unique and that if $\tau'_S > 0$ it corresponds to a maximum (to a minimum if $\tau'_S < 0$). Q.E.D.

Remark 7. From

$$d\tau = \tau'_S dS + \tau'_p dp,$$

it follows that, along Λ,

$$\tau'_S \frac{dS}{d\tau} = 1 - \tau'_p M = \tau'_p(p'_\tau - M).$$

Hence, at Z_+,

$$M < p'_\tau < 0$$

(p'_τ is the slope of the isentropic adiabat).

Remark 8. From equation (8.14) one has

$$P(l_\alpha)|_{Z_+} > 0, \tag{8.19a}$$

$$P(l_\alpha)|_{\hat{Z}} < 0. \tag{8.19b}$$

Now,

$$P(l_\alpha) = (1 - c_s^2) l_\alpha l^\alpha (v_\Sigma^2 \Gamma_\Sigma^2 - c_s^2 \Gamma_s^2), \tag{8.20}$$

where v_Σ is the normal speed of propagation of Σ with respect to the fluid, $\Gamma_\Sigma^2 = (1 - v_\Sigma^2)^{-1}$, $\Gamma_s^2 = (1 - c_s^2)^{-1}$.

Hence, from (8.19a), because $l_\alpha l^\alpha > 0$, one finds

$$(v_\Sigma)_+ > (c_s)_+ \text{ at } Z_+ \tag{8.21a}$$

(the shock speed is supersonic with respect to the fluid ahead), from (8.19b), applied to the shock state Z_-,

$$(v_\Sigma)_- < (c_s)_- \text{ at } Z_- \tag{8.21b}$$

(the shock speed is subsonic with respect to the fluid behind).

PROPOSITION 8.8. *Under the assumptions* (8.10a)–(8.10b) *and* (8.17a) *the entropy S is a strictly monotonic function of τ on each connected component of the Taub adiabat $\mathscr{H}$.*

Proof. Suppose that there exists $\hat{Z} \in \mathscr{H}$ such that $\left(\dfrac{dS}{d\tau}\right)_{\mathscr{H}}(\hat{Z}) = 0$. The isentropic adiabat $\mathscr{C}$ at $\hat{Z}$ is tangent to $\mathscr{H}$.

Let $\Lambda: p - p_+ = M(\tau - \tau_+)$ be the straight line passing through $\hat{Z}$ and Z_+. Now, from (8.9) it follows that Λ is tangent to $\mathscr{H}$ at $\hat{Z}$. But at $\hat{Z}$, $\left(\tau'_s \frac{dS}{d\tau}\right)_\Lambda (\hat{Z}) > 0$ by Proposition 8.7, which is a contradiction. Q.E.D.

Remark 9. One has

$$\left(\frac{dS}{dp}\right)_{\mathscr{H}} = \left(\frac{dS}{d\tau}\right)_{\mathscr{H}} \left(\frac{dp}{d\tau}\right)_{\mathscr{H}}$$

and, if at some point $Z^* \in \mathscr{H}, Z^* \neq Z_+$, along $\mathscr{H}$, $(dS/dp)(Z^*) = 0$, then one must necessarily have $(d\tau/dp)(Z^*) = 0$. But this contradicts equation (8.9), hence S is also a strictly monotonic function of p on each connected component of $\mathscr{H}$.

Until now the existence of shock waves has been postulated and their general properties have been investigated.

In order to prove the existence of shock wave solutions more definite assumptions are needed on the state equation.

PROPOSITION 8.9. *Along the isentropic adiabat $\mathscr{C}$, corresponding to $S = S_+$, in the region $\tau \leq \tau_+$, one has*

$$\left(\frac{d\mathscr{H}}{d\tau}\right)_{\mathscr{C}} > 0, \quad \mathscr{H}(Z_+, Z) < 0, \quad \forall Z \in \mathscr{C}.$$

Proof. From (8.9), along $\mathscr{C}$, one finds

$$\left(\frac{d\mathscr{H}}{d\tau}\right)_{\mathscr{C}} = (\tau - \tau_+)\left(\frac{dp}{d\tau}\right)_{\mathscr{C}} - (p - p_+) > 0$$

(from the concavity of $\mathscr{C}$)

$$\left(\frac{d^2\mathscr{H}}{d\tau^2}\right)_{\mathscr{C}} = (\tau - \tau_+)\left(\frac{d^2 p}{d\tau^2}\right)_{\mathscr{C}} < 0. \qquad \text{Q.E.D.}$$

PROPOSITION 8.10. *Under the assumptions* (8.10a)–(8.10b) *and* (8.17b) *and, furthermore, from*

$$\tau'_S > 0, \tag{8.22a}$$

$$\lim_{\tau \to 0} p(\tau, S) = +\infty, \tag{8.22b}$$

the Taub adiabat is a connected curve. Furthermore, for any $M < p'_\tau$ there is a *corresponding unique point* $Z_- \neq Z_+$ *of* $\mathscr{H}$ *such that*

$$M = \frac{p_- - p_+}{\tau_- - \tau_+}.$$

Moreover, on $\mathscr{H}$ *one has*

$$\left(\frac{dS}{d\tau}\right)_{\mathscr{H}} < 0, \quad \left(\frac{dS}{dp}\right)_{\mathscr{H}} > 0. \tag{8.22c}$$

Proof. Let $M < p'_\tau$ and Λ be the straight line

$$p - p_+ = M(\tau - \tau_+).$$

The isentropic adiabat $\mathscr{C}$ through Z_+ is strictly concave and therefore Λ meets $\mathscr{C}$ at a unique point Z^*, $Z^* \neq Z_+$. From Proposition 8.9,

$$\mathscr{H}(Z_+, Z^*) < 0.$$

Because $S(Z_+) = S(Z^*)$, S is stationary on Λ (between Z_+ and Z^*) at a unique point Z_s which is a strict maximum for S (Proposition 8.7).

Now, since $\tau'_S > 0$,

$$\left(\frac{dS}{d\tau}\right)_\Lambda (Z_+) < 0,$$

hence also

$$\left(\frac{d\mathscr{H}}{d\tau}\right)_\Lambda (Z_+) < 0.$$

Therefore, when Z ranges from Z_+ to Z^* along Λ, $\mathscr{H}(Z_+, Z)$ is at first positive, hence, because $\mathscr{H}(Z_+, Z^*) < 0$, Z must vanish at some Z_- of Λ,

$$\mathscr{H}(Z_+, Z_-) = 0.$$

The point Z_- is necessarily unique because, by Proposition 8.7,

$$\left(\frac{d\mathscr{H}}{d\tau}\right)_\Lambda (Z_-) > 0.$$

In order to prove that equations (8.22c) hold on $\mathscr{H}$ then it is sufficient to check that they hold in a neighborhood of Z_+ (weak shocks, Proposition 8.3). Q.E.D.

Remark 10. The point Z_- on $\mathscr{H}$ corresponds to the jump conditions, once M (and hence the shock speed) has been fixed.

8.2. Solutions of the jump conditions

First of all, some general formulas will be derived, independent of the state equation.

The energy-momentum flux continuity, equation (8.3), with

$$T^{\mu\nu} = (e+p)u^\mu u^\nu + pg^{\mu\nu},$$

can be written as

$$(e+p)_- u^\mu_- a_- + p_- l^\mu = (e+p)_+ u^\mu_+ a_+ + p_+ l^\mu, \tag{8.23a}$$

where

$$a_\pm = u^\mu_\pm l_\mu.$$

Contracting equation (8.23a) with l_μ yields

$$(e+p)_+ a_+^2 - (e+p)_- a_-^2 = G(p_- - p_+), \tag{8.23b}$$

where

$$G = l_\mu l^\mu.$$

Now, let the hypersurface Σ be timelike, that is, $G > 0$. From equation (8.17j) one has for the proper speeds u_+, u_-,

$$u_\pm^2 = a_\pm^2/G$$

and therefore equation (8.23b) is written as

$$(e+p)_+ u_+^2 - (e+p)_- u_-^2 = p_- - p_+. \tag{8.24a}$$

Contracting equation (8.23a) with $u_{\pm\mu}$, respectively, yields

$$(e+p)_+ a_+ u^\mu_+ u_{-\mu} = -a_-(e_- + p_+),$$
$$(e+p)_- a_- u^\mu_- u_{+\mu} = -a_+(e_+ + p_-).$$

Now, $u^\mu_+ u_{-\mu} \neq 0$ because both vectors are timelike.

Hence, a_+ and a_- are both nonzero for a nonvanishing shock ($p_- \neq p_+$). It follows that

$$\frac{u_+^2}{u_-^2} = \frac{a_+^2}{a_-^2} = \frac{e_- + p_+}{e_+ + p_-} \cdot \frac{(e+p)_-}{(e+p)_+}. \tag{8.24b}$$

From equations (8.24a)–(8.24b) one obtains

$$u_-^2 = \frac{(p_- - p_+)(e_+ + p_-)}{(e+p)_-(e_- - e_+ + p_+ - p_-)}, \tag{8.25a}$$

$$u_+^2 = \frac{(p_- - p_+)(e_- + p_+)}{(e+p)_+(e_- - e_+ + p_+ - p_-)} \tag{8.25b}$$

and for the three-velocities v_+, v_-, defined by $u_+^2 = \Gamma_+^2 v_+^2, \Gamma_+^2 = 1 + u_+^2$ and similarly for v_-,

$$v_-^2 = \frac{(p_- - p_+)(p_- + e_+)}{(e_- + p_+)(e_- - e_+)}, \tag{8.26a}$$

$$v_+^2 = \frac{(p_- - p_+)(p_+ + e_-)}{(e_+ + p_-)(e_- - e_+)}. \tag{8.26b}$$

Notice that one has

$$\frac{p_- - p_+}{e_- - e_+} > 0$$

and therefore

$$v_- v_+ = \frac{p_- - p_+}{e_- - e_+}. \tag{8.26c}$$

For a compressive shock, $e_- > e_+, p_- > p_+$ and therefore $u_-^2 < u_+^2$, hence $a_-^2 < a_+^2$. Also, for the three-velocities, $v_-^2 < v_+^2$.

Having established these general results, we now investigate some special state equations which are physically relevant.

The first case to be considered is that of barotropic fluids. These fluids must be treated separately because their state equation depends only on one parameter and does not fit into the general discussion of Section 8.1.

PROPOSITION 8.11. *For a barotropic fluid described by an equation of state $p = p(e)$, let η be the quantity defined by equation* (2.17),

$$\eta = \exp\left(\int \frac{de}{e+p}\right).$$

Then under the assumptions that

$$0 < p'_e < 1,$$

and that η is increasing across the shock, the shock hypersurface Σ is timelike.

Furthermore, if $p'_e = 1$, the hypersurface Σ is null (this latter result holds irrespective of the type of barotropic fluid).

Proof. From the definition of η for a barotropic fluid

$$\eta = \exp\left(\int \frac{de}{(e+p)}\right)$$

it follows that

$$\eta_- - \eta_+ = \int_{p_+}^{p_-} \frac{e'_p \eta}{e+p} \mathrm{d}p,$$

and therefore $\eta_- - \eta_+ > 0$ implies $p_- - p_+ > 0$ (compressive shock).

From equations (8.23b) and (8.24b) one obtains

$$G(p_- - p_+) = \frac{a_-^2 (e+p)_-}{e_+ + p_-}(e_- - e_+ + p_+ - p_-)$$

(notice that equation (8.24b) for a_+^2/a_-^2 holds also for an arbitrary G).

Now

$$p_- - p_+ = \int_{e_+}^{e_-} p'_e \mathrm{d}e < e_- - e_+,$$

which, together with the above equation, implies $G > 0$.

It is apparent that, for $p'_e = 1$, the above equation implies (for a nonvanishing shock) $G = 0$. Q.E.D.

Remark 11. For a barotropic fluid consisting of ultrarelativistic particles in thermal equilibrium,

$$\frac{\mathrm{d}p}{\mathrm{d}T} = \frac{1}{T}(e+p) = \eta,$$

hence

$$p_- - p_+ = \int_{T_+}^{T_-} \eta \mathrm{d}T.$$

Therefore, under the assumptions of Proposition (8.11) one has also

$$T_- > T_+.$$

Remark 12. For a barotropic fluid corresponding to "cold matter" the previous arguments cannot be applied, because they correspond to zero entropy, $S = 0$, hence η cannot be interpreted as entropy density. Therefore, in order to prove that, for such a fluid, rarefaction shocks cannot occur, it is necessary to resort to different arguments, based on stability. This will be seen in Chapter 10. The following proposition is the equivalent of Proposition 8.3 for barotropic fluids.

PROPOSITION 8.12. *For a barotropic fluid with state equation $p = p(e)$ let*

$$W = 2p'_e(1 - p'_e) + (e+p)p''_e. \tag{8.27a}$$

Then if

$$W > 0 \qquad \text{(the Weyl condition)},$$

for weak compressive shock waves one has

$$|v_+| > (p'_e)_+^{1/2}, \tag{8.27b}$$

$$|v_-| < (p'_e)_-^{1/2}. \tag{8.27c}$$

Vice versa, if equations (8.27b)–(8.27c) *hold for shocks of arbitrary strength the Weyl condition holds.*

Proof. Let

$$A(e_-) = \frac{(p_- - p_+)(p_+ + e_-)}{(e_+ + p_-)(e_- - e_+)}.$$

For weak shocks, $e_- = e_+ + [[e]]$, $[[e]] \ll e_+$,

$$p_- = p_+ + (p'_e)_+[[e]] + \tfrac{1}{2}(p''_e)_+([[e]])^2 + O(|[[e]]|^2).$$

Then

$$A(e_-) = (p'_e)_+ + \frac{[[\mathrm{e}]]}{2(e_+ + p_+)} W.$$

Hence, since $[[e]] > 0$, equation (8.27b) immediately follows.

In a similar way, one proves equation (8.27c) by writing

$$e_+ = e_- - [[e]], p_+ = p_- - (p'_e)_-[[e]] \\ + \tfrac{1}{2}(p''_e)_-([[e]])^2 + O(|[[e]]|^2).$$

Obviously, if equations (8.27b)–(8.27c) hold for shocks of arbitrary strength, they will hold also for weak shocks and by using the above expansions one sees that one must have $W_+ > 0$, $W_- > 0$. Q.E.D.

Remark 13. The compressibility assumption $W > 0$ (Weyl condition) has already been introduced in Section 5.9, Proposition 5.5, when the steepening of magnetoacoustic waves was discussed.

The simplest example of barotropic fluids is that which has a constant speed of sound c_s,

$$p = c_s^2 e, \quad c_s = \text{constant}. \tag{8.28}$$

In this case it is possible to solve the jump conditions explicitly.

It is convenient to work in the inertial frame $\mathscr{I}$ in which the fluid ahead of the shock is at rest.

The results obtained in this frame will be useful in later chapters.

In this frame one has for the components of l_μ,

$$l_\mu = (-V_\Sigma, \nu^i),$$

where ν^i is a unit three-vector, $\delta^{ij}\nu_i\nu_j = 1$, (representing the normal to the wave front's two-surfaces, $\Sigma \cap (t = \text{const.})$, where t is the time in the inertial frame $\mathscr{I}$) and v_Σ is the normal speed of propagation of Σ with respect to the fluid ahead ($v_\Sigma > 0$).

Also, in this frame one has

$$u_+^\mu = (1, 0, 0, 0)$$

and

$$u_-^\mu = \Gamma_b(1, v_b^i), \quad \Gamma_b = (1 - v_b^i v_b^i)^{-1/2}.$$

Hence,

$$a_+ = -v_\Sigma,$$
$$a_- = \Gamma_b(v_b - v_\Sigma),$$

where $v_b = v_b^i \nu^i$ is the normal fluid speed behind the shock. We define the velocity jump as $[[v]] = v_b$. Finally, one has $G = l^\mu l_\mu = 1 - v_\Sigma^2$.

From the definition of u_+^2 and u_-^2 and the three-velocities we have

$$v_+^2 = v_\Sigma^2, \quad v_-^2 = \frac{(v_b - v_\Sigma)^2}{(1 - v_b v_\Sigma)^2}.$$

By substituting into equation (8.26c) one finds two equations

$$v_\Sigma^2 - (1 + c_s^2)v_\Sigma v_b + c_s^2 = 0,$$
$$v_\Sigma^2 - (1 - c_s^2)v_\Sigma v_b - c_s^2 = 0.$$

The first equation has real solutions only if $v_b > 2c_s/(1 + c_s^2)$ and therefore this choice is not consistent with the weak shock limit. The second equation has the only positive solution

$$v_\Sigma = \tfrac{1}{2}\{(1 - c_s^2)[[v]] + [(1 - c_s^2)^2[[v]]^2 + 4c_s^2]^{1/2}\} \tag{8.29}$$

and from equation (8.26b)

$$\frac{e_-}{e_+} = \frac{v_\Sigma(1 - [[v]]^2)}{v_\Sigma(1 + c_s^2[[v]]^2) - (1 + c_s^2)[[v]]}. \tag{8.30}$$

Notice that when $[[v]] \to 1$, $v_\Sigma \to 1$, $v_- \to c_s^2$, and $\dfrac{e_-}{e_+} \to \infty$ (Figs 8.2a–8.2b).

The second case that will be considered is that of polytropic fluids,

Fig. 8.2a. v_Σ as a function of $[[v]]$ for a barotropic fluid with $c_s = (1/3)^{1/2}$.

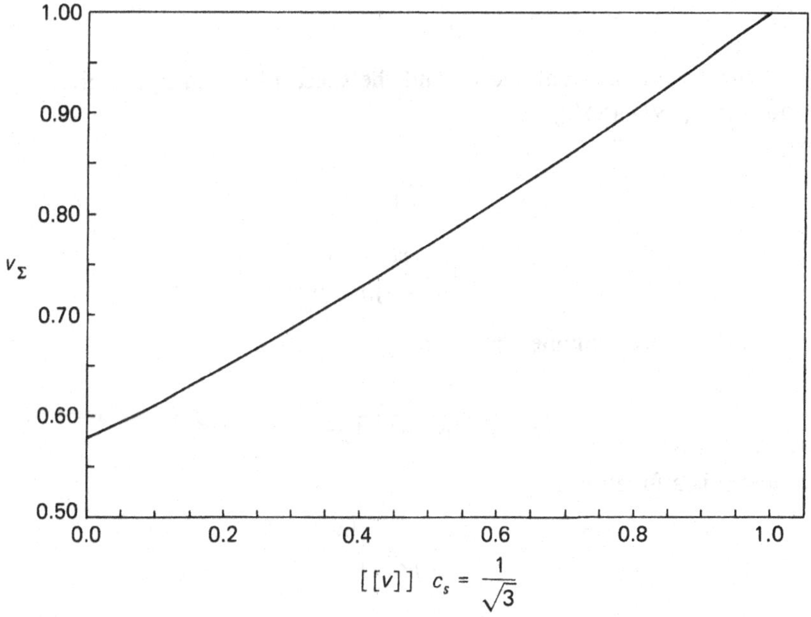

Fig. 8.2b. e_-/e_+ as a function of $[[v]]$ for a barotropic fluid as in Fig. 8.2a.

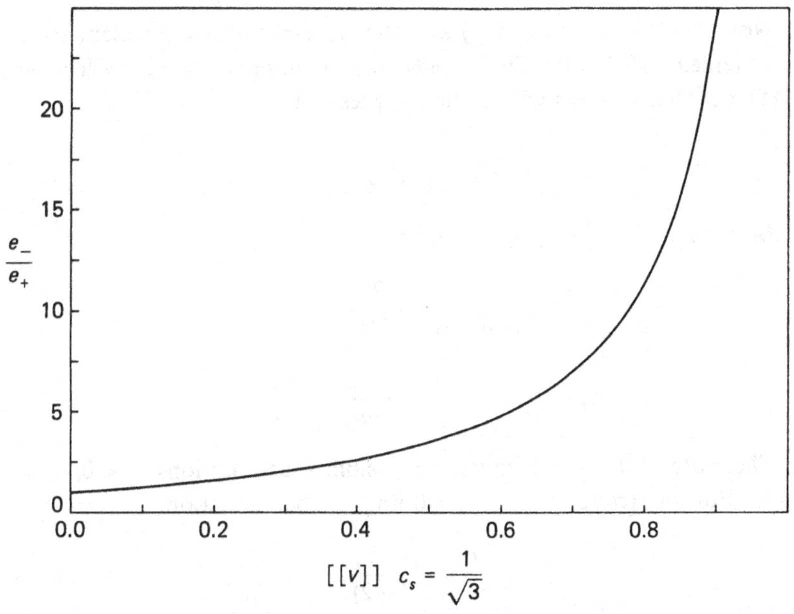

described by equation (2.19)

$$p = k(S)\rho^{\gamma},$$

and the specific internal energy and the speed of sound c_s are given by equations (5.91)–(5.92),

$$\varepsilon = \frac{1}{\gamma - 1}\frac{p}{\rho},$$

$$c_s^2 = \frac{\gamma p(\gamma - 1)}{(\gamma - 1)\rho + \gamma p}.$$

The dynamical volume τ is given by

$$\tau = \frac{1}{\rho} + \frac{\gamma}{\gamma - 1}\frac{p}{\rho^2},$$

where ρ is a function of p, S,

$$\rho = \left(\frac{p}{k(S)}\right)^{1/\gamma}.$$

Then

$$\tau'_S = \frac{1}{\rho}\left(1 + \frac{2\gamma p}{\gamma - 1}\frac{1}{\gamma}\right)\frac{k'(S)}{\gamma k(S)}.$$

Now $\gamma - 1 > 0$ (from $c_s^2 > 0$) and therefore $\tau'_S > 0$ is equivalent to the requirement $k'(S) > 0$. This holds for a nonrelativistic perfect gas ($k(S) = e^{S/c_v}$) and a gas with radiative pressure

$$\left(k(S) = \frac{1}{3}a_R\left(\frac{3S}{4a_R}\right)^{4/3}\right).$$

Moreover,

$$\tau'_p = -\frac{1}{\rho\gamma p} + \frac{\gamma^2 - 2\gamma}{\gamma(\gamma - 1)\rho^2},$$

$$\tau''_p = \frac{1}{\rho p^2}\left(\frac{1}{\gamma} + \frac{1}{\gamma^2}\right) - \frac{2}{\gamma p \rho^2}\frac{\gamma - 2}{\gamma - 1}.$$

Therefore, if $1 < \gamma \le 2$, the compressibility assumptions $\tau'_p < 0$, $\tau''_p > 0$ hold. For an arbitrary $\gamma > 1$, $\tau'_p < 0$ implies the restriction

$$\frac{p}{\rho} < \frac{\gamma - 1}{\gamma(\gamma - 2)},$$

whereas $\tau_p'' > 0$ implies

$$\frac{p}{\rho} < \frac{\gamma^2 - 1}{2\gamma(\gamma - 2)}.$$

It follows immediately that if $\gamma > 1$ then $\tau_p' < 0$ implies $\tau_p'' > 0$.

The general method for solving the jump conditions is based on the study of the Taub adiabat.

However, in the case of a polytropic fluid, it is simpler to resort to a more direct method.

From equation (8.23a) one finds, from the space components,

$$\Gamma_b^2(e+p)_-(v_\Sigma - v_b)v_b^i = (p_- - p_+)v^i \tag{8.31a}$$

and, from the time component,

$$\Gamma_b^2(e+p)_-(v_\Sigma - v_b) = (p_- + e_+)v_\Sigma, \tag{8.31b}$$

where

$$v_b = v_b^i \nu_i \tag{8.32}$$

is the fluid's normal velocity behind the shock, as measured in the frame $\mathscr{I}$.

From (8.31a) one finds

$$v_b^i = v_b \nu^i, \tag{8.33}$$

which states that the tangential components of the fluid's velocity are continuous across Σ.

Then equation (8.31a) is equivalent to

$$\Gamma_b^2(e+p)_-(v_\Sigma - v_b)v_b = p_- - p_+. \tag{8.34}$$

Finally, in the frame $\mathscr{I}$, equation (8.5) is written

$$\rho_-\Gamma_b(v_\Sigma - v_b) = \rho_+ v_\Sigma. \tag{8.35}$$

From equations (8.34)–(8.35) one obtains immediately

$$p_- = \frac{p_+ + e_+ v_\Sigma v_b}{1 - v_\Sigma v_b}, \tag{8.36}$$

$$\frac{1+\varepsilon_-}{\Gamma_b} = \frac{p_+}{\rho_+}\frac{v_b}{v_\Sigma} + 1 + \varepsilon_+. \tag{8.37}$$

It is convenient to introduce the following dimensionless variables

$$q = \frac{p_-}{p_+}, \quad r = \frac{\rho_-}{\rho_+}, \quad \mu = \frac{p_+}{\rho_+}, \quad \xi = \frac{v_b}{v_\Sigma}. \tag{8.38}$$

Then equation (8.36) reads,

$$\xi = \frac{\left(1 - \frac{1}{\Gamma_b^2}\right)}{q-1}\left(q + \frac{1}{\mu} + \frac{1}{\gamma - 1}\right) \tag{8.39}$$

and equation (8.37) reads

$$\frac{1}{\Gamma_b}\left\{1 + \frac{\Gamma_b \mu q}{(\gamma - 1)}(1 - \xi)\right\} = \mu\xi + 1 + \frac{\mu}{\gamma - 1}. \tag{8.40}$$

By substituting equation (8.39) into equation (8.40) one finds

$$A\frac{1}{\Gamma^2} + B\frac{1}{\Gamma_b} + C = 0, \tag{8.41}$$

with

$$A = \left(1 + \frac{q}{\gamma - 1}\right)\left\{1 + \mu\left(q + \frac{1}{\gamma - 1}\right)\right\}, \tag{8.42}$$

$$B = q - 1, \tag{8.43}$$

$$C = -\frac{q\gamma}{\gamma - 1}\left(1 + \mu\frac{\gamma}{\gamma - 1}\right). \tag{8.44}$$

Because $A > 0$, $B > 0$ (for a compressive shock), and $C > 0$, equation (8.41) has only one positive root for $1/\Gamma_b$, given q and μ. From this root, by using equation (8.39) one determines ξ (hence v_Σ), and from equation (8.35), which reads

$$r = \frac{1}{\Gamma_b(1 - \xi)}, \tag{8.45}$$

one obtains r.

This procedure is an unambiguous method for solving parametrically the jump conditions for a polytropic fluid.

Other choices for the parameters (e.g., r instead of q) lead to more complicated expressions (which, moreover, also have some sign ambiguity).

Notice that the parameter μ is related to the nonrelativistic speed of sound in the gas.

In Figs. 8.3 one plots the speed v_b versus v_Σ for various values of γ and μ.

In Figs. 8.4 the ratios r and q are plotted as functions of $[[v]]$.

Notice that, for very strong shocks, in the nonrelativistic theory, $r \to \frac{\gamma + 1}{\gamma - 1}$, whereas in special relativity this limit is removed.

Fig. 8.3. (a) Gas velocity as a function of v_Σ for $\gamma = 4/3$ for $\mu = 0.001, 0.01, 0.1, 1, 10$. (b) Same as in (a), but with $\gamma = 2.5$ and $\mu = 0.001, 0.01, 0.1, 0.4, 1$.

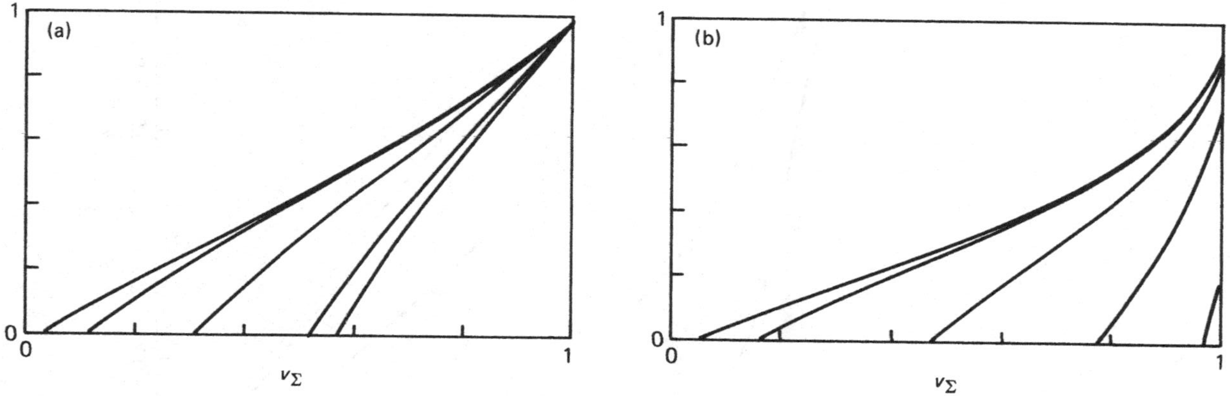

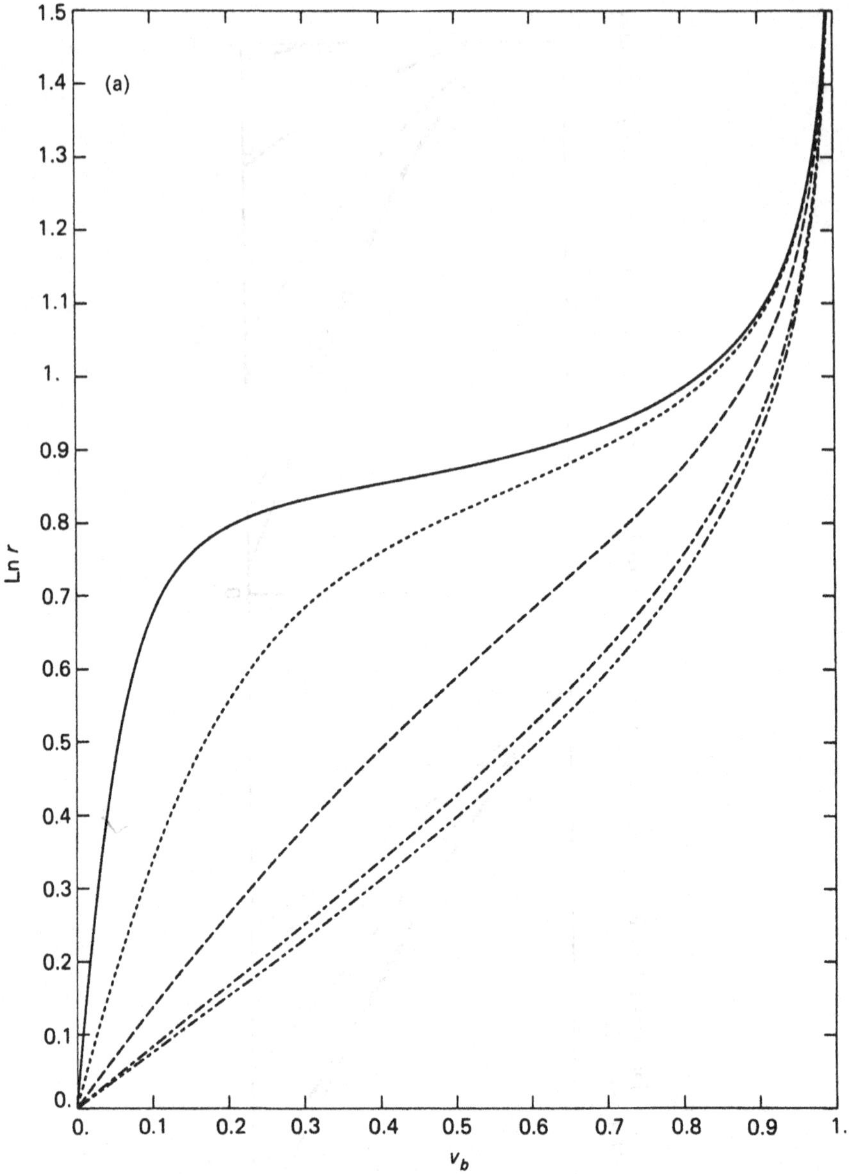
(a)
Ln r
v_b
0.
0.1
0.2
0.3
0.4
0.5
0.6
0.7
0.8
0.9
1.
1.1
1.2
1.3
1.4
1.5

Fig. 8.4. Solution of the jump conditions for a polytropic gas. Density (a) and pressure (b) ratios as functions of the fluid velocity v_b. $\gamma = 4/3$, $\mu = 0.001, 0.01, 0.1, 1, 10$ (left to right).

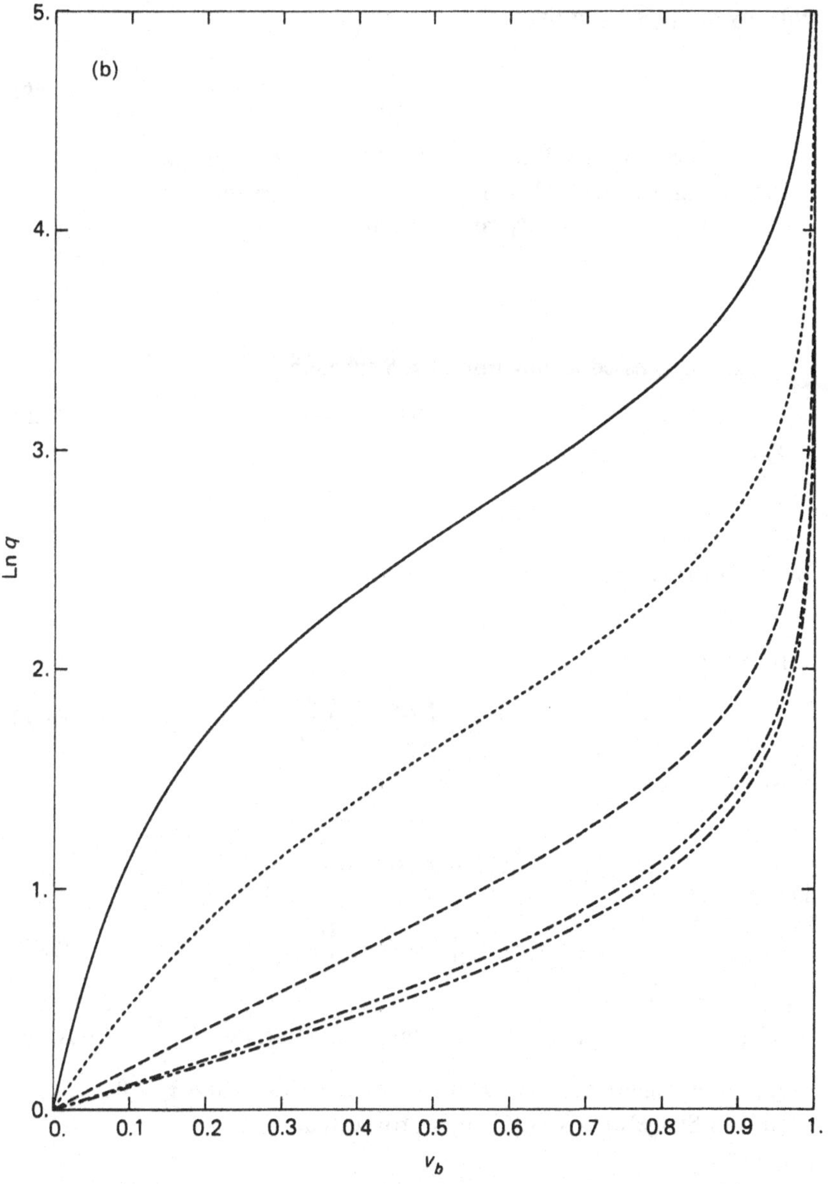

The specific entropy jump $[[S]]$ cannot be calculated unambiguously unless the physical model for the polytropic fluid is specified.

In particular, for a radiation dominated fluid, with $\gamma = 4/3$, the specific entropy jump is given by

$$\frac{S_-}{S_+} = \left(\frac{q}{r}\right)^{3/4}. \tag{8.46}$$

The final case that will be treated is that of the Synge gas.

First of all, we check that the compressibility assumptions hold.

The expression for the dynamical volume is

$$\tau = \frac{G(z)}{zp}, \tag{8.47}$$

where z is expressed as function of p, S through

$$e^{-m/k_B(S-S_0)} = pzL(z). \tag{8.48}$$

Then

$$z'_S = -\frac{m}{k_B}\frac{zL}{L+zL'} = \frac{m}{k_B}\frac{1}{zG'}$$

by using the equality

$$L + zL' = -z^2 LG'.$$

It follows that

$$\tau'_S = \frac{m}{k_B}\frac{1}{pz}\left(G' - \frac{G}{z}\right)\frac{1}{zG'}. \tag{8.48'}$$

Also,

$$z'_p = \frac{-zL}{p(L+zL')} = \frac{1}{pzG'},$$

hence

$$\tau'_p = \frac{1}{p^2 z}\left(\frac{1}{z} - G - \frac{G}{z^2 G'}\right). \tag{8.49}$$

Synge (1957) has proved that the functions G and $G^2 - \dfrac{2G}{z}$ are positive and monotonically decreasing. It follows that $\tau'_S > 0$ and $\tau'_p < 0$.

Also in Synge's book (1957) it is proved that

$$\frac{d^2(GL)}{d\left(\frac{1}{zL}\right)^2} > 0,$$

which, because

$$p = \frac{\mathrm{e}^{-m(S-S_0)/K_B}}{zL(z)}$$

and

$$\tau = G(z)L(z)\mathrm{e}^{m(S-S_0)/K_B},$$

is equivalent to $\tau''_p > 0$.

In order to solve the jump conditions we proceed as follows. Since $\tau'_p < 0$, the shock hypersurface Σ is timelike.

Therefore, by using the definitions (8.17j), the mass conservation equation (8.5) is written

$$\rho_- u_- = \rho_+ u_+. \tag{8.50}$$

From the equation of the Taub adiabat (8.7) we have

$$G_-^2 - G_+^2 + \left(\frac{\rho_+}{z_+} - \frac{\rho_-}{z_-}\right)\left(\frac{G_+}{\rho_+} - \frac{G_-}{\rho_-}\right) = 0,$$

which, using equation (8.50), yields

$$F(\eta) \equiv \frac{G_-}{z_+}\eta^2 - A\eta - \frac{G_+}{z_-} = 0, \tag{8.51}$$

where

$$\eta = \frac{u_-}{u_+}, \quad A = G_+^2 - G_-^2 + \frac{G_-}{z_-} - \frac{G_+}{z_+}. \tag{8.52}$$

Now, from Proposition (8.4) we have that $S_- - S_+ > 0$ implies $f_- > f_+$, that is, $G(z_-) > G(z_+)$. Since $G(z)$ is a decreasing function of z, it follows that

$$z_- < z_+. \tag{8.53}$$

Equation (8.51) admits only a positive root given by

$$\eta = \frac{z_+}{2G_-}\left\{A + \sqrt{A^2 + \frac{4G_+G_-}{z_+z_-}}\right\}. \tag{8.54}$$

From equations (8.25a)–(8.26b) one obtains

$$\frac{u_-^2}{u_+^2} = \frac{(e+p)_+ v_-}{(e+p)_- v_+},$$

hence

$$\eta = \frac{G_- v_-}{G_+ v_+}. \tag{8.55}$$

Therefore

$$v_- = B(z_-, z_+)v_+, \tag{8.56}$$

with

$$B(z_-, z_+) = \frac{z_+}{2G_+}\left\{A + \sqrt{A^2 + \frac{4G_+G_-}{z_+z_-}}\right\}.$$

Furthermore, from equations (8.25a)–(8.26b) we obtain easily

$$\Gamma_- G_- = \Gamma_+ G_+, \tag{8.57}$$

whence

$$|v_+| = \sqrt{\frac{g^2 - 1}{g^2 - B^2}}, \tag{8.58}$$

with

$$g = \frac{G_-}{G_+}.$$

In this way the solutions of the jump conditions have been obtained explicitly once the temperatures $T_\pm$ ahead and behind the shock have been assigned.

In Figs. 8.5–8.8 the quantities v_-, v_+ are plotted as functions of z_- for $z_+ = .01, .1, 1, 10$.

When the fluid ahead of the shock is cold it is possible to obtain simpler expressions.

In this case, $T_+ \to 0$, $z_+ \to 1$, and $G_+ \to 1$. Then equation (8.57) becomes

$$G_- \Gamma_- = \Gamma_+. \tag{8.59}$$

Fig. 8.5. The velocities $|v_-|$, $|v_+|$ as functions of z_-, for $z_+ = 0.01$.

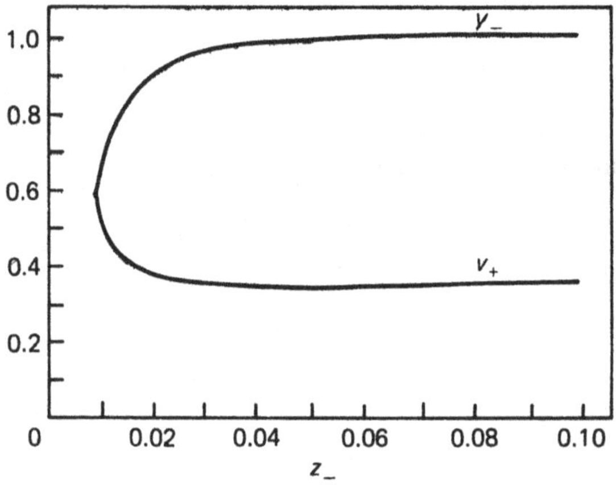

Fig. 8.6. As in Fig. 8.5 with $z_+ = 0.1$.

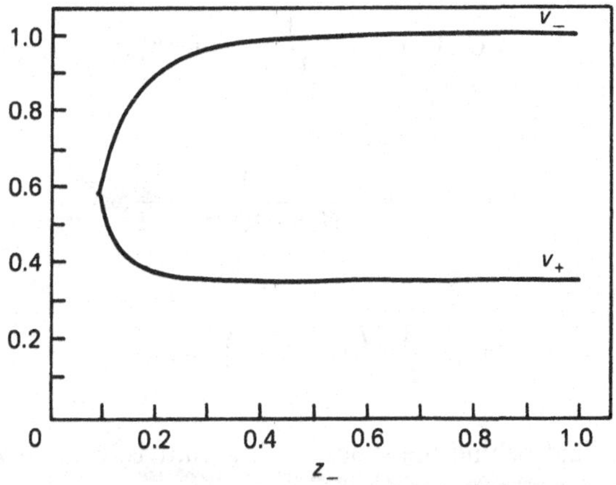

Fig. 8.7. As in Fig. 8.6 with $z_+ = 1$.

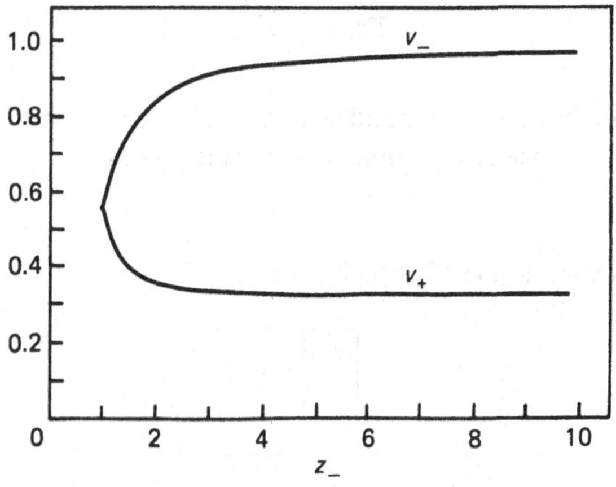

Fig. 8.8. As in Fig. 8.7 with $z_+ = 10$.

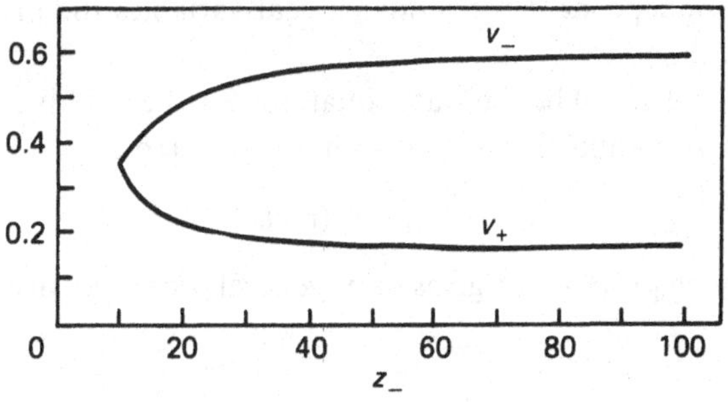

Furthermore, equation (8.16) yields

$$G_-\Gamma_-v_- + \frac{1}{z_-\Gamma_-v_-} = \Gamma_+v_+, \tag{8.60}$$

whence

$$\Gamma_-^2 = 1 + \frac{1}{z_-^2G_- - 2G_-z_- - z_-^2} \tag{8.61}$$

and

$$|v_-| = \frac{1}{z_-}\left\{\left(G_- - \frac{1}{z_-}\right)^2 - 1\right\}^{1/2}. \tag{8.62}$$

When the fluid behind the shock is ultrarelativistic $T_- \to \infty$, $G_- \to \dfrac{4}{z_-}$, and from equation (8.61) it follows that

$$\Gamma_-^2 = \frac{9 - z_-^2}{8 - z_-^2}. \tag{8.63}$$

When both the fluids ahead and behind the shock are ultrarelativistic, it is convenient to consider equation (8.26c), which gives

$$3v_+v_- = 1 \tag{8.64}$$

jointly with equation (8.57) which gives

$$\left[\frac{\Gamma}{z}\right] = 0. \tag{8.65}$$

8.3. Relativistic detonation and deflagration waves

The concept of combustion is introduced as follows (Coll, 1976).

We use as independent thermodynamical variables the pressure p and dynamical volume τ.

Let $e(\tau, p)$ and $\tilde{e}(\tau, p)$ be the state equations for the fluid before and after combustion. A combustion is said to have occurred if

$$e(\tau, p) > \tilde{e}(\tau, p). \tag{8.66}$$

The following proposition gives some general properties of a combustion process.

PROPOSITION 8.13. *The following inequalities are all equivalent:*

$$\begin{aligned}
&\text{(a)} \quad e(\tau, p) > \tilde{e}(\tau, p),\\
&\text{(b)} \quad \rho(\tau, p) > \tilde{\rho}(\tau, p),\\
&\text{(c)} \quad f(\tau, p) > \tilde{f}(\tau, p),\\
&\text{(d)} \quad \varepsilon(\tau, p) > \tilde{\varepsilon}(\tau, p).
\end{aligned} \tag{8.67}$$

Proof.

$$\begin{aligned}
\text{(a)} &\Rightarrow e(\tau, p) + p > \tilde{e}(\tau, p) + p\\
&\Rightarrow \rho(\tau, p) f(\tau, p) > \tilde{\rho}(\tau, p) \tilde{g}(\tau, p)\\
&\Rightarrow \rho^2(\tau, p)\tau > \tilde{\rho}^2(\tau, p)\tau,
\end{aligned}$$

whence (b) and (c) follow.

Furthermore, since $f = 1 + \varepsilon + \dfrac{p}{\rho}$, (c) implies

$$\varepsilon(\tau, p) + \frac{p}{\rho(\tau, p)} > \varepsilon(\tau, p) + \frac{p}{\tilde{\rho}(\tau, p)},$$

whence (d) follows.

Therefore the sequence (a)⇒(b)⇒(c)⇒(d) has been proved.

It is immediately seen that (c)⇒(b), hence (b) and (c) are equivalent.

Now, from (c) it follows that

$$f(\tau, p)\rho(\tau, p) > \tilde{f}(\tau, p)\tilde{\rho}(\tau, p),$$

hence

$$e(\tau, p) - p > \tilde{e}(\tau, p) - p,$$

which implies (a). Therefore (a) and (c) are equivalent.

From (d) one obtains

$$f(\tau, p) - \frac{p}{\rho(\tau, p)} > \tilde{f}(\tau, p) - \frac{p}{\tilde{\rho}(\tau, p)},$$

whence

$$\begin{aligned}
f(\tau, p) - \tilde{f}(\tau, p) &> p\left[\frac{1}{\rho(\tau, p)} - \frac{1}{\tilde{\rho}(\tau, p)}\right]\\
&= p\left[\frac{\tau}{f(\tau, p)} - \frac{\tau}{\tilde{f}(\tau, p)}\right],
\end{aligned}$$

which implies

$$[f(\tau,p)-\tilde{f}(\tau,p)]\left(1+\frac{\tau p}{f(\tau,p)\tilde{f}(\tau,p)}\right)>0,$$

which proves the equivalence between (d) and (c). Q.E.D.

The nonrelativistic definition of a combustion process is

$$\varepsilon(V,p)>\tilde{\varepsilon}(V,p), \tag{8.68}$$

where $V=1/\rho$ is the specific volume.

PROPOSITION 8.14. *Let $T(\tau,p)$ be the temperature. Under the assumptions $\tau'_S>0$ and*

$$T'_p>0 \tag{8.69}$$

the nonrelativistic and relativistic definitions of combustion coincide.

Proof. By Proposition 8.13 it is sufficient to prove that equation (8.68) is equivalent to (b), that is,

$$V(\tau,p)<\tilde{V}(\tau,p). \tag{8.70}$$

Let

$$\hat{V}(S,p)=V(\tau(S,p),p).$$

Then

$$\hat{V}'_S=V_\tau\tau'_S.$$

From the first law of thermodynamics,

$$d\varepsilon=T\,dS-p\,dV=T\,dS-d(pV)+\hat{V}\,dp,$$
$$d(\varepsilon+pV)=T\,dS+\hat{V}\,dp,$$

which yields the integrability condition

$$T'_p=\hat{V}'_S=V'_\tau\tau'_S,$$

whence

$$V'_\tau=\frac{T'_p}{\tau'_S}>0.$$

It follows that $V(\tau,p)$ is monotonically increasing with τ. Let $\tau=\tau(V,p)$ be the inverse function.

The inequality $V(\tau,p)<\tilde{V}(\tau,p)$ is equivalent to

$$\tau(V,p)>\tilde{\tau}(V,p),$$

where $\tilde{\tau}(V,p)$ is the inverse function of $\tilde{V}(\tau,p)$.

But $\tau = fV = V(1 + \varepsilon + pV)$, and therefore

$$V(1 + \varepsilon(V, p) + pV) > (1 + \tilde{\varepsilon}(V, p) + pV)V$$

and equation (8.68) is proved. Q.E.D.

The combustion process is idealized as follows.

A given fluid element in the unburnt fluid crosses a surface of discontinuity and at the same time, instantaneously, liberates energy (thereby changing its equation of state).

The burnt and unburnt fluid are separated by a moving surface of discontinuity, which in space-time is represented by a hypersurface Σ, which, locally, is given by the equations

$$\varphi(x^\alpha) = 0, \quad d\varphi|_\Sigma \neq 0.$$

Unlike a shock wave, a combustion front separates two different media and surface effects need to be taken into account. The required junction conditions have been derived by Israel (1966) with the aid of the Gauss-Codazzi formalism. However, for nearly plane fronts, we can neglect the curvature of the separating surface and subsequently we shall restrict ourselves to this situation.

Then the jump conditions (8.3)–(8.4) apply across Σ, the only difference with the case of shock waves being the change in the equation of state (8.66).

The jump condition (8.4) yields, as in the case of shock waves, since $l_\alpha = \nabla_\alpha \varphi$,

$$m = \rho_+ u_+^\mu l_\mu = \rho_- u_-^\mu l_\mu. \tag{8.71}$$

From the jump condition (8.3), proceeding as in Section 8.1 one has

$$f_+^2 - (\tilde{f}_-^2) + (p_- - p_+)(\tau_+ + \tau_-) = 0 \tag{8.72}$$

and also

$$m^2(\tau_- - \tau_+) = l^\mu l_\mu (p_+ - p_-). \tag{8.73}$$

Let Π be the half-plane (τ, p), $\tau > 0$, with $Z \equiv (\tau, p)$ its generic point. The detonation function $\tilde{\mathscr{H}}(Z, Z_+)$ is defined as

$$\tilde{\mathscr{H}}(Z, Z_+) = -f_+^2 + \tilde{f}^2 - (p - p_+)(\tau_+ + \tau). \tag{8.74}$$

The detonation adiabat $\mathscr{C}$ relative to the point Z_+ is defined as

$$\tilde{\mathscr{H}}(Z, Z_+) = 0. \tag{8.75}$$

Remark 14. Unlike the case of shock waves, the initial point Z_+ does not belong to the detonation adiabat. In fact,

$$\tilde{\mathscr{H}}(Z_+, Z_+) < 0.$$

Remark 15. One also has, in Π,

$$d\tilde{\mathscr{H}} = 2\tilde{f}T\,dS + (\tau - \tau_+)dp - (p - p_+)d\tau. \tag{8.76}$$

PROPOSITION 8.15. *Under the assumptions $\tau'_p < 0$, $\tau'_S > 0$ the detonation adiabat $\mathscr{C}$:*

$$\tilde{\mathscr{H}}(Z, Z_+) = 0$$

meets the straight line $\mathscr{L}': \tau - \tau_+ = 0$ at a unique point (τ_+, p') such that $p' > p_+$. Similarly, it meets the line $\mathscr{L}'': p - p_+ = 0$ at a unique point (τ'', p_+) such that $\tau'' > \tau_+$.

Proof. On $\mathscr{L}': \tau - \tau_+ = 0$ one has $\tau'_S dS = -\tau'_p dp$, hence

$$\frac{d\tilde{\mathscr{H}}}{dp} = -2\tilde{f}T\frac{\tau'_p}{\tau'_S} > 0.$$

Because $\tilde{\mathscr{H}}(Z_+, Z_+) < 0$, it follows that $\tilde{\mathscr{H}}(Z, Z_+)$ must vanish at a point $Z' \equiv (\tau_+, p')$ on $\mathscr{L}'$, with $p' > p_+$.

Similarly, on $\mathscr{L}'': p - p_+ = 0$, one has

$$\frac{d\tilde{\mathscr{H}}}{d\tau} = 2\tilde{f}\frac{T}{\tau'_S} > 0$$

and it follows that $\tilde{\mathscr{H}}(Z, Z_+)$ must vanish at a point $Z'' \equiv (\tau'', p_+)$ on $\mathscr{L}''$, with $\tau'' > \tau_+$. The uniqueness of the points Z', Z'' follows from the monotonicity of $\tilde{\mathscr{H}}(Z, Z_+)$ on $\mathscr{L}'$ and on $\mathscr{L}''$.

Q.E.D.

Remark 16. The tangent to the detonation adiabat $\mathscr{C}$ is found in the following way. From equation (8.76), on $\mathscr{C}$,

$$2\tilde{f}T\left(\frac{dS}{d\tau}\right)_{\mathscr{C}} + (\tau - \tau_+)\left(\frac{dp}{d\tau}\right)_{\mathscr{C}} - (p - p_+) = 0.$$

Now, for $S = S(\tau, p)$,

$$\frac{dS}{d\tau} = S'_\tau + S'_p\frac{dp}{d\tau}.$$

Hence

$$(\tau - \tau_+ + 2\tilde{f}TS'_p)\left(\frac{\mathrm{d}p}{\mathrm{d}\tau}\right)_{\mathscr{C}} = p - p_+ - 2\tilde{f}TS'_\tau. \tag{8.77}$$

From $\mathrm{d}\tau = \tau'_S \mathrm{d}S + \tau'_p \mathrm{d}p$ one has

$$S'_p = -\frac{\tau'_p}{\tau'_S}, \quad S'_\tau = \frac{1}{\tau'_S}.$$

Under the assumptions $\tau'_p < 0$, $\tau'_S > 0$ one has $S'_p > 0$, $S'_\tau > 0$.

Then $(\mathrm{d}p/\mathrm{d}\tau)_{\mathscr{C}}$ is finite at any point with $\tau \geq \tau_+$. Furthermore, $\left(\frac{\mathrm{d}p}{\mathrm{d}\tau}\right)_{\mathscr{C}} \neq 0$ at all points at which $\tau \geq \tau_+$, $p \leq p_+$.

PROPOSITION 8.16. *Under the assumptions $\tau'_p < 0$, $\tau'_S > 0$, $\tau''_p > 0$, a straight line $\mathscr{L}$ from $Z_+ \equiv (\tau_+, p_+)$ intersects the detonation adiabat $\mathscr{C}$ at most in two points.*

Proof. If $\mathscr{L}$ intersects $\mathscr{C}$ in two points, A and B say, then $\tilde{\mathscr{H}}$ (and S) must have a stationary point Z^* on the segment AB in $\mathscr{L}$. From equation (8.76) it follows that

$$\mathrm{d}S|_{Z^*} = 0.$$

Let

$$\mathscr{L}: p - p_+ = M(\tau - \tau_+).$$

Then

$$M = \left(\frac{\mathrm{d}p}{\mathrm{d}\tau}\right)_{\mathscr{L}} = p'_\tau + p'_S\left(\frac{\mathrm{d}S}{\mathrm{d}\tau}\right)_{\mathscr{L}},$$

hence

$$0 = \left(\frac{\mathrm{d}^2p}{\mathrm{d}\tau^2}\right)_{\mathscr{L}} = p''_\tau + 2p''_{\tau S}\left(\frac{\mathrm{d}S}{\mathrm{d}\tau}\right)_{\mathscr{L}} + p''_S\left(\frac{\mathrm{d}S}{\mathrm{d}\tau}\right)^2_{\mathscr{L}} + p'_S\left(\frac{\mathrm{d}^2S}{\mathrm{d}\tau^2}\right)_{\mathscr{L}}.$$

It follows that at a stationary point Z^*,

$$p''_\tau \geqq p'_S\left(\frac{\mathrm{d}^2S}{\mathrm{d}\tau^2}\right)_{\mathscr{L}}\bigg|_{Z^*} = 0,$$

whence

$$\left(\frac{\mathrm{d}^2S}{\mathrm{d}\tau^2}\right)_{\mathscr{L}}\bigg|_{Z^*} = -\frac{p''_\tau}{p'_S}.$$

Now

$$p'_S = -\frac{\tau'_S}{\tau'_p} > 0,$$

$$p''_\tau = -\frac{1}{(\tau'_p)^3}\tau''_p > 0$$

imply

$$\left.\left(\frac{d^2S}{d\tau^2}\right)_{\mathscr{L}}\right|_{Z^*} < 0,$$

which corresponds to a maximum of S on $\mathscr{L}$, which is obviously unique.
Q.E.D.

Remark 17. Henceforth we shall assume that Σ is timelike, an assumption which is certainly consistent with causality.

Then, from equation (8.73), since $l_\alpha l^\alpha > 0$, the state after combustion must lie on the straight line $\mathscr{L}$,

$$p - p_+ = -\frac{m^2}{l_\alpha l^\alpha}(\tau - \tau_+), \tag{8.78}$$

which has a negative slope. The qualitative features of the detonation adiabat can be seen from Fig. 8.9. One notices that the detonation adiabat $\mathscr{C}$ splits into three parts,

$$\mathscr{C} = \mathscr{C}_D \cup \mathscr{C}_I \cup \mathscr{C}_d,$$

where the region $\mathscr{C}_I$ is inaccessible, and the regions $\mathscr{C}_D$, $\mathscr{C}_d$ correspond to detonations and deflagrations, respectively. The points DJ and dJ where the straight line $\mathscr{L}$ from Z_+ is tangent to $\mathscr{C}$ are the Jouguet detonation and deflagration points, respectively. They are defined by the condition

$$\left(\frac{dp}{d\tau}\right)_{\mathscr{C}} = \left(\frac{dp}{d\tau}\right)_{\mathscr{L}}. \tag{8.79}$$

A property distinguishing between detonation and deflagration waves which can be obtained immediately is the following.

From the mass conservation law equation (8.71) one has

$$\frac{u_+^2}{u_-^2} = \frac{\tilde{\rho}_-^2}{\rho_+^2}.$$

From the detonation adiabat equation (8.72) one obtains that for detonations ($p_- > p_+, \tau_- < \tau_+$) one has $\tilde{f}_- > f_+$, that is, $\tau_- \tilde{\rho}_- > \tau_+ \rho_+$, which implies $\dfrac{\tilde{\rho}_-}{\rho_+} > \dfrac{\tau_+}{\tau_-} > 1$. Therefore, for detonation waves $u_+^2 > u_-^2$.

In a similar way one proves that, for deflagrations, $\tilde{\rho}_- < \rho_+$ and $u_+^2 < u_-^2$ (expansive nature of deflagrations).

From equation (8.76), along $\mathscr{C}$, one has

$$2\tilde{f}T\left(\frac{\mathrm{d}S}{\mathrm{d}\tau}\right)_{\mathscr{C}}+(\tau-\tau_+)\left(\frac{\mathrm{d}p}{\mathrm{d}\tau}\right)_{\mathscr{C}}-(p-p_+)=0. \tag{8.80}$$

Also, on $\mathscr{L}$,

$$p-p_+=\left(\frac{\mathrm{d}p}{\mathrm{d}\tau}\right)_{\mathscr{L}}(\tau-\tau_+)$$

and therefore at the Jouguet points

$$\left(\frac{\mathrm{d}S}{\mathrm{d}\tau}\right)_{\mathscr{C}}=\left(\frac{\mathrm{d}S}{\mathrm{d}\tau}\right)_{\mathscr{L}}=0, \tag{8.81}$$

and at these points the isentropic curves are tangent to $\mathscr{C}$.

Remark 18. From equation (8.14) one sees that at the Jouguet points

$$\tau'_S\left(\frac{\mathrm{d}S}{\mathrm{d}p}\right)_J=\left(\frac{P(l_\alpha)e'_p}{m^2}\right)_J=0$$

and therefore *the Jouguet points correspond to acoustic wave fronts,*

$$P(l_\alpha)|_J=0. \tag{8.82}$$

PROPOSITION 8.17. *Under the assumptions $\tau'_p<0$, $\tau'_S>0$, $\tau''_p>0$, for the Jouguet wavefronts the entropy after combustion is stationary. It is a minimum for detonations and a maximum for deflagrations, compared to all other states with the same thermodynamic state before combustion.*

Proof. By differentiating equation (8.80) along $\mathscr{C}$ yields

$$2\frac{\mathrm{d}}{\mathrm{d}\tau}(\tilde{f}T)\frac{\mathrm{d}S}{\mathrm{d}\tau}+2\tilde{f}T\frac{\mathrm{d}^2S}{\mathrm{d}\tau^2}+(\tau-\tau_+)\frac{\mathrm{d}^2p}{\mathrm{d}\tau^2}=0,$$

which, at the Jouguet points, reduces to

$$2\tilde{f}T\left(\frac{\mathrm{d}^2S}{\mathrm{d}\tau^2}\right)_J=-(\tau-\tau_+)\left(\frac{\mathrm{d}^2p}{\mathrm{d}\tau^2}\right)_J. \tag{8.83}$$

Now, along $\mathscr{C}$,

$$\frac{\mathrm{d}p}{\mathrm{d}\tau}=\frac{\partial p}{\partial\tau}+\frac{\partial p}{\partial S}\frac{\mathrm{d}S}{\mathrm{d}\tau},$$

$$\frac{\mathrm{d}^2p}{\mathrm{d}\tau^2}=\frac{\partial^2p}{\partial\tau^2}+2\frac{\partial^2p}{\partial\tau\partial S}\frac{\mathrm{d}S}{\mathrm{d}\tau}+\frac{\partial^2p}{\partial S^2}\left(\frac{\mathrm{d}S}{\mathrm{d}\tau}\right)^2+\frac{\partial p}{\partial S}\frac{\mathrm{d}^2S}{\mathrm{d}\tau^2}$$

and at the Jouguet points

$$\left(\frac{\mathrm{d}^2 p}{\mathrm{d}\tau^2}\right)_J = \left(\frac{\partial^2 p}{\partial \tau^2}\right)_J + \left(\frac{\partial p}{\partial S}\right)_J \left(\frac{\mathrm{d}^2 S}{\mathrm{d}\tau^2}\right)_J. \tag{8.84}$$

Also,

$$\frac{\partial p}{\partial S} = -\frac{\tau'_S}{\tau'_p}, \quad \frac{\partial p}{\partial \tau} = \frac{1}{\tau'_p}, \quad \frac{\partial^2 p}{\partial \tau^2} = -\frac{\tau''_p}{(\tau'_p)^3},$$

hence

$$\frac{\partial p}{\partial S} > 0, \quad \frac{\partial^2 p}{\partial \tau^2} > 0.$$

It follows that $\left(\frac{\mathrm{d}^2 p}{\mathrm{d}\tau^2}\right)_J$ and $\left(\frac{\mathrm{d}^2 S}{\mathrm{d}\tau^2}\right)_J$ cannot vanish simultaneously, and, furthermore, that they are both nonzero.

From Proposition (8.16) one has that $\mathscr{C}$ must be concave upward. In fact, if $\mathscr{C}$ were concave downward any straight line $\mathscr{L}$ from Z_+ would intersect $\mathscr{C}$ at most at one point. Hence, S would not have a stationary point on $\mathscr{L}$ and from equation (8.11) this would mean $l_\alpha l^\alpha < 0$, which contradicts the assumption that Σ is timelike. It follows that

$$\left(\frac{\mathrm{d}^2 p}{\mathrm{d}\tau^2}\right)_J > 0,$$

hence, for $\tau > \tau_+$, $\left(\frac{\mathrm{d}^2 S}{\mathrm{d}\tau^2}\right)_{dJ} < 0$ (corresponding to a maximum entropy), and for $\tau < \tau_+$, $\left(\frac{\mathrm{d}^2 S}{\mathrm{d}\tau^2}\right)_{DJ} > 0$ (corresponding to a minimum entropy).

Q.E.D.

PROPOSITION 8.18. *For the Jouguet wavefronts the speed v_Σ relative to the medium ahead is stationary. It is a minimum for a detonation and a maximum for a deflagration, with respect to the other states with the same thermodynamical state before combustion.*

Proof. From

$$V_\Sigma^2 = \frac{(u^\mu l_\mu)^2}{l^\mu l_\mu + (u^\mu l_\mu)^2}$$

one has

$$V_\Sigma^2 = \frac{M}{M - \dot{\rho}^2},$$

where

$$M = -\frac{m^2}{l_\mu l^\mu}.$$

From

$$M = \frac{[[p]]}{[[\tau]]},$$

it follows that

$$(V_\Sigma^2)_+ = \frac{[[p]]}{[[p]] - \rho_+^2 [[\tau]]}.$$

Then

$$\left(\frac{\mathrm{d}(V_\Sigma^2)_+}{\mathrm{d}\tau}\right)_{\mathscr{C}} = \frac{\rho_+^2}{([[p]] - \rho_+^2[[\tau]])^2}\left([[p]] - [[\tau]]\left(\frac{\mathrm{d}p}{\mathrm{d}\tau}\right)_{\mathscr{C}}\right),$$

which vanishes at the Jouguet points.

Also,

$$\left(\frac{\mathrm{d}^2(V_\Sigma^2)_+}{\mathrm{d}\tau^2}\right)_J = -\frac{-\rho_+^2[[\tau]]}{([[p]] - \rho_+^2[[\tau]])^2}\left(\frac{\mathrm{d}^2 p}{\mathrm{d}\tau^2}\right)_J. \qquad \text{Q.E.D.}$$

Fig. 8.9. Sketch of the detonation adiabat.

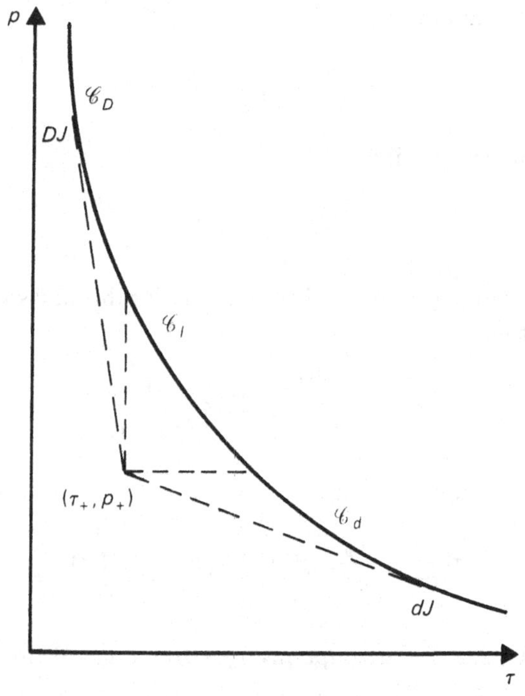

Let $\mathscr{L}$ be straight line from Z_+. Along $\mathscr{L}$ the entropy S has a unique maximum at $Z_* \equiv (\tau_*, p_*)$.

Hence $\dfrac{dS}{d\tau} < 0$ for $\tau > \tau_*$ and $\dfrac{dS}{d\tau} > 0$ for $\tau < \tau_*$.

Hence, from Fig. 8.9, in the case of detonations, one has

$$\frac{dS}{d\tau}(\tau_+, p_+) < 0 \tag{8.85a}$$

and in the case of deflagrations

$$\frac{dS}{d\tau}(\tau_+, p_+) > 0. \tag{8.85b}$$

Also, for strong detonations,

$$\frac{dS}{d\tau}(\tau_-, p_-) > 0 \tag{8.85c}$$

and for weak detonations

$$\frac{dS}{d\tau}(\tau_-, p_-) < 0. \tag{8.85d}$$

For weak deflagrations

$$\frac{dS}{d\tau}(\tau_-, p_-) > 0 \tag{8.85e}$$

and for strong deflagrations

$$\frac{dS}{d\tau}(\tau_-, p_-) < 0. \tag{8.85f}$$

By using equation (8.17j) and the definitions for the fluid's speed relative to the wavefront,

$$u_\pm = \frac{|v_\pm|}{\sqrt{1 - V_\Sigma^2}}, \quad \Gamma_s = \frac{1}{\sqrt{1 - c_s^2}},$$

equation (8.14) can be written as

$$\tau'_S \frac{dS}{d\tau} = \frac{e'_p}{\tau'_p m^2}(l_\alpha l^\alpha)(1 - c_s^2)(u^2 - c_s^2 \Gamma_s^2). \tag{8.86}$$

Remark 19. Under the assumptions $\tau'_S > 0$ and $\tau'_p < 0$, from equations

(8.85a)–(8.85b) it follows that

$$|v_+| > (c_s)_+$$

for detonations and

$$|v_+| < (c_s)_+$$

for deflagrations.

Also, $|v_-| < (c_s)_-$ for strong detonations and $|v_-| > (c_s)_-$ for weak detonations.

Finally, $|v_-| < (c_s)_-$ for weak deflagrations and $|v_-| > (c_s)_-$ for strong deflagrations.

The stability analysis for combustion fronts is a rather difficult subject (Courant and Friedrichs, 1976; Liberman and Velikovich, 1985, and references therein) and in a relativistic framework it has not been attempted yet. A weak form of stability is the evolutionary condition which will be briefly touched upon in Section 8.5 for shock fronts. An extension of such an analysis to combustion fronts (Coll, 1976) leads to the conclusion that for evolutionary waves the interface always moves subsonically (or sonically) relative to the medium behind but, relative to the medium ahead, it may be either supersonic (giving a strong detonation) or subsonic (giving a weak deflagration). By using the method of characteristics (Whitham, 1974; Courant and Friedrichs, 1976) it is possible to see that, whereas the evolution of a strong detonation is determined uniquely by the fluid dynamical equations and the jump conditions, this is not the case for weak deflagrations. In fact, for deflagrations, the details of the combustion process regulate the development of the fluid dynamical flow.

Now a concrete example will be discussed.

This simple model, which may be relevant for cosmology and high-energy nuclear collisions, is based on the following assumptions:

(i) both the fluid before combustion and the fluid after combustion are in thermal equilibrium;
(ii) the temperature is so high that the fluids are highly relativistic.

This simple model can be applied in the following situations:

(a) Bubble growth during false vacuum decay at finite nonzero temperatures. In some scenarios for the early universe, the universe is trapped in a metastable phase. This occurs as a first order phase transition through the nucleation of bubbles in the metastable phase, inside of which is the stable phase. If the bubbles that are nucleated are above a critical size, they begin to grow, accelerating very rapidly. The force causing the acceleration originates in the process of converting the metastable phase to the stable one.

Therefore, there is a strong analogy between a combustion wave and the bubble growth. In both cases the front acceleration is due to the energy derived from converting the fluid outside the bubble into the fluid inside the bubble (Steinhardt, 1982).

(b) Phase transition from a quark-gluon plasma to hadron matter through nucleation of hadron bubbles. This is usually thought to be the last of the transitions and would have occurred when the temperature of the universe was around 200 MeV. At very high temperatures, strongly interacting matter behaves as a plasma of massless free quarks and gluons but as the temperature is gradually decreased below 1 GeV, a critical temperature is reached below which the confined hadron state is energetically favored. It has been speculated that the cosmological quark-hadron transition (assumed to be first order) could have led to formation of inhomogeneities of various kinds, which could be significant for dark matter problems, the formation of pregalactic structure, and possible subsequent effects on nucleosynthesis (Miller and Pantano, 1987, and references therein).

In an ultrarelativistic nuclear collision ($El_{ab} > 1$ Tev per nucleon) a transient state of a quark-gluon plasma could be produced. The phase transition to hadron matter could have observable consequences (Clare and Strottman, 1986; Cleymans et al., 1986).

Let B be the additional stored energy in the fluid ahead. Then

$$e = a_R T^4 + B, \tag{8.87}$$

$$\tilde{e} = a_R T^4. \tag{8.88}$$

The pressure p can be determined as follows.

The integrability condition for the first law of thermodynamics

$$T\,\mathrm{d}H = (e+p)\mathrm{d}V + V\,\mathrm{d}e$$

(where V is the volume and H the entropy) is

$$\frac{\partial}{\partial T}\left(\frac{e+p}{T}\right) = 4a_R T^2 \tag{8.89}$$

and inserting equation (8.87) into equation (8.89) gives

$$p = \tfrac{1}{3}a_R T^4 - B. \tag{8.90}$$

The state equation for the fluid before combustion is then

$$e = 3p + 4B \tag{8.91}$$

and after combustion

$$\tilde{e} = 3p. \tag{8.92}$$

In the examples discussed above these state equations have the following interpretation.

In the case of false vacuum decay, the quantity B represents the false vacuum energy.

In the case of the quark-gluon plasma phase transition to hadron matter, these state equations (apart from the degeneracy factors) are analogous to the bag model equation of state with B the bag constant (Clare and Strottman, 1986).

In order to discuss detonation and deflagration waves it is necessary to consider the jump conditions.

For the model under consideration it is not necessary to consider the mass conservation law (8.4).

However, if in the fluid there is a conserved charge such as baryon number (this would be the case if in the fluid there is a small amount of baryons which do not contribute in a significant way to the energy density and pressure), then equation (8.4) must be included. Also, it is necessary to include the mass conservation law (or the generalized conservation of a quantum number) if we want to specify the position of the combustion front with respect to "Eulerian observers." This is the case of Lagrangian coordinates calculations used by several authors (Miller and Pantano, 1987).

Notice that, in this case, for the fluid after combustion one can define $\tau(p, s)$ in the following way. The entropy density is [equation (2.17)]

$$\eta = \frac{4}{3} a_R \left(\frac{3p}{a_R} \right)^{3/4},$$

hence the specific entropy is η / ρ.

It follows that

$$\tau = \frac{4p}{\tilde{\rho}^2} = \frac{9}{4} \left(\frac{1}{a_R} \right)^2 S^2 \left(\frac{3p}{a_R} \right)^{-3/2}.$$

Therefore the conditions $\tau'_S > 0$, $\tau'_p < 0$, $\tau''_p > 0$ hold.

From equation (8.72), since

$$\tilde{j}^2 = 4p\tau, \quad \tilde{j}^2_+ = 4(p_+ + B)\tau_+,$$

one obtains the equation for the detonation adiabat $\mathscr{C}$

$$4p\tau - 4p_+\tau_+ - (p - p_+)(\tau_+ + \tau) = 4B\tau_+. \tag{8.93}$$

This equation can also be put in the form

$$(p + \tfrac{1}{3}p_+)(\tau - \tfrac{1}{3}\tau_+) = \tfrac{8}{9}\tau_+ p_+ + \tfrac{4}{3}B\tau_+. \tag{8.94}$$

It is apparent that $\mathscr{C}$ does not pass through the given initial point (τ_+, p_+). The Taub adiabat is obtained by putting $B = 0$ in equation (8.94). In Fig. 8.10 both the Taub and the detonation adiabat are plotted.

8.4. The jump conditions in relativistic magneto-fluid dynamics

The field equations of relativistic magneto-fluid dynamics in the form of conservation laws have been discussed already and are

$$\nabla_\mu(\rho u^\mu) = 0, \tag{8.95}$$

$$\nabla_\mu(b^\mu u^\nu - u^\mu b^\nu) = 0, \tag{8.96}$$

$$\nabla_\mu T^{\mu\nu} = 0, \tag{8.97}$$

with

$$T^{\mu\nu} = (e + p + |b|^2)u^\mu u^\nu + (p + \tfrac{1}{2}|b|^2)g^{\mu\nu} - b^\mu b^\nu.$$

The state equation is given in the form $\tau = \tau(p, S)$.

Proceeding as in Section 8.1, one obtains the following shock invariants (Lichnerowicz, 1967, 1971, 1976).

Fig. 8.10. Various detonation adiabats as functions of the "bag constant" B. Plot of p/p_+ versus τ/τ_+.

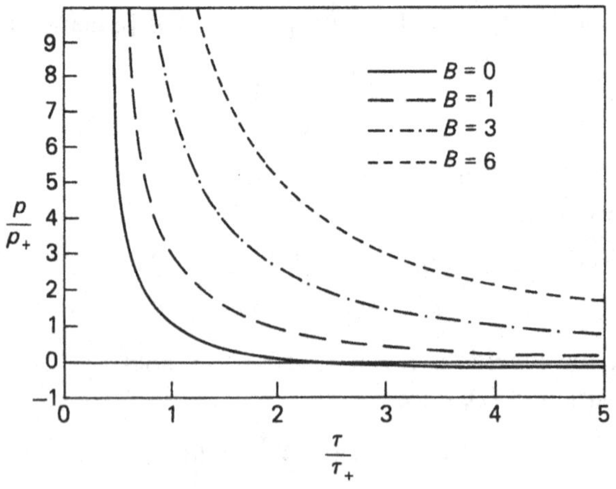

From equation (8.95) one has, as in Section 8.1,

$$m = \rho_- u^\mu_- l_\mu = \rho_+ u^\mu_+ l_\mu, \tag{8.98}$$

and from (8.96)

$$V^\mu = B_- u^\mu_- - \frac{m}{\rho_-} b^\mu_- = B_+ u^\mu_+ - \frac{m}{\rho_+} b^\mu_+, \tag{8.99}$$

where $B = b^\mu l_\mu$. Notice that $V^\mu l_\mu = 0$, that is, V^μ is tangent to the hypersurface Σ.

Finally, from equation (8.97) it follows that

$$\begin{aligned} W^\mu &= \left(\frac{\tau + |b|^2}{\rho^2}\right)_- m\rho_- u^\mu_- + q_- l^\mu - B_- b^\mu_- \\ &= \left(\frac{\tau + |b|^2}{\rho^2}\right)_+ m\rho_+ u^\mu_+ + q_+ l^\mu - B_+ b^\mu_+, \end{aligned} \tag{8.100}$$

where $q = p + \frac{1}{2}|b|^2$.

The contact discontinuity corresponds to $m = 0$, which yields

$$u^\mu_+ l_\mu = u^\mu_- l_\mu = 0. \tag{8.101}$$

Then from (8.99) it follows that

$$B_+ u^\mu_+ = B_- u^\mu_- \tag{8.102}$$

and from (8.100),

$$(q_+ - q_-) l^\mu - (B_+ b^\mu_+ - B_- b^\mu_-) = 0. \tag{8.103}$$

If $B_+ \neq 0$, then $B_- \neq 0$ and equation (8.102) implies that u^μ_+ and u^μ_- are collinear. Because they are both unitary and future directed vectors, they must coincide, that is,

$$[[u^\mu]] = 0, \tag{8.104}$$

hence also $[[B]] = 0$, and from (8.103),

$$[[q]] l^\mu - B_+ [[b^\mu]] = 0.$$

By contracting this latter equation with $[[b_\mu]]$, since $l^\mu [[b_\mu]] = [[B]] = 0$, it follows that

$$[[b^\mu]] = 0, \tag{8.105}$$

and from (8.103),

$$[[q]] = 0,$$

hence

$$[[p]] = 0. \tag{8.106}$$

Therefore, in this case, only $[[\rho]]$ is undetermined.

In the case $B_+ = 0$, then $B_- = 0$ and from equation (8.103) it follows that

$$[[p + \tfrac{1}{2}|b|^2]] = 0, \tag{8.107}$$

the other discontinuities being undetermined.

The case $m \neq 0$ corresponds to shock waves and will be treated in the following.

PROPOSITION 8.19. *The following four scalars are shock invariants,*

(i) $m = l_\alpha u^\alpha \rho$,

(ii) $H = \dfrac{B^2}{m^2} - \dfrac{|b|^2}{\rho^2}$,

(iii) $\mathscr{B} = fB$,

(iv) $\mathscr{I} = \alpha + \dfrac{q l_\mu l^\mu}{m^2}$,

with $\alpha = \tau - H$.

Proof. Since V^μ is invariant, so is $H = -\dfrac{1}{m^2} V^\mu V_\mu$.

Also, one has

$$b^\mu = -\frac{\rho}{m} V^\mu + \frac{B\rho}{m} u^\mu$$

and substituting into equation (8.100) yields

$$W^\mu = m\alpha\rho u^\mu + q l^\mu + \frac{B\rho}{m} V^\mu.$$

Then it is easy to check that $\mathscr{B} = fB = -\dfrac{1}{m} V_\mu W^\mu$ and that

$$\mathscr{I} = \frac{1}{m^2} l_\mu W^\mu. \qquad \text{Q.E.D.}$$

PROPOSITION 8.20. *Under the assumption* $l^\mu l_\mu \neq 0$ *the following scalars are also shock invariants:*

(v) $\mathscr{E} = q + \dfrac{\alpha m^2}{l_\mu l^\mu}$,

(vi) $K = f^2 + \dfrac{m^2\tau^2}{l_\alpha l^\alpha} + 2\tau\chi - H\chi$,

(vii) $\mathscr{L} = \chi\alpha^2$,

where $\chi = |b|^2 - \dfrac{m^2 H}{l_\alpha l^\alpha}$.

Proof. $\mathscr{E} = \dfrac{\mathscr{I} m^2}{l_\mu l^\mu}$ proves that $\mathscr{E}$ is invariant.

Let $H_{\mu\nu} = g_{\mu\nu} - \dfrac{l_\mu l_\nu}{l_\alpha l^\alpha}$ be the projection tensor onto Σ, and set

$$s_\mu = H_{\mu\nu}\rho u^\nu = \rho u_\mu - \frac{m l_\mu}{l_\alpha l^\alpha}, \quad s_\mu l^\mu = 0.$$

If we define

$$X^\mu = W^\mu - \frac{(W_\nu l^\nu) l^\mu}{l_\alpha l^\alpha},$$

then

$$X^\mu = \alpha m s^\mu + \frac{B\rho}{m} V^\mu.$$

Also,

$$s_\mu s^\mu = -\left(\rho^2 + \frac{m^2}{l_\alpha l^\alpha}\right)$$

and

$$s_\mu V^\mu = -\rho B,$$

whence $K = -\dfrac{1}{m^2} X_\mu X^\mu = \alpha^2\left(\rho^2 + \dfrac{m^2}{l_\alpha l^\alpha}\right) + \dfrac{2B^2\rho^2\alpha}{m^2} + \dfrac{B^2\rho^2 H}{m^2}$.

It remains to be shown that K coincides with the expression (vi). In fact,

$$\begin{aligned} K &= (\tau - H)^2\left(\rho^2 + \frac{m^2}{l_\alpha l^\alpha}\right) + \frac{2B^2\rho^2}{m^2}(\tau - H) + \frac{B^2\rho^2 H}{m^2} \\ &= \tau^2\left(\rho^2 + \frac{m^2}{l_\alpha l^\alpha}\right) + 2\tau\left(\frac{B^2\rho^2}{m^2} - H\left(\rho^2 + \frac{m^2}{l_\alpha l^\alpha}\right)\right) \\ &\quad - H\left(\frac{B^2\rho^2}{m^2} - H\left(\rho^2 + \frac{m^2}{l_\alpha l^\alpha}\right)\right). \end{aligned}$$

Finally,

$$\frac{B^2\rho^2}{m^2} - \left(\rho^2 + \frac{m^2}{l_\alpha l^\alpha}\right)\left(\frac{B^2}{m^2} - \frac{|b|^2}{\rho^2}\right) = |b|^2 - \frac{m^2}{l_\alpha l^\alpha} H = \chi.$$

In order to prove the invariance of $\mathscr{L}$ one notices that

$$\begin{aligned} KH &= H\left(\rho^2 + \frac{m^2}{l_\alpha l^\alpha}\right)\alpha^2 + \frac{B^2\rho^2}{m^2}H(2\alpha + H) \\ &= H\left(\rho^2 + \frac{m^2}{l_\alpha l^\alpha}\right)\alpha^2 + \frac{B^2\rho^2}{m^2}(\tau^2 - \alpha^2) \\ &= \left\{H\left(\rho^2 + \frac{m^2}{l_\alpha l^\alpha}\right) - \frac{B^2\rho^2}{m^2}\right\}\alpha^2 + \frac{f^2B^2}{m^2} \\ &= -\chi\alpha^2 + \frac{f^2B^2}{m^2}. \end{aligned}$$

Q.E.D.

Remark 20. It can be immediately checked that

$$K = f^2 + |b|^2\tau + |b|^2\alpha + \frac{m^2\alpha^2}{l_\alpha l^\alpha}. \tag{8.108}$$

In fact,

$$\begin{aligned} K &= f^2 + \frac{\tau^2 m^2}{l_\alpha l^\alpha} + 2\tau\chi - H\chi \\ &= f^2 + \frac{m^2}{l_\alpha l^\alpha}\tau^2 + 2\tau\left(|b|^2 - \frac{m^2}{l_\alpha l^\alpha}H\right) - H\left(|b|^2 - \frac{m^2 H}{l_\alpha l^\alpha}\right) \\ &= f^2 + |b|^2(2\tau - H) + \frac{m^2}{l_\alpha l^\alpha}(\tau^2 - 2\tau H + H^2) \\ &= f^2 + |b|^2\tau + |b|^2\alpha + \frac{m^2}{l_\alpha l^\alpha}\alpha^2. \end{aligned}$$

PROPOSITION 8.21. *One has*

(i) $(l^\alpha l_\alpha)\chi \geq 0$;
(ii) *if* $l^\alpha l_\alpha = 0$, $\chi = 0$, *then* l_α *belongs to the two-plane spanned by* u^α, b^α, *and vice versa;*
(iii) $l^\alpha l_\alpha \leq 0$ *implies* $H \leq 0$, *hence* $\alpha > 0$.

Proof. One has

$$(l^\alpha l_\alpha)\chi = (l^\alpha l_\alpha)|b|^2 - B^2 + \frac{|b|^2 m^2}{\rho^2}.$$

In an orthonormal frame $\{u^\mu, \hat{b}^\mu, \underset{(2)}{e}{}^\mu, \underset{(3)}{e}{}^\mu\}$, as defined in the proof of

Proposition 2.2,

$$l^\mu = -\frac{m}{\rho}u^\mu + \frac{B}{|b|}\hat{b}^\mu + c_2 \underset{(2)}{e}{}^\mu + c_3 \underset{(3)}{e}{}^\mu,$$

hence

$$(l^\alpha l_\alpha)\chi = |b|^2(c_2^2 + c_3^2) \geq 0$$

and (i) is proved.

Also, if $l^\alpha l_\alpha \neq 0$ and $\chi = 0$, then $c_2 = c_3 = 0$ and $l^\mu = -\frac{m}{\rho}u^\mu + \frac{B}{|b|}\hat{b}^\mu$, and viceversa.

Finally, if $l^\alpha l_\alpha \leq 0$, then $\frac{m^2}{\rho^2} \geq \frac{B^2}{|b|^2} + c_2^2 + c_3^2 \geq \frac{B^2}{|b|^2}$, whence

$H \leq 0$, and since $\tau > 0$, (iii) follows. Q.E.D.

Remark 21. The two thermodynamical variables (p, S) and the three scalars $|b|^2, B, u^\alpha l_\alpha$, obey the five scalar laws (i), (ii) (of Proposition 8.19), and (v), (vi), (vii) (of Proposition 8.20). The other shock invariants (iii) and (iv) (of Proposition 8.19) are obviously functionally dependent on these.

From the shock invariants (v), (vi), (vii) it is possible to obtain χ (hence $|b|^2$) and the two thermodynamic variables, whereas from (i), (ii) the normal components of u^α and b^α are obtained, that is, $u^\alpha l_\alpha$, $b^\alpha l_\alpha$. In order to get the tangential components of u^α and b^α, one proceeds as follows. From the invariance of V^μ and W^μ [equations (8.99)–(8.100)] one obtains for the tangential components $''u^\mu$, $''b^\mu$ of u^μ and b^μ,

$$B_- {''u^\mu_-} - \frac{m}{\rho_+}{''b^\mu_-} = B_+ {''u^\mu_+} - \frac{m}{\rho_+}{''b^\mu_+}, \tag{8.109}$$

$$\left(\tau + \frac{|b|^2}{\rho^2}\right)_- m\rho_- {''u^\mu_-} - B_- {''b^\mu_-} = \left(\tau + \frac{|b|^2}{\rho^2}\right)_+ m\rho_+ {''u^\mu_+} - B_+ {''b^\mu_+}. \tag{8.110}$$

The determinant of this linear system in the unknowns $''u^\mu_-$, $''b^\mu_-$ is $m^2\alpha_-$. Therefore, if $\alpha_- \neq 0$, equations (8.109)–(8.110) determine $''u^\mu_-$, $''b^\mu_-$ as functions of the state ahead.

Remark 22. The Alfvén waves are characterized by $A^2 = 0$, $A = Ea^2 - B^2$. It is immediately seen that

$$A = m^2\alpha. \tag{8.111}$$

Remark 23. If at some point $x \in \Sigma$, one has $\alpha_- = 0$, from the invariant $\mathscr{L}$ one has

$$\chi_+\alpha_+^2 = 0,$$

whence

$$\alpha_+ = 0 \quad \text{or} \quad \chi_+ = 0.$$

The *Alfvén shock* is defined to be the case $\alpha_- = \alpha_+ = 0$.

When $\alpha_+ \neq 0$, $\chi_+ = 0$, $\alpha_- = 0$, or $\alpha_+ = 0$, $\chi_- = 0$, $\alpha_- \neq 0$ one has the *singular shocks*. Alfvén and singular shocks are discussed by Lichnerowicz (1967, 1971).

For the rest of this section it will be assumed that $\alpha_-\alpha_+ \neq 0$.

PROPOSITION 8.22. *Let $Y_\pm = (\tau_\pm, p_\pm, u^\mu_\pm, b^\mu_\pm)$ be the states ahead and behind the shock, respectively. Then, under the assumption $\alpha_-\alpha_+ \neq 0$, the following relationship holds:*

$$f_-^2 - f_+^2 - (\tau_- + \tau_+)(p_- - p_+) + \tfrac{1}{2}(\tau_- - \tau_+)\left(\chi_+ + \chi_- - 2\chi_+\frac{\alpha_+}{\alpha_-}\right) = 0. \tag{8.112}$$

This equation defines the Lichnerowicz adiabat $\mathscr{C}$ which is the extension to relativistic magneto-fluid dynamics of the classical Hugoniot adiabat.

Proof. From Proposition (8.20), the invariance of K and $\mathscr{E}$, it follows that

$$f_-^2 - f_+^2 - \frac{(\tau_+ + \tau_-)(\tau_+ - \tau_-)}{\alpha_- - \alpha_+}(q_+ - q_-) + (\tau_- + \alpha_-)\chi_-$$
$$- (\tau_+ + \alpha_+)\chi_+ = 0.$$

Also, from the invariance of H and the definition of χ,

$$|b|_-^2 - |b|_+^2 = \chi_- - \chi_+,$$

hence

$$f_-^2 - f_+^2 - (\tau_+ + \tau_-)(p_- - p_+) - \tfrac{1}{2}(\tau_+ + \tau_-)(\chi_- - \chi_+) + \tau_-\chi_-$$
$$+ \alpha_-\chi_- - \tau_+\chi_+ + \alpha_+\chi_+ = 0,$$

which also is written as

$$f_-^2 - f_+^2 - (\tau_- + \tau_+)(p_- - p_+) + \tfrac{1}{2}(\tau_- - \tau_+)(\chi_- + \chi_+)$$
$$+ \chi_-\alpha_- - \chi_+\alpha_+ = 0.$$

From the invariance of $\mathscr{L}$ one has

$$\chi_-\alpha_- = (\chi_+\alpha_+)\frac{\alpha_+}{\alpha_-}$$

and equation (8.112) follows by using the identity

$$(\tau_+ - \tau_-)\frac{\chi_+\alpha_+}{\alpha_-} = (\alpha_+ - \alpha_-)\frac{\chi_+\alpha_+}{\alpha_-}. \qquad \text{Q.E.D.}$$

Let $\mathbf{Y}_+ = (\tau_+, p_+, u^\mu_+, b^\mu_+)$ and $\mathbf{Y} = (\tau, p, u^\mu, b^\mu)$.

In the space of the vectors $\mathbf{Y}$, let N denote all possible states satisfying the conditions

$$H(\mathbf{Y}) = H(\mathbf{Y}_+) = H = \text{const.}, \quad \mathscr{L}(\mathbf{Y}) = \mathscr{L}(\mathbf{Y}_+) = \mathscr{L} = \text{const.}, \tag{8.113}$$

with

$$(b^\alpha l_\alpha)(b^\alpha_+ l_\alpha) \geq 0.$$

In the set N, for $\alpha \neq 0$, one has

$$\chi = \frac{\mathscr{L}}{(\tau - H)^2} = \frac{\chi_+\alpha_+}{(\tau - H)^2}.$$

Let

$$\bar{q} = p + \tfrac{1}{2}\chi. \tag{8.114}$$

Then, in N,

$$\bar{q} = p(\tau, S) + \frac{\mathscr{L}}{2(\tau - H)^2}, \tag{8.115}$$

where $p(\tau, S) \geq 0$ is obtained by inverting the state equation $\tau = \tau(p, S)$.

Let $\Pi = \{(\tau, \bar{q}),\ \tau \geq 0\}$. In the set N, a thermodynamical state of the medium, defined by (τ, p), determines a point $(\tau, \bar{q}) \in \Pi$ such that

$$\bar{q} \geq \frac{\mathscr{L}}{2(\tau - H)^2}. \tag{8.116}$$

If $H > 0$, this inequality defines two connected convex regions $\mathscr{R}$ of the half plane Π. If $H \leq 0$, since $\tau \geq 0$, there is only one connected convex region $\mathscr{R}$.

In $\mathscr{R}$ a point $Z = (\tau, \bar{q})$ determines a thermodynamical state of the fluid.

On $\mathscr{R}$ one defines the Lichnerowicz function $\mathscr{H}(Z_+, Z)$ as

$$\mathscr{H}(Z_+, Z) = f^2 - f_+^2 - (\tau + \tau_+)(p - p_+) + \tfrac{1}{2}(\tau - \tau_+)\left(\chi_- + \chi_+ - \frac{2\chi_+\alpha_+}{\alpha}\right). \tag{8.117}$$

Obviously, $\mathscr{H}(Z_+, Z_+) = 0$ and the equation of the Lichnerowicz adiabat $\mathscr{C}$ reads

$$\mathscr{H}(Z_+, Z_-) = 0.$$

Remark 24. The differential of $\mathscr{H}$ in $\mathscr{R}$ is given by

$$\mathrm{d}\mathscr{H} = 2fT\,\mathrm{d}S + (\tau - \tau_+)\mathrm{d}\bar{q} - (\bar{q} - \bar{q}_+)\mathrm{d}\tau, \tag{8.117'}$$

and can be checked by direct calculation using the fact that, on $\mathscr{R}$, $\mathrm{d}\alpha = \mathrm{d}\tau$.

Remark 25. Let Λ be the straight line of $\mathscr{R}$ joining Z_+ to Z,

$$\bar{q} - \bar{q}_+ = M(\tau - \tau_+). \tag{8.118}$$

Then, one has

$$\mathrm{d}\mathscr{H} = 2fT\,\mathrm{d}S + (\tau - \tau_+)^2\mathrm{d}M. \tag{8.119}$$

PROPOSITION 8.23. *Under the following assumptions on the state equation* $\tau = \tau(p, S)$,

$$\tau'_S \neq 0, \tag{8.120}$$

$$\tau'_p < 0, \tag{8.121}$$

the hypersurface Σ *is timelike.*

Proof. This proposition is the exact analog of Proposition (8.1).

The proof runs along similar lines. It is based on the following equation which holds on the straight line Λ joining Z_+ and Z_-,

$$\bar{q}_- - \bar{q}_+ = \frac{m^2}{l_\alpha l^\alpha}(\tau_- - \tau_+). \tag{8.122}$$

Then

$$\alpha\tau'_p\mathrm{d}S = -\frac{N_4(l^\mu)}{m^2 l^\mu l_\mu}\mathrm{d}\tau = \frac{N_4(l^\mu)}{m^2 l^\mu l_\mu}\mathrm{d}q, \tag{8.123}$$

where, from equation (2.82),

$$N_4(l^\mu) = \eta(e'_p - 1)\frac{m^4}{\rho^4} - (\eta + e'_p|b|^2)\frac{m^2}{\rho^2}l^\mu l_\mu + B^2 l^\mu l_\mu. \qquad \text{Q.E.D.}$$

The theory of shock waves in relativistic magneto-fluid dynamics can now be developed in close parallelism with that of Section 8.1. Only the main results will be stated here. The proofs run along the same lines as the analogous proofs of Section 8.1. The details of the calculations can be found in Lichnerowicz (1967, 1971, 1976).

In particular, Proposition 8.4 holds also for shock waves in relativistic magneto-fluid dynamics (provided that one avoids Alfvén and singular shocks, which is already assumed in this section).

One can also prove the following proposition, which closely parallels Remark 8.

PROPOSITION 8.24. *Under the assumptions:*

(i) $\tau_S' > 0$,
(ii) $\tau_p' < 0$,
(iii) $\tau_p'' > 0$ (the Weyl condition),

the normal speeds $|v_+|, |v_-|$ *of the fluid with respect to* Σ *are such that:*
(1) *if* $0 < \alpha_- < \alpha_+$, *then*

$$\begin{aligned} &(v_s)_+ < (v_A)_+ < (v_f)_+ < |v_+|, \\ &(v_s)_- < (v_A)_- < |v_-| < (v_f)_-, \end{aligned} \tag{8.124a}$$

corresponding to the fast magnetoacoustic shock.
(2) *if* $\alpha_- < \alpha_+ < 0$, *then*

$$\begin{aligned} &(v_s)_+ < |v_+| < (v_A)_+ < (v_f)_+, \\ &|v_-| < (v_s)_- < (v_A)_- < (v_f)_-, \end{aligned} \tag{8.124b}$$

corresponding to the slow magnetoacoustic shock.

Remark 26. Along the lines of Proposition 8.4 it can be proved (Lichnerowicz, 1976) that, under the assumptions $\tau_S' > 0$, $\tau_p' < 0$, $\tau_p'' > 0$ and the entropy growth condition $S_- - S_+ > 0$, one has

$$p_- > p_+, \quad f_- > f_+, \quad \tau_- < \tau_+, \quad \rho_- > \rho_+.$$

The existence of the fast and slow magnetoacoustic shocks is established by the following proposition.

PROPOSITION 8.25. *Under the assumptions* (i), (ii), (iii) *of Proposition* 8.24, *if*:

(i) $\tau = \tau(p, S)$ *is invertible giving* $S = S(p, \tau)$, *and*
(ii) *there exists* $\bar{\tau} \geq 0$ *such that* $\lim_{\tau \to \bar{\tau}} p(\tau, s) = \infty$, *then to any state* $\mathbf{Y}_+$ *such that* $\alpha_+ N_4(l^\mu)_+ > 0$ *there corresponds a unique nontrivial solution* $\underset{\sim}{\mathbf{Y}}_-$ *of the jump conditions such that* $\alpha_+\alpha_- > 0$.

Remark 27. As will be seen in the next section, the condition $\alpha_+\alpha_- > 0$ is necessary in order for the shock to be evolutionary, that is, such that the initial value problem for the linearized perturbations is well posed.

In the physically important case of a Synge gas, explicit solutions of the jump conditions can be obtained as follows (Majorana and Anile, 1987).

We shall consider propagating shock waves, for which $\alpha_+ > 0$. It is convenient to introduce the angle φ between the magnetic field and the wavefront direction of propagation, as measured in the local rest frame of the fluid ahead of the shock. In this frame the components of b^α_+ and l_α are

$$|b|_+(0, \cos\varphi, \sin\varphi, 0)$$

and

$$\Gamma_+(|v_+|, 1, 0, 0),$$

where $|v_+|$ is the magnitude of the normal velocity of the shock front with respect to the fluid ahead. Then

$$B = \Gamma_+|b|\cos\varphi.$$

It is convenient to introduce a frame $\mathscr{I}$ such that

$$l^\mu = (0, 1, 0, 0), \quad u^\mu_+ = \Gamma_+(1, |v_+|, 0, 0)$$

in which the shock front is momentarily at rest. By an appropriate rotation of the 1-axis we can set $b^3_+ = 0$, and then the invariance of V^α, W^α, with $\alpha = 3$, yields

$$B_- u^3_- - ab^3_- = 0, \tag{8.125}$$

$$n_- G_- a_- u^3_- + |b|^2_- a_- u^3_- - B_- b^3_- = 0, \tag{8.126}$$

where $a = u_\mu l^\mu$, $n = \rho/\rho_+$.

We shall assume that the wavefront is not Alfvénic with respect to the fluid behind the shock, that is, $\alpha_- \neq 0$. Then, from equations (8.125)–(8.126) it follows that

$$u^3_- = b^3_- = 0.$$

The case of normal propagation, $\sin\varphi = 0$, is easy to study. In this case, in the local rest frame of the fluid ahead of the shock, the 2-component of b^α_+ is vanishing. The transformation to the frame $\mathscr{I}$ is achieved by a rotation of the 0- and 1-axis and therefore, also in this frame the 2-component of b^α_+ vanishes. Therefore, in the frame $\mathscr{I}$, the invariance of V^α, W^α with $\alpha = 2$, gives, as above,

$$u^2_- = b^2_- = 0.$$

Hence, in the frame $\mathscr{I}$ we have

$$b^\alpha_\pm = b^1_\pm(|v_\pm|, 1, 0, 0), \quad u^\alpha_\pm = \Gamma_\pm(1, |v_\pm|, 0, 0).$$

Then, from the invariants H and $\mathscr{B}$ of Proposition 8.19 we have

$$[[|b|^2]] = 0, \tag{8.127}$$

$$[[\Gamma G]] = 0. \tag{8.128}$$

Finally, the invariant $\mathscr{E}$ of Proposition 8.20 gives

$$\left[\left[na^2G + \frac{n}{z}\right]\right] = 0. \tag{8.129}$$

Remark 28. Equations (8.128)–(8.129), together with the mass conservation law

$$[[n\Gamma v]] = 0,$$

are the same jump conditions as in the pure gas dynamical case. Therefore the magnetic field decouples from the fluid quantities and one can then use the procedure expounded in Section 8.2.

In the general case one must resort to a numerical solution of the jump conditions. A detailed analysis has been performed by Majorana and Anile (1987). The results are plotted in Figs. 8.11 – 8.17, for $\rho_- = 1$.

Fig. 8.11. The quantity z_+ as a function of $|v_-|$ for $z_- = 10$, $|b_-| = 1, \varphi = 0°$.

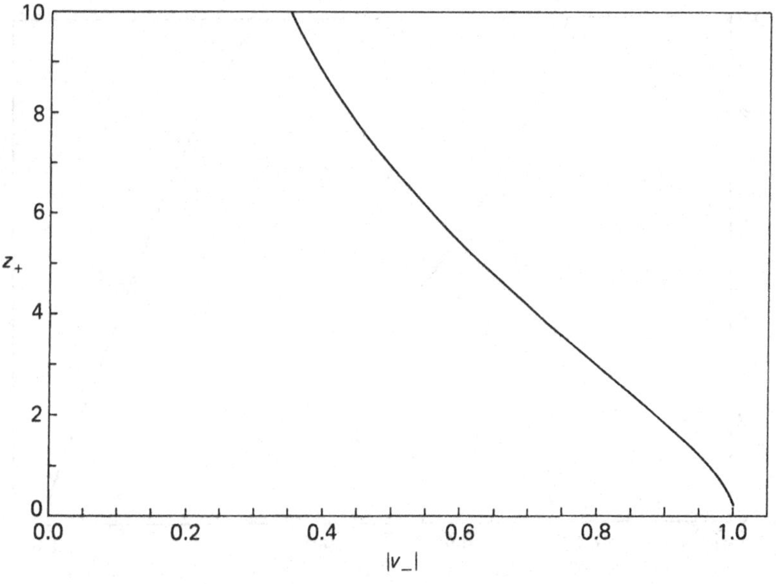

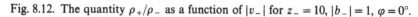

Fig. 8.12. The quantity ρ_+/ρ_- as a function of $|v_-|$ for $z_- = 10$, $|b_-| = 1$, $\varphi = 0°$.

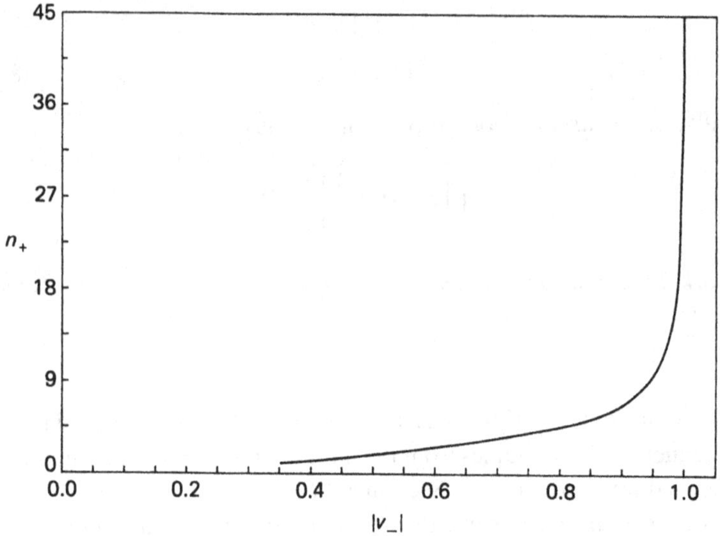

Fig. 8.13. The quantity z_+ as function of $|v_-|$ for $z_- = 10$, $|b_-| = 1$, $\varphi = 45°$. The two branches correspond to fast and slow magnetoacoustic shocks.

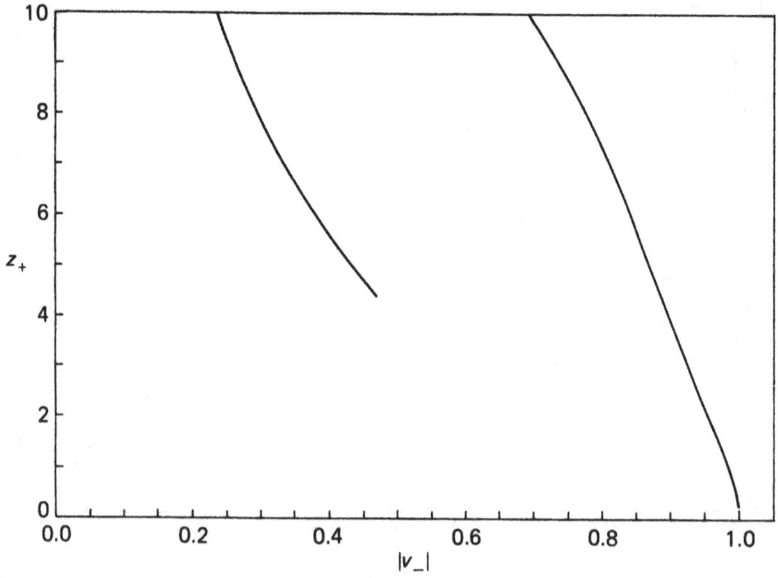

Fig. 8.14. The quantity ρ_+/ρ_- as a function of $|v_-|$ for $z_- = 10$, $|b_-| = 1$, and $\varphi = 45°$.

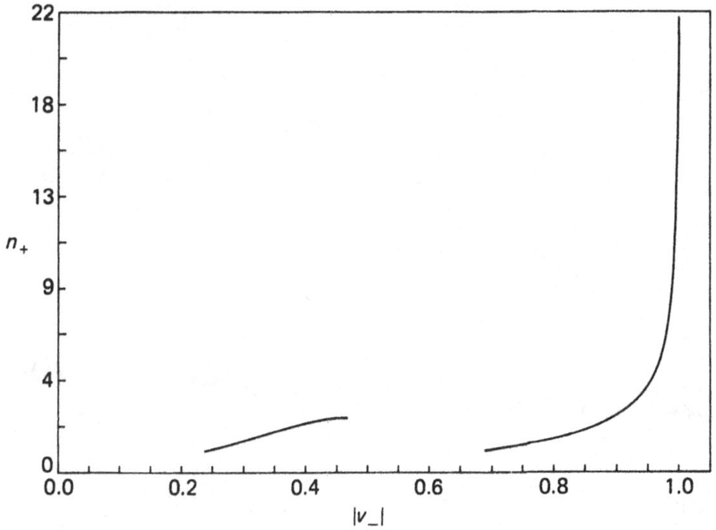

Fig. 8.15. The quantity $|b_+|$ as a function of $|v|$ for $z_- = 10$, $|b_-| = 1$, $\varphi = 45°$.

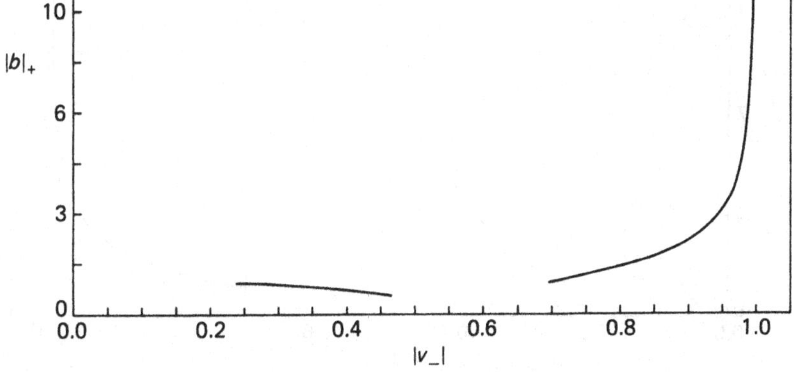

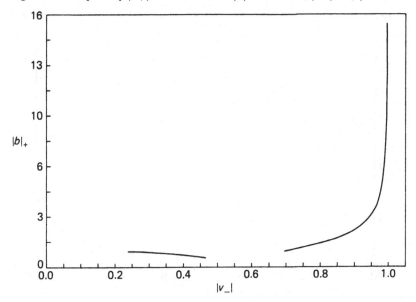

Fig. 8.16. The quantity z_+ as a function of $|v_-|$ for $z_- = 10$, $|b_-| = 1$, $\varphi = 90°$.

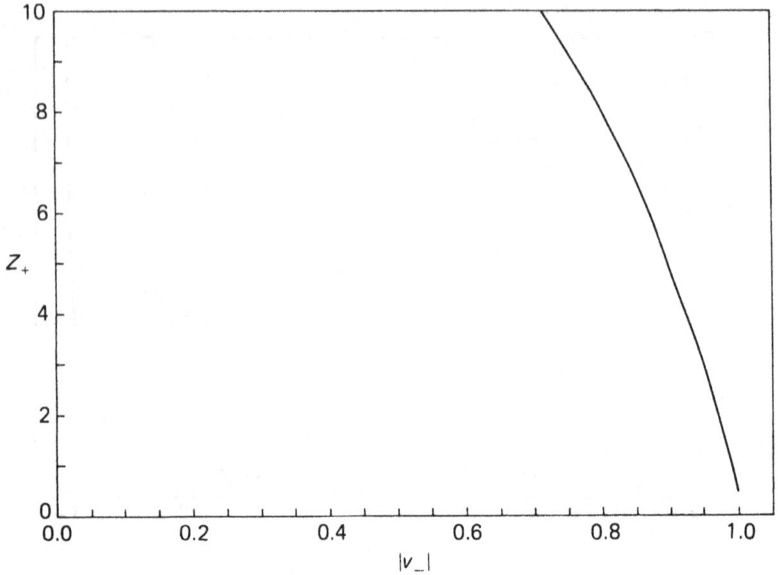

Fig. 8.17. The quantity ρ_+/ρ_- as a function of $|v_-|$ for $z_- = 10$, $|b_-| = 1$, $\varphi = 90°$.

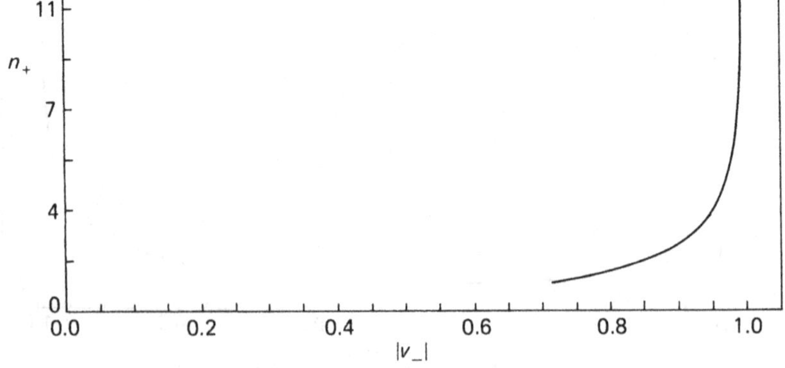

Remark 29. We notice that for $\varphi = 0$ a discontinuity appears in the quantities behind the shock as a function of the angle. In fact, since we are in the case $\varphi = 0$, the fluid quantities obey the purely gas-dynamical jump relations. Then a fast magnetoacoustic shock that for $\varphi \neq 0$ is weak (meaning that its propagation speed v_Σ differs slightly from the fast magnetoacoustic speed) can become strong for $\varphi = 0$, because in this case v_Σ can be much larger than the speed of sound c_s. In fact, in the neighborhood of $\varphi = 0$ there exist three solutions of the jump conditions, but two of them do not satisfy the evolutionary condition $\alpha_+ \alpha_- > 0$. For $\varphi \neq 0$ the shock corresponding to the admissible solution is weak. If this solution is followed for $\varphi \to 0^+$, we obtain a solution satisfying the jump conditions but which is not evolutionary (Majorana and Anile, 1987).

8.5. Evolutionary conditions for relativistic shock waves

The shock waves which have been discussed in the previous sections, in order to be physically admissible, must have some property of structural stability.

A very weak form of structural stability for a shock wave is to be evolutionary.

A shock will be said to be evolutionary if the resolution of the interaction of an incident small disturbance wave and the shock into outgoing disturbance waves and a disturbed shock motion comprise a well-posed problem (in the sense of Hadamard) (Jeffrey, 1976).

For a shock wave of arbitrary profile, in general, it is very difficult to check the evolutionary condition.

This problem has been solved in the case of step-shocks, that is, shock waves connecting two constant states; and for one dimensional perturbations.

A necessary condition in order for a step-shock to be evolutionary is the Lax condition (Jeffrey, 1976) that the number of outgoing waves be $N - 1$ (where N is the order of the system). The equations of one dimensional relativistic fluid dynamics can be deduced from equations (2.4) and (2.9)–(2.10). Let $\mathcal{M}$ denote Minkowski space and (x, y, z, t) denote inertial coordinates. The four-velocity of the fluid is $u^\alpha = \Gamma(1, v, 0, 0)$, $\Gamma = (1 - v^2)^{-1/2}$. Let

$$\mathbf{U} = \begin{bmatrix} \rho \\ u \\ p \end{bmatrix}, \tag{8.130}$$

where $u = \Gamma v$. By introducing the field vector $\mathbf{U}$ the equations of one dimensional relativistic fluid dynamics can be taken to be

$$\mathscr{A}^0 \partial_t \mathbf{U} + \mathscr{A} \partial_x \mathbf{U} = 0, \tag{8.131}$$

where

$$\mathscr{A}^0 = \begin{bmatrix} \Gamma & \frac{\rho}{\Gamma} u & 0 \\ 0 & \Gamma(e+p) & \Gamma \\ -\Gamma a^2 & 0 & \Gamma \end{bmatrix}, \quad \mathscr{A} = \begin{bmatrix} u & \rho & 0 \\ 0 & u(e+p) & 1+u^2 \\ -a^2 u & 0 & \Gamma \end{bmatrix},$$

with

$$a^2 = \left(\frac{\partial p}{\partial \rho}\right)_S = \frac{c_s^2(e+p)}{\rho}.$$

In this case $N = 3$ and therefore the number of outgoing waves must be 2.

The characteristic speeds are

$$\lambda_1 = \frac{v - c_s}{1 - v c_s},$$

$$\lambda_2 = v,$$

$$\lambda_3 = \frac{v + c_s}{1 + v c_s},$$

$$\lambda_1 < \lambda_2 < \lambda_3.$$

Simple reasoning, paralleling classical reasoning (Jeffrey, 1976), then shows that, for a right propagating shock, the Lax condition selects the acoustic shock, satisfying the supersonic-subsonic relationship

$$\begin{aligned} &v_b < v_\Sigma < \frac{v_b + (c_s)_-}{1 + v_b (c_s)_-}, \\ &v_\Sigma > (c_s)_+ . \end{aligned} \tag{8.132}$$

Under the assumptions on the state equation discussed in Section 8.1 (Proposition 8.4 and Remark 8) the conditions (8.132) are automatically satisfied.

A similar analysis can be performed in the case of relativistic magneto-fluid dynamics.

Lichnerowicz (1967, 1971) has proved that *the evolutionary condition selects only the fast and slow magnetoacoustic shocks.*

In particular, the singular shocks are not evolutionary.

However, care must be exercised when drawing the conclusion that singular shocks are not physically realizable. In fact, when the shock speed coincides with one of the characteristic speeds, taking into account dissipative effects (heat conduction, viscosity, resistivity, which hitherto have been neglected) can change significantly the stability analysis (Liberman and Velikovich, 1985).

Hitherto, this approach has not been developed in a relativistic framework.

The evolutionary condition selects a subclass of discontinuity waves (shocks and contact discontinuities). These, together with simple waves, form the basis for the solution of the shock-tube problem. In relativistic fluid dynamics the shock-tube problem can be easily solved in a manner exactly analogous to the nonrelativistic case (Thompson, 1986) once explicit expressions for the simple wave solutions and the shock solutions are available. As such the shock-tube problem solution provides an excellent benchmark against which to test numerical codes (Hawley et al., 1984a). The shock-tube problem for relativistic magneto-fluid dynamics has not been tackled yet.

9

Propagation of relativistic shock waves

9.0. Introduction

The propagation of relativistic shocks is of great relevance to several astrophysical problems. Obviously, it is a central feature of the standard model of supernova explosions (Chevalier, 1981; Shapiro and Teukolsky, 1983). Also, in some models of jet emission in extragalactic radio sources and quasars, relativistic shocks propagate along the jet (Appl and Camezind, 1988, and references therein). In theories of galaxy formation relativistic shocks may occur either as nonlinear developments of adiabatic perturbations (Peebles, 1980) or, being caused by some explosion in the pregalactic material, as seeds for the formation of protogalactic structure (Ikeuchi, Tomisaka, and Ostriker, 1983).

The evolution of relativistic shock waves hitherto has not been studied as extensively as in the nonrelativistic case. Some results have been obtained for the case of self-similar solutions representing blast waves by, among others, Johnson and McKee (1971), Eltgroth (1971, 1972), Blandford and McKee (1976), Deb Ray and Chakraborty (1978), Bogoyavlenski (1978), Moschetti (1987), and Anile, Moschetti, and Bogoyavlenski (1987), and, by using approximate methods, by Ishizuka (1980) and Ishizuka and Sakashita (1980).

For an arbitrary shock wave the problem of its evolution is a fundamental one. The damping of a plane relativistic shock was first investigated by Liang (1977a) and Liang and Baker (1977), who suggested the interesting effect that, at variance with the nonrelativistic case, the damping time would tend to infinity the stronger the shock and this could have important consequences for the nonlinear damping of primeval adiabatic perturbations. However, their treatment of the shock damping was rather crude and unsatisfactory. The problem was taken up again by Anile, Miller, and Motta (1980), who proved the Liang and Baker conjecture rigorously using methods of the theory of singular surfaces. Subsequently, Anile et al. (1983) and Lanza et al. (1982, 1985) performed numerical calculations confirming the previous results. Independent calculations have also been done by Granik (1983). The propagation of a curved relativistic shock has been investigated by Singh (1984) and by Anile (1984) who found that, under certain circumstances, a similar effect on the damping time is also present.

For weak shocks a theory which closely parallels the nonrelativistic one can be developed (Liang, 1977a; Anile, 1983).

The plan of the chapter is the following. In Section 9.1 we investigate the damping of a plane relativistic shock propagating into a constant state. The initial damping time is defined and it is proved that, in the limit of an extremely relativistic shock, it tends to infinity. In Section 9.2 the evolution of a curved shock propagating into an arbitrary state is first investigated and some general results are given. When the shock propagates into an arbitrary static medium in special relativity, an analogous effect on the damping time is found. In Section 9.3 the propagation of relativistic weak shocks is studied by employing the method of Hunter and Keller (1983).

9.1. Damping of plane one dimensional shock waves in a relativistic fluid

The damping of relativistic shock waves presents unexpected features which are at striking variance with the nonrelativistic case. In order to elucidate these novel features it is appropriate to begin with the simplest possible case, that is, with a barotropic fluid and a plane one dimensional shock propagating into a constant state.

Let $\mathscr{M}$ be Minkowski space-time, (t, x, y, z) be inertial coordinates, and $u^\mu = \Gamma(1, v, 0, 0)$ be the fluid's four-velocity. Then, for a barotropic fluid for which $p = c_s^2 e$ ($c_s^2 = \text{const.}$), the conservation of energy (2.9) and conservation of momentum (2.10) yield

$$\partial_t \mathbf{U} + \mathscr{A}\,\partial_x \mathbf{U} = 0, \tag{9.1}$$

where

$$\mathbf{U} = \begin{pmatrix} v \\ p \end{pmatrix}$$

and

$$\mathscr{A} = \frac{1}{c_s^2 v^2 - 1}\begin{bmatrix} v(c_s^2 - 1) & -\dfrac{1}{(p+e)\Gamma^4} \\ -(p+e)c_s^2 & v(c_s^2 - 1) \end{bmatrix}.$$

Let $\Sigma: x = x_s(t)$ be the path of a shock wave, and we assume the fluid ahead to be at rest and uniform. Taking the jump of equation (9.1) across Σ

and using the kinematic compatibility relations in the form (3.23) gives

$$\mathscr{D}[[\mathbf{U}]] + (\mathscr{A}_- - v_\Sigma I)\partial_x[[\mathbf{U}]] = 0, \tag{9.2}$$

where $\mathscr{D}$ is the Thomas derivative, $v_\Sigma = \dfrac{\mathrm{d}x_s}{\mathrm{d}t}$ is the shock speed and I is the identity matrix.

By using the solutions of the jump conditions for v_Σ [equation (8.29)] and for $[[p]]$ [equation (8.30)], and substituting into equation (9.2), one finds after some manipulations

$$\mathscr{D}[[v]] + \frac{\Phi^2([[v]]) - c_s^2(1-[[v]]^2)^2}{\Phi([[v]]) + \left(\dfrac{c_s^2}{1+c_s^2}\right)H([[v]])}(c_s[[v]]^2 - 1)^2[[\partial_x v]] = 0, \tag{9.3}$$

where

$$\Phi([[v]]) = v_\Sigma(1 - c_s^2[[v]]^2) - [[v]](1 - c_s^2),$$

$$H([[v]]) = \frac{1-[[v]]^2}{r([[v]])v_\Sigma}\{-2[[v]]r([[v]])v_\Sigma + (1-[[v]]^2) \times (v_\Sigma' r([[v]]) - v_\Sigma r'([[v]])\},$$

with

$$r([[v]]) = v_\Sigma(1 + c_s^2[[v]]^2) - [[v]](1 + c_s^2)$$

and a prime denoting the derivative with respect to the argument $[[v]]$.

It can be immediately checked that, when $c_s = 1$, one has $v_\Sigma = 1$ and $\mathscr{D}[[v]] = 0$. Therefore, in an extreme fluid the shock amplitude is constant along the shock line.

For a given shock the *initial damping time* τ_D can be defined as

$$\tau_D = \left|\frac{[[v]]}{\mathscr{D}[[v]]}\right|_{t=0}. \tag{9.4}$$

From equation (9.3) it is possible to compute τ_D exactly once the initial velocity profile is assigned.

For a given initial $[[\partial_x v]]$, and an extremely relativistic shock, $v \to 1$, one has

$$\Phi([[v]]) = \left(\frac{4c_s^2}{1+c_s^2}\right)(1-[[v]]) + O(1-[[v]]^2),$$

$$H([[v]]) = 2(1-[[v]]) + O(1-[[v]]^2).$$

Hence

$$\tau_D = O((1-[[v]]^{-1})) \quad \text{as} \quad [[v]] \to 1, \tag{9.5}$$

which shows that for an extreme relativistic fluid, $\tau_D \to \infty$ as $v \to 1$.

A more quantitative result can be obtained by choosing a specific initial velocity profile, for example, a triangular one,

$$v(0,x) = \begin{cases} x[[v]]_{t=0}, & x \in [0,1], \\ 0, & x \notin [0,1], \end{cases}$$

in a radiation fluid, $c_s = \dfrac{1}{\sqrt{3}}$.

In Fig. 9.1 the quantity $Q = \tau_D^{-1}$ (the shock damping rate) is plotted as a function of $[[v]]$.

Notice that the maximum value of Q occurs at the mildly relativistic $[[v]] \simeq 0.4$.

The phenomenon which we have just investigated is a novel feature of relativistic fluid dynamics (Anile et al., 1980).

In classical fluid dynamics it can be easily checked that Q increases with the shock amplitude.

A tentative explanation of the relativistic feature could be the following. The condition for the shock to damp is that the shock speed v_Σ is less than the speed of sound c_b behind the shock (each measured in the rest frame ahead of the shock). As $[[v]] \to 1$ both v_Σ and c_b tend to the speed of light and therefore the damping will tend to vanish. In order to assess the importance of such an effect on the nonlinear damping of primeval adiabatic perturbations in the radiation era of the universe it is necessary to investigate the influence of an expanding background medium.

Fig. 9.1. The quantity $Q \equiv \tau_D^{-1}$ (the shock damping rate) is plotted as a function of $[[v]]$.

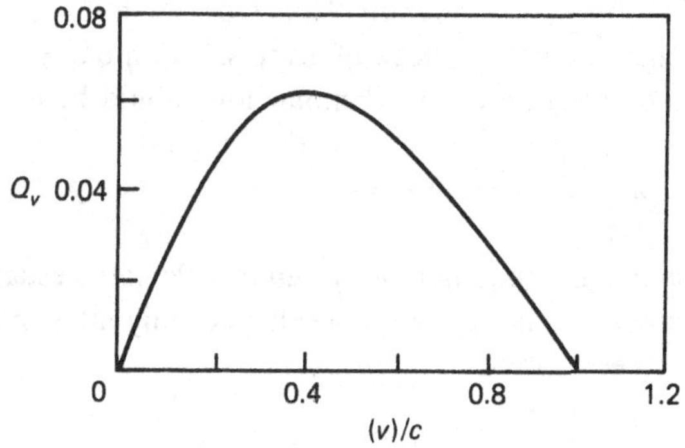

This can be done for a radiation dominated fluid $\left(c_s = \frac{1}{\sqrt{3}}\right)$ in a spatially flat Robertson-Walker Universe as in Section 5.5, by using the transformation defined in Proposition 5.2.

It is convenient to work with the quantity

$$Q_p = \left| \frac{1}{[[p]]} \mathscr{D}[[p]] \right|_{t_0}, \tag{9.6}$$

where t_0 is a fiducial initial time, rather than with Q.

The quantity Q_p is to be distinguished from Q, but they have similar asymptotic properties.

By applying the transformations (5.86) it is easy to see that

$$Q_p = \left| \frac{4}{R} \frac{\mathrm{d}R}{\mathrm{d}t} + \frac{\tilde{Q}_p}{R} \right|_{t=t_0},$$

where

$$\tilde{Q}_p = \left| \frac{1}{[[\tilde{p}]]} \mathscr{D}[[\tilde{p}]] \right|_{\tau=\tau_0}.$$

Since $\tilde{Q}_p \geq 0$ it follows that $Q_p \geq \frac{4}{R} \frac{\mathrm{d}R}{\mathrm{d}t}$ if $\frac{\mathrm{d}R}{\mathrm{d}t} > 0$, with the equality holding only in the limit $[[v]] \to 1$. Therefore, in an expanding spatially flat Robertson-Walker Universe the damping time never exceeds the cosmic expansion time even for extremely relativistic shocks (Anile et al., 1980) and therefore the "antidamping phenomenon" we have seen operating in the propagation into a stationary state does not seem important in this case.

This result will not change much in nonspatially flat models, if the pulse wavelength is small compared with the radius of curvature of the spatial sections.

The phenomenon which we have discussed is based on the analysis of the initial damping time τ_D, equation (9.4). This quantity gives only an indication for how quickly a shock of given strength damps. However, a more accurate description of the phenomenon should be given in terms of the actual damping time $\hat{\tau}_D$, that is the time taken from starting the initial shock profile to the point at which $[[v]]$ has damped to 1/2.7 times the initial $[[v]]_{t=0}$.

Whereas the initial damping time τ_D can be calculated exactly from the given initial shock profile, $\hat{\tau}_D$ can be evaluated only after the dynamical equations have been solved.

Anile et al. (1983) have performed a detailed study of the formation of strong relativistic shocks from simple waves and their subsequent damping, by numerically integrating the equations of special relativistic fluid dynamics in one dimension. The equations are written in Lagrangian mass coordinates (t, μ, y, z) and have been deduced in Section 5.3, equations (5.62)–(5.65). In this calculation the fluid was assumed to have radiation pressure, hence

$$p = \tfrac{1}{3}\rho\varepsilon.$$

In the framework of nondissipative fluid dynamics which we adopt in this book, shock waves are described as mathematical discontinuities. From the numerical viewpoint this represents an awkward problem, because, as energy is channeled into higher wavenumbers (due to nonlinearity), a limit is eventually reached when the wavelength is equal to two zones of the grid. Since higher wavenumbers cannot be attained, energy accumulates at the highest allowed wavenumber causing large oscillations to appear, which corrupt the solution and often destroy it completely.

In reality, dissipative effects convert the kinetic energy of the oscillations into internal gas energy, but in a purely nondissipative system this cannot occur.

There are several ways to get around these difficulties by introducing implicit numerical diffusion or explicit artificial viscosity, or by employing the characteristic or flux-corrected transport methods.

The method employed in this calculation was the well-established one of the artificial viscosity of Von Neumann and Richtmyer, which amounts to adding to the pressure a viscous scalar stress with a scale-dependent coefficient which spreads the shock over several zones while automatically ensuring that the jump relations are satisfied.

In particular, the following relativistic generalization of this procedure, due to May and White (1967), was adopted

$$q = \begin{cases} k\rho(\Delta\mu)^2 \dfrac{(U_\mu)^2}{\Gamma}, & \rho_t > 0, \\ 0, & \rho_t \le 0. \end{cases}$$

The initial waveforms for the calculations were taken to be right-propagating simple waves. A value was chosen for $\bar{c}_s$ (the speed of sound in the unperturbed medium) and the profile of the three-velocity v was specified as a function of μ. The local speed of sound c_s was then calculated

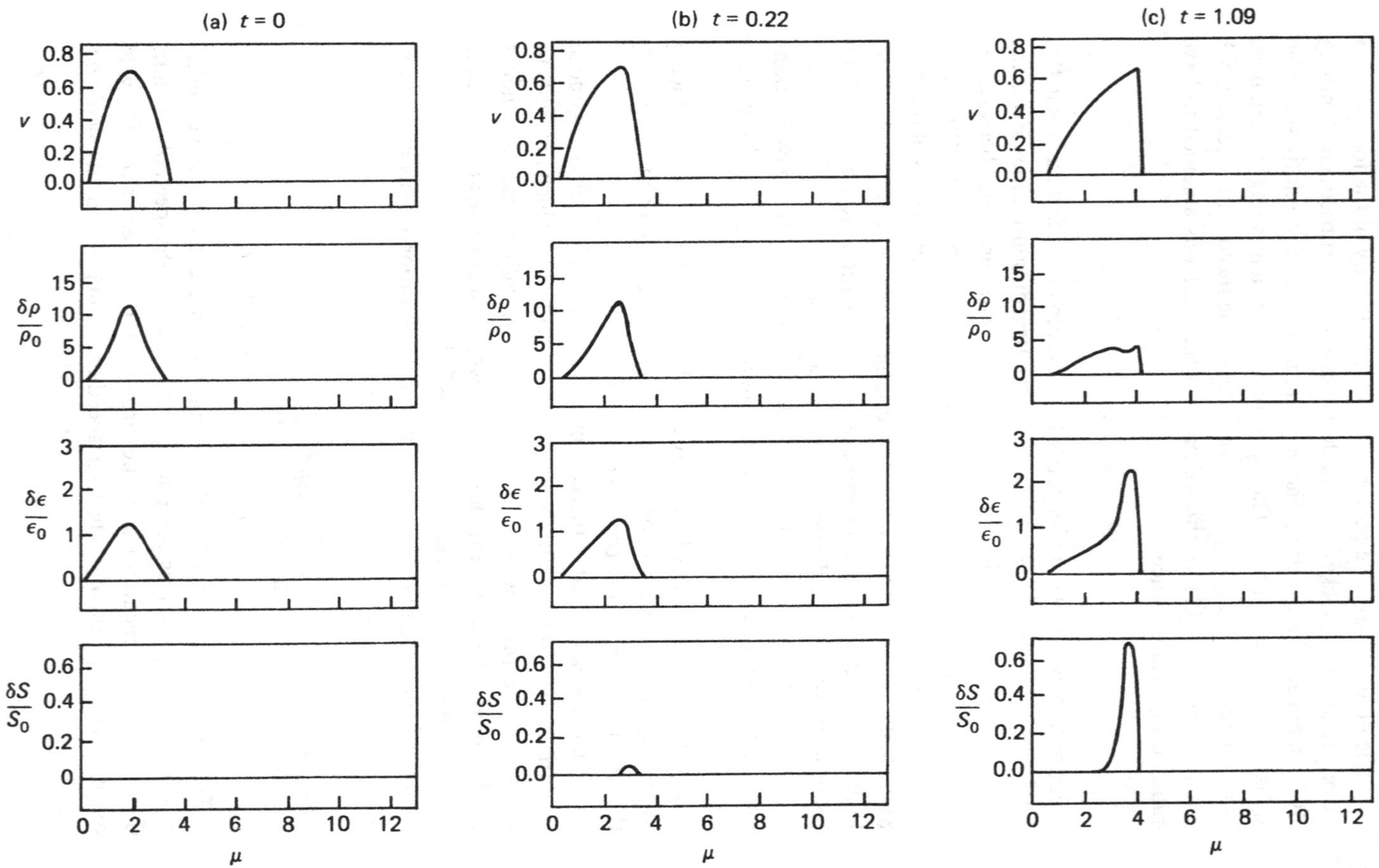

(a) t = 0
(b) t = 0.22
(c) t = 1.09
v
δρ/ρ0
δε/ε0
δS/S0
μ
0.8
0.6
0.4
0.2
0.0
15
10
5
0
3
2
1
4
6
8
12

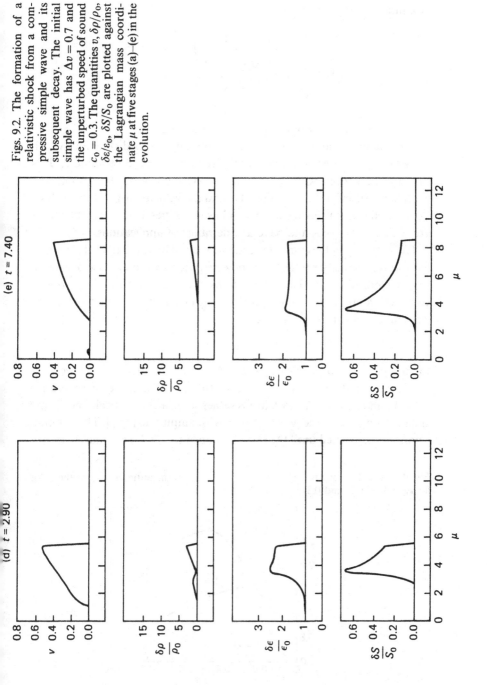

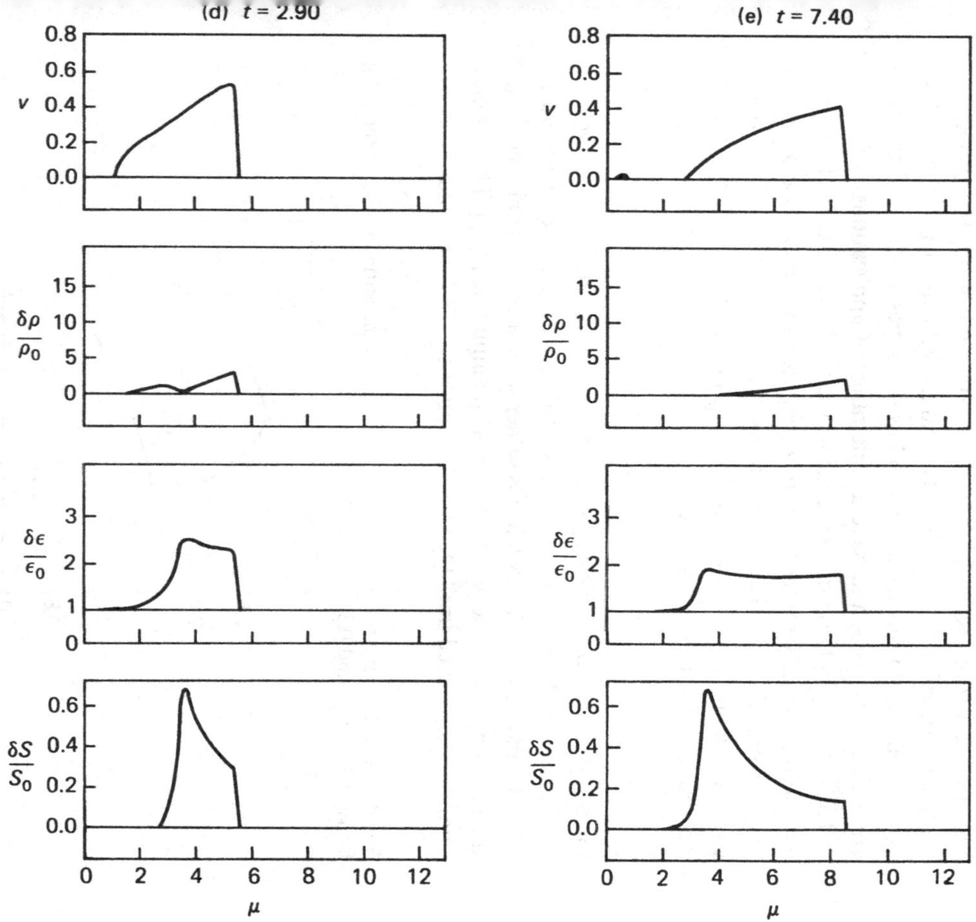

Figs. 9.2. The formation of a relativistic shock from a compressive simple wave and its subsequent decay. The initial simple wave has $\Delta v = 0.7$ and the unperturbed speed of sound $c_0 = 0.3$. The quantities v, $\delta\rho/\rho_0$, $\delta\varepsilon/\varepsilon_0$, $\delta S/S_0$ are plotted against the Lagrangian mass coordinate μ at five stages (a)–(e) in the evolution.

as a function of μ by

$$c_s = \frac{\dfrac{1+\bar{c}_s\sqrt{3}}{1-\bar{c}_s\sqrt{3}}\left(\dfrac{1+v}{1-v}\right)^{\frac{1}{2}\sqrt{3}} - 1}{\sqrt{3}\left[\dfrac{1+\bar{c}_s\sqrt{3}}{1-\bar{c}_s\sqrt{3}}\left(\dfrac{1+v}{1-v}\right)^{\frac{1}{2}\sqrt{3}} - 1\right]}.$$

From this one can obtain the initial value of ε.

The reference state, corresponding to the unperturbed state into which the wave propagates, will be characterized by the speed of sound $\bar{c}_s$. It is easy to check that, with the polytropic state equation, the problem is completely specified by the value of $\bar{c}_s$ and the velocity distribution as long as one then works using the dimensionless quantities. In this way the results are scale independent and have a wide range of applications.

The accuracy of the computer code was tested by running trial problems with square velocity profiles. Errors were generally found to be of order a few percent or less. The calculations were performed for a purely compressive initial sine pulse propagating into a static uniform medium. The initial velocity v is specified as

$$v = \Delta v \sin \mu$$

for $0 \leq \mu \leq \pi$, zero otherwise, where Δv is the initial maximum velocity amplitude. In Figs. 9.2 the results of the calculation are presented for $\Delta v = 0.7$ and $c_0 = 0.3$. A shock forms rather soon, after a critical time T_c, and then extends across the wave's profile with amplitude $[[v]]$. The computation was continued until $[[v]]$ had damped to 1/2.7 the initial Δv for the

Fig. 9.3. The damping rate Q plotted as a function of the initial Δv across the wave for $c_0 = 0.01$, 0.1, and 0.3.

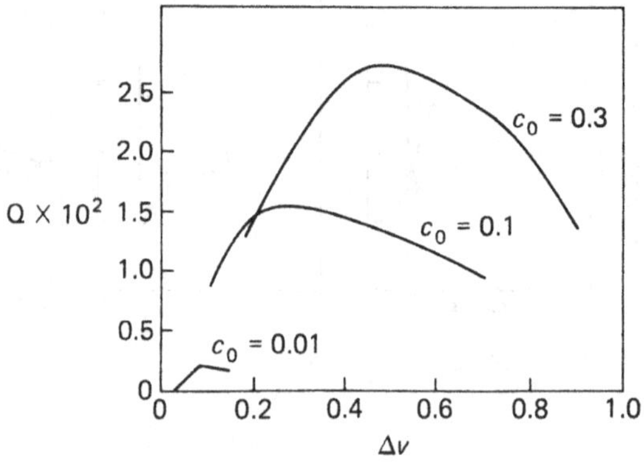

wave. The value of Eulerian time at the shock front at this stage is taken as the shock damping time $\hat{t}_D$. This quantity differs very little from the $\hat{t}_D$ previously defined, because the critical time for shock formation is very small $T_c \ll \hat{t}_D$.

In Fig. 9.3 the damping rate Q is plotted as a function of the initial Δv across the wave, for $c_0 = 0.01$, 0.1, and 0.3. The curves terminate at the points where the computations cease to be accurate.

All of them exhibit a maximum and have the appearance of tending toward zero as $\Delta v \to 1$. Therefore, the peculiar phenomenon, which was discussed analytically above, is confirmed by the numerical calculation.

For all cases which have been treated numerically, the maximum occurs at medium or weakly relativistic Δv. Since very extreme conditions are not required, it is possible that this effect might be important in realistic physical situations.

The investigation of curved relativistic shocks requires considerably more labor. By using the covariant theory of singular hypersurfaces developed in Chapter 3 it is possible to prove that the same qualitative effects on the damping are present, under appropriate conditions (Anile, 1984).

9.2. Propagation of weak shock waves

In the previous section we have exploited the compatibility relations in order to obtain analytic information on the propagation of relativistic shock waves of arbitrary strength. In general, it is not possible to obtain a more detailed knowledge without integrating numerically the full set of partial differential equations, because the jumps in the field's derivatives are known only at the initial time. Some analytical progress can be made only if one makes some assumptions about these parameters, as in the case of self-similar solutions. This situation is to be contrasted with that of the ordinary discontinuity waves, where, as shown in Chapter 4, one can construct an exact theory of the propagation of the wavefront based on the methods of bicharacteristics.

In this section we show that, if one restricts oneself to the case of *weak shocks* occurring within the profile of an asymptotic wave, it is indeed possible to develop a theory for the propagation of the wavefront. As in Section 2.1 we shall deal with a system of conservation laws

$$\nabla_\alpha F^\alpha(\mathbf{U}) = f(\mathbf{U}), \tag{9.7}$$

where $F^\alpha(\mathbf{U})$ and $f(\mathbf{U})$ are analytic functions of $\mathbf{U}$ in a suitable domain $\mathscr{D}$ of $\mathbb{R}^N$.

By introducing the matrixes

$$A_B^{\alpha A}(\mathbf{U}) = \frac{\partial F^{\alpha A}}{\partial U^B}$$

the system (9.7) is rewritten

$$A_B^{\alpha A}(\mathbf{U})\nabla_\alpha U^B = f^A(\mathbf{U}). \tag{9.8}$$

We shall assume that the system (9.8) is hyperbolic in the time-direction defined by the timelike vector field ξ, according to the Definition 2.1.

Let Ω be a coordinate neighborhood in M and $\psi \in D(\Omega)$. Let Σ be the hypersurface of Ω described by

$$\psi(x^\alpha) = 0, \quad \mathrm{d}\psi|_\Sigma \neq 0.$$

We assume that Σ is a shock wave for the system (9.7) and let $l_\alpha = \nabla_\alpha \psi$. The jump conditions for the system (9.7) can be obtained as in Section 8.1 (where the case of relativistic fluid dynamics is treated), and are

$$[[F^\alpha]]l_\alpha = 0. \tag{9.9}$$

Let us assume that equation (9.9) admits a solution of the kind

$$\mathbf{U}_- = \mathbf{U}_-(\mathbf{U}_+, l_\alpha) \tag{9.10}$$

that is nontrivial, that is, $\mathbf{U}_- \neq \mathbf{U}_+$, for an assigned unperturbed state $\mathbf{U}_+$.

Given the timelike vector ξ_α, one can always decompose

$$l_\alpha = -\sigma\xi_\alpha + \zeta_\alpha, \tag{9.11}$$

with

$$\zeta_\alpha\zeta^\alpha = 1, \quad \zeta_\alpha\xi^\alpha = 0.$$

Now let us consider the eigenvalue problem

$$\det\{(-\mu\xi_\alpha + \zeta_\alpha)A^\alpha\}|_{\mathbf{U}_+} = 0.$$

By the assumptions of hyperbolicity the above equation has N real solutions $\mu_+^{(1)}, \ldots, \mu_+^{(N)}$. Then we shall assume that there exists an integer k such that

$$\lim_{\sigma \to \mu_+^{(k)}} \mathbf{U}_- = \mathbf{U}_+. \tag{9.12}$$

This expresses the existence of the weak shock limit, that is, when the shock speed σ (relative to the timelike congruence defined by ξ) tends to one of the characteristic speeds, the shock strength tends to zero.

Now let

$$\Omega_\pm = (\psi(x^\alpha) \gtrless 0).$$

We shall extend the treatment of asymptotic waves given in Section 7.1 by looking for solutions of (9.93) of the kind

$$\mathbf{U} = \mathbf{U}_0(x^\mu), \quad x \in \overset{0}{\Omega}_+,$$
$$\mathbf{U} = \mathbf{U}_0 + \frac{1}{\omega}\mathbf{U}_1(x^\mu, \xi) + \frac{1}{\omega^2}\mathbf{U}_2(x^\mu, \xi) + O(\omega^{-2}), \tag{9.13}$$

with $\xi = \omega\phi(x^\mu)$, ϕ a phase function, and $\mathbf{U}_0$ continuous across Σ. We shall assume that Σ is influenced by the wave in the sense that $\psi(x^\mu)$ can be written as

$$\psi(x^\mu) = \tilde{\psi}(x^\mu, \xi)$$

and, therefore, the equation describing Σ is

$$\tilde{\psi}(x^\mu, \xi) = 0. \tag{9.14}$$

We have

$$F^\mu(\mathbf{U}_-) = F(\mathbf{U}_{0+}) + \frac{1}{\omega}\underset{0}{A}{}^{A\mu}_{+B}U_1^B + \frac{1}{2\omega^2}N^{A\mu}_{+BC}U_1^B U_1^C$$
$$+ O(\omega^{-2}), \tag{9.15}$$

with

$$N^{A\mu}_{BC} = \left(\frac{\partial^2 F^{A\mu}}{\partial U^B \partial U^C}\right)_{\mathbf{U}_0}.$$

By using the expansion (9.15) in the jump conditions (9.9), to the leading order we obtain

$$\underset{0}{A}{}^{A\mu}_{B}[[U^B]]\phi_\mu \frac{\partial\psi}{\partial\xi} = 0,$$

which gives, for ϕ a simple root of the characteristic equation,

$$[[\mathbf{U}]] = [[\Pi(x^\mu, \xi)]]\mathbf{R}_+(x^\mu), \tag{9.16}$$

with $\mathbf{R}_+$ the right eigenvector of the matrix $\underset{0}{A}{}^{A\mu}_{+B}\phi_\mu$ and $\Pi(x^\mu, \xi)$ defined as in Section 7.1.

To the next order we obtain

$$\underset{0}{A}{}^{A\mu}_{+B}\psi_\mu R_+^B[[\Pi]] + \tfrac{1}{2}N^{A\mu}_{+BC}R_+^B R_+^C[[\Pi]]^2\phi_\mu \frac{\partial\psi}{\partial\xi} = 0. \tag{9.17}$$

In Ω_- the transport equation for $\Pi(x^\mu, \xi)$ is

$$\frac{\partial\Pi}{\partial\sigma} + N\Pi\frac{\partial\Pi}{\partial\xi} + M\Pi = 0, \tag{9.18}$$

where the symbols have the same meaning as in Section 7.1. By multiplying equation (9.17) with L_{+A} we get

$$\frac{\partial\psi}{\partial\sigma}[[\Pi]] + \tfrac{1}{2}N[[\Pi]]^2\frac{\partial\psi}{\partial\xi} = 0. \tag{9.19}$$

It is apparent that equation (9.19) can be interpreted as providing the jump conditions for equation (9.18) across Σ, in the plane (σ, ξ). Therefore, one can apply the well-known methods of shock fitting to equation (9.18) in order to study the evolution of the shock (Whitham, 1974; Hunter and Keller, 1983).

This method could be applied in order to study the propagation of a relativistic plane weak shock propagating into a constant state (in special relativity). By using the results of the Section 7.3 it is a simple exercise to show that the damping law of the shock is

$$[[v]](t) = \frac{[[v]]_0}{\left(1 + \alpha\frac{[[v]]_0}{L}t\right)^{1/2}}, \tag{9.20}$$

where

$$\alpha = 1 - c_s^2 + \frac{(e+p)}{2c_s^2}p''_e, \tag{9.21}$$

a formula that coincides with the result of Liang (1977a) obtained by standard methods.

10

Stability of relativistic shock waves

10.0. Introduction

The stability of relativistic shock waves is a fundamental problem of relativistic fluid dynamics and its resolution might have interesting consequences for astrophysics, plasma physics, and nuclear physics. In particular, in models of gravitational collapse and supernova explosions, the bounce shock traverses regions of varying state equations. The stability of this shock is essential for the overall validity of the model. Present numerical codes, being restricted to spherical symmetry, are not capable of detecting these kinds of instabilities. In high-energy heavy ion collisions, the formation of a quark-gluon plasma might occur through a relativistic shock and its instability might signal the transition (e.g., by observing shock splitting; Barz et al., 1985).

In nonrelativistic fluid dynamics the stability of plane shock fronts has been investigated by D'Yakov (1956) and Erpenbeck (1962), using linear stability methods.

The problem has also been studied by Gardner (1963) in the framework of shock splitting methods and more recently by Fowels (1981, and references therein) who considered the stability of the front against an impinging acoustic wave from behind. By requiring that the perturbations decay sufficiently fast at infinity these authors obtained restrictions on the equation of state of a fluid in order for it to sustain plane fronted shock waves.

The linear stability analysis has been extended by Gardner and Kruskal (1964) to the fast magnetoacoustic shock and by Lessen and Deshpande (1967) to the slow magnetoacoustic shock.

In the framework of linear stability analysis, a mathematically very general approach has been lately introduced by Majda (1984), which is applicable also to curved shock fronts.

Another approach, which differs significantly from the linear stability one, is based on an intuitive definition of stability due to Whitham (1974).

According to Whitham we say that a plane shock front is stable if, by perturbing the shock front, the shock velocity decreases where the front is expanding (convex perturbation) and increases where the front is converging (concave perturbation) (refer to Fig. 10.1, p. 304).

By using such an intuitive definition one could obtain necessary and sufficient stability criteria if the relationship between the shock speed and what we might call, losely speaking, the shock front rate of deformation, were known. Unfortunately, usually this is not the case, except in special circumstances (characteristic shocks, or approximation methods). By applying Whitham's geometric shock dynamics approximation (Whitham, 1974) in order to obtain such a relationship in the case of nonrelativistic fluid dynamics it is possible to recover the same restrictions as obtained with linear stability theory (Liberman and Velikovich, 1985). The problem of obtaining a relationship between the shock speed and the front deformation rate can also be treated within other approximation methods. In particular, Anile and Russo (1986a, 1988) have resorted to the generalized wavefront expansion method in order to study the propagation of step-shocks.

If one is interested in obtaining only necessary stability conditions it is possible to avoid resorting to approximation schemes. This requires that Whitham's intuitive definition be first put into a rigorous form. Anile and Russo (1986b) have achieved this by giving a precise definition of *corrugation stability* and applying it to the case of relativistic fluid dynamics. Such a concept of corrugation stability is purely local (on the wavefront) and is nonlinear (it does not require linearizing the equations).

Subsequently, Anile and Russo (1987) have performed a linear stability analysis for plane fronted relativistic shock waves and their results coincide with those obtained in the framework of corrugation stability.

In Section 10.1 we shall treat the linear stability of a plane relativistic shock front.

In Section 10.2 the definition of corrugation stability is given and is applied to relativistic fluid dynamics.

In Section 10.3 the interaction of a relativistic shock wave with an acoustic weak discontinuity wave is considered and conclusions for stability are drawn.

In Section 10.4 the restrictions on the equation of state of a relativistic fluid in order for it to sustain a stable plane fronted shock wave are discussed from the viewpoint of thermodynamics. Examples of interest for nuclear matter theory and for astrophysics are treated.

Throughout this chapter a coordinate dependent formulation in the framework of special relativity will be adopted. Such a choice has the advantage of making the treatment closer to that of nonrelativistic fluid dynamics so that the results can be immediately compared. Furthermore, it brings to light some invariance properties related to a change of field variables. A covariant approach is really needed when investigating the stability of curved shock fronts in a gravitational field, a problem that might be of considerable interest in astrophysics.

10.1. Linear stability of a relativistic shock

Let $\mathscr{M}$ denote Minkowki space-time and (x, y, z, t) be inertial coordinates. The four-velocity of the fluid has components

$$u^\alpha = \Gamma(1, v^i), \quad \Gamma = (1 - v^2)^{-1/2}, \quad v^2 = v^i v^i \delta_{ij}.$$

By introducing the field vector

$$\mathbf{U} = \begin{pmatrix} \rho \\ u^i \\ p \end{pmatrix}, \quad i = 1, 2, 3$$

the field equations of relativistic fluid dynamics (2.4)–(2.5) are written

$$\mathscr{A}^0 \partial_t \mathbf{U} + \mathscr{A}^k \partial_k \mathbf{U} = 0, \tag{10.1}$$

with

$$\mathscr{A}^0 = \begin{bmatrix} \Gamma & \frac{\rho}{\Gamma} u^i & 0 \\ 0 & \Gamma(e+p)\delta_{ij} & \Gamma u^j \\ -\Gamma a^2 & 0 & \Gamma \end{bmatrix},$$

$$\mathscr{A}^k = \begin{bmatrix} u^k & \rho\delta_i^k & 0 \\ 0 & u^k(e+p)\delta_{ij} & \delta_j^k + u^k u_j \\ -a^2 u^k & 0 & u^k \end{bmatrix},$$

where $a^2 = \left(\frac{\partial p}{\partial \rho}\right)_S = \frac{c_s^2(e+p)}{\rho}$.

Let the equation of the shock hypersurface Σ be

$$\varphi(x, y, z, t) = 0,$$

and write

$$M^i = \rho u^i, \quad W^j = (e+p)\Gamma u^j.$$

Then the Rankine-Hugoniot relations (8.3)–(8.4) can be written in the form

$$\begin{aligned} \varphi_t[[\rho\Gamma]] + \varphi_i[[M^i]] &= 0, \\ \varphi_t[[W^i]] + \varphi_i[[T^{ij}]] &= 0, \\ \varphi_t[[(e+p)\Gamma^2]] + \varphi_i[[W^i]] &= 0. \end{aligned} \tag{10.2}$$

A plane fronted step-shock consists of a one dimensional flow in which two half spaces of constant flow are separated by a shock moving with a constant speed $\overset{0}{v}_\Sigma$, that is,

$$\mathbf{U}(x,y,z,t) = \begin{cases} \mathbf{U}_1 = \text{const. in the region } \Pi_- : x - \overset{0}{v}_\Sigma t < 0, \\ \mathbf{U}_0 = \text{const. in the region } \Pi_+ : x - \overset{0}{v}_\Sigma t \geq 0. \end{cases} \tag{10.3}$$

By a Lorentz transformation the shock front can be taken at rest, that is, $\overset{0}{v}_\Sigma = 0$, hence

$$\varphi(x,y,z,t) = x.$$

Then the Rankine-Hugoniot relations (10.2) yield

$$\begin{aligned} [[M^1]] &= 0, \\ [[T^{1j}]] &= 0, \\ [[W^1]] &= 0. \end{aligned} \tag{10.4}$$

From these equations it is easy to see that one can always make $u^2 = u^3 = 0$, hence, for the unperturbed flow, $u^\alpha = \Gamma(1, v, 0, 0)$.

The linear stability of the solution (10.3) is studied as follows. By linearizing equation (10.1) around the constant state (ahead or behind the shock) one obtains

$$\mathscr{A}^0 \partial_t \delta\mathbf{U} + \mathscr{A}^i \partial_i \delta\mathbf{U} = 0, \tag{10.5}$$

where $\delta\mathbf{U}$ is the perturbation and $\mathscr{A}^0$, $\mathscr{A}^i$ are constant matrixes [they are different behind and ahead of the shock, i.e., $\mathscr{A}^\alpha = \mathscr{A}^\alpha(\mathbf{U}_1)$ in Π_- and $\mathscr{A}^\alpha = \mathscr{A}^\alpha(\mathbf{U}_0)$ in Π_+].

Also, the shock wavefront is perturbed according to

$$\varphi = x + \delta\varphi(t, y, z). \tag{10.6}$$

One seeks special classes of solutions of the linearized problem of the kind (normal mode analysis)

$$\delta\mathbf{U}(t,x,y,z) = \mathbf{Y}(x) e^{-i(\omega t + ly + mz)}, \tag{10.7}$$

with l, m real, $\omega \in \mathbb{C}$, and

$$\lim_{x \to \pm\infty} |\mathbf{Y}(x)| = 0. \tag{10.8}$$

Also,

$$\delta\varphi(t, y, z) = \delta\bar{\varphi}\mathrm{e}^{-i(\omega t + ly + mz)}, \tag{10.9}$$

with $\delta\bar{\varphi}$ constant.

The requirement (10.8) that $\mathbf{Y}(x)$ vanish at infinity expresses the physical condition that the perturbation is not driven by external sources.

Linearizing the jump conditions (10.2) yields

$$\begin{aligned}
&[[\delta M^1]] = i\delta\varphi\omega[[\rho\,\Gamma]],\\
&[[\delta T^{11}]] = i\delta\varphi\omega[[W^1]],\\
&[[\delta T^{12}]] = i\delta\varphi l[[p]],\\
&[[\delta T^{13}]] = i\delta\varphi m[[p]],\\
&[[\delta W^1]] = i\delta\varphi\omega[[\Gamma^2(e+p) - p]],
\end{aligned} \tag{10.10}$$

where δM^1, δT^{11}, etc. denote the variations in M^1, T^{11}, etc. due to the perturbation $\delta\mathbf{U}$.

Substituting (10.7) into (10.5) gives

$$\left(\omega\mathscr{A}^0 + i\mathscr{A}^1\frac{\mathrm{d}}{\mathrm{d}x} + l\mathscr{A}^2 + m\mathscr{A}^3\right)\mathbf{Y}(x) = 0. \tag{10.11}$$

By further assuming that $\mathbf{Y}(x)$ admits a Laplace transform, from equation (10.11) one obtains

$$(\omega\mathscr{A}^0 \pm iq\mathscr{A}^1 + \mathscr{A}^2 + m\mathscr{A}^3)\hat{\mathbf{Y}}(q) \mp i\mathscr{A}^1\mathbf{Y}(0) = 0 \tag{10.12}$$

in $\Pi_\pm$, respectively, where

$$\hat{\mathbf{Y}}(q) = \int_0^\infty \mathrm{e}^{-qx}\mathbf{Y}(x)\mathrm{d}x \qquad \text{in } \Pi_+,$$

$$\hat{\mathbf{Y}}(q) = \int_0^\infty \mathrm{e}^{-qx}\mathbf{Y}(-x)\mathrm{d}x \quad \text{in } \Pi_-.$$

Let $q = \mp ik$, $k \in \mathbb{C}$, in $\Pi_\pm$, respectively. Then equation (10.12) is rewritten

$$\mathscr{A}\hat{\mathbf{Y}} \mp i\mathscr{A}^1\mathbf{Y}(0) = 0, \tag{10.13}$$

where $\mathscr{A}$ is the matrix

$$\mathscr{A} = \omega\mathscr{A}^0 + k\mathscr{A}^1 + l\mathscr{A}^2 + m\mathscr{A}^3 \tag{10.14}$$

and $\hat{\mathbf{Y}}$ stands for $\hat{\mathbf{Y}}(\mp ik)$.

When ω and k are both real, $\mathscr{A}$ is the characteristic matrix for the system

(10.1) (evaluated in $\mathbf{U}_0$ and $\mathbf{U}_1$), with the identification

$$n^1 = \frac{k}{\sqrt{k^2+l^2+m^2}}, \quad n^2 = \frac{l}{\sqrt{k^2+l^2+m^2}}, \quad n^3 = \frac{m}{\sqrt{k^2+l^2+m^2}},$$
$$\lambda = \frac{-\omega}{\sqrt{k^2+l^2+m^2}}.$$

It immediately checks that the characteristic speeds are the following

$$\begin{aligned} \lambda_1 &= \frac{\Gamma c_s - u_n\sqrt{\Delta}}{u_n c_s - \Gamma\sqrt{\Delta}}, \\ \lambda_2 &= v_n \text{ (triple)}, \\ \lambda_1 &= \frac{\Gamma c_s + u_n\sqrt{\Delta}}{u_n c_s - \Gamma\sqrt{\Delta}}, \end{aligned} \tag{10.15}$$

where

$$v_n = v^i n_i, \quad u_n = u^i n_i, \quad \Delta = 1 + (u_{\mathrm{T}})^2(1-c_s^2),$$

with $u_{\mathrm{T}}^i = u^i - n^i u_n$ being the tangential components of u^i.

The shock under consideration will be assumed to satisfy the Lax conditions (8.132). In particular, the unperturbed step-shock will be a right propagating acoustic shock, for which, in the chosen inertial frame,

$$\begin{aligned} &\lambda_3^{(+)} < 0, \quad \lambda_1^{(+)} < 0, \quad \lambda_2^{(+)} < 0, \\ &\lambda_3^{(-)} > 0, \quad \lambda_1^{(-)} < 0, \quad \lambda_2^{(-)} < 0, \end{aligned} \tag{10.16}$$

where the eigenvalues correspond to the unperturbed shock wavefront $x = 0$, that is, $l = m = 0$.

From (10.15)–(10.16) it immediately follows that,

$$\begin{aligned} &v_+ < 0, \quad v_- < 0, \\ &(c_s - |v|_+) < 0, \quad (c_s - |v|_-) > 0, \end{aligned} \tag{10.17}$$

the latter inequalities expressing the well-known result that the flow (with respect to the shock) is supersonic ahead of the front and subsonic behind it.

In the general case, a straightforward calculation yields

$$\det \mathscr{A} = \Gamma^3\Omega^3\{\Gamma^2\Omega^2 - c_s^2((\Gamma^2\Omega v + k)(\omega v + k) + l^2 + m^2)\}, \tag{10.18}$$

where $\Omega = \omega + kv$.

The solutions of $\det \mathscr{A} = 0$ are the following.

The root $\Omega = 0$ is triple and in the case where k and ω are both real it corresponds to the eigenvalue λ_2.

The corresponding value for k, denoted by k_2, is given by

$$k_2 = -\frac{\omega}{v}, \tag{10.19}$$

and for such a root one has

$$\operatorname{Im}\omega = -v \operatorname{Im} k. \tag{10.20}$$

The other two roots of $\det \mathscr{A} = 0$ are given by

$$k_{3,1} = \frac{-\Gamma^2 v(1-c_s^2)\omega \pm c_s\sqrt{\omega^2 + \Gamma^2(v^2 - c_s^2)(l^2 + m^2)}}{\Gamma^2(v^2 - c_s^2)}, \tag{10.21}$$

where $\sqrt{\ }$ denotes the principal determination of the square root.

When l, m are real these roots correspond to the eigenvalues $\lambda_{3,1}$, respectively (according to the choice $\pm$).

In the following, for the sake of definiteness, it will be assumed $\operatorname{Re}\omega \geq 0$.

One can state the following proposition.

PROPOSITION 10.1 *Ahead of the shock, the following relationships hold:*
(A1) $\operatorname{Im}\omega > 0$ iff $\operatorname{Im} k_1 > 0$,
(A2) $\operatorname{Im}\omega > 0$ iff $\operatorname{Im} k_2 > 0$,
(A3) $\operatorname{Im}\omega > 0$ iff $\operatorname{Im} k_3 > 0$, in Π_+;
whereas, behind the shock,
(B1) $\operatorname{Im}\omega > 0$ iff $\operatorname{Im} k_1 > 0$,
(B2) $\operatorname{Im}\omega > 0$ iff $\operatorname{Im} k_2 > 0$,
(B3) $\operatorname{Im}\omega > 0$ iff $\operatorname{Im} k_3 < 0$, in Π_-.

Proof. The relationships (A2) and (B2) are immediate consequences of equation (10.20).

Let

$$z = \omega^2 + \Gamma^2(v^2 - c_s^2)(l^2 + m^2).$$

The principal square root of z is given by

$$\sqrt{z} = \mathscr{L}_+(z) + i(\operatorname{sign} \operatorname{Im} z)\mathscr{L}_-(z),$$

where

$$\mathscr{L}_\pm(z) = \frac{1}{\sqrt{2}}\{((\operatorname{Re} z)^2 + (\operatorname{Im} z)^2)^{1/2} \pm \operatorname{Re} z\}^{1/2}.$$

For the root k_1 one has

$$\Gamma^2(v^2 - c_s^2)\operatorname{Im} k_1 = -\Gamma^2 v(1-c_s^2)\operatorname{Im}\omega - c_s\frac{\operatorname{Im}\omega}{|\operatorname{Im}\omega|}\mathscr{L}_-(z).$$

Ahead of the shock, $v^2 - c_s^2 > 0$, and it is immediately seen that $\dfrac{\mathscr{L}_-(z)}{|\mathrm{Im}\,\omega|} < 1$, hence, for $\mathrm{Im}\,\omega > 0$,

$$\Gamma^2(v^2 - c_s^2)\,\mathrm{Im}\,k_1 > \Gamma^2(1 - c_s v)(|v| - c_s)\,\mathrm{Im}\,\omega,$$

which proves (A1).

Behind the shock, $v^2 - c_s^2 < 0$ and $\dfrac{\mathscr{L}_-(z)}{|\mathrm{Im}\,\omega|} > 1$, hence

$$\Gamma^2(c_s^2 - v^2)\,\mathrm{Im}\,k_1 > \Gamma^2(c_s + v)(1 - c_s v)\,\mathrm{Im}\,\omega,$$

which proves (B1).

Finally,

$$\Gamma^2(v^2 - c_s^2)\,\mathrm{Im}\,k_3 = -\Gamma^2 v(1 - c_s v)\,\mathrm{Im}\,\omega + c_s\frac{\mathrm{Im}\,\omega}{|\mathrm{Im}\,\omega|}\mathscr{L}_-(z).$$

Ahead of the shock, $v^2 - c_s^2 > 0, v < 0$, hence (A3) follows. Behind the shock, $v^2 - c_s^2 < 0$, $\dfrac{\mathscr{L}_-(z)}{|\mathrm{Im}\,\omega|} > 1$, hence

$$\Gamma^2(v^2 - c_s^2)\,\mathrm{Im}\,k_3 > \Gamma^2(c_s - v)(1 + vc_s)\,\mathrm{Im}\,\omega$$

and (B3) follows. Q.E.D.

Now let $\mathbf{R}_1$, $\mathbf{R}_3$, $\mathbf{R}_2^{(i)}$, $\mathbf{L}_1$, $\mathbf{L}_3$, $\mathbf{L}_2^{(i)}$ be the right and left eigenvectors of the matrix $\mathscr{A}$ corresponding to the roots k_1, k_3, k_2 (triple; in this case the eigenvectors are chosen such that ${}^{\mathrm{T}}\mathbf{L}_2^{(i)}\mathscr{A}^1\mathbf{R}_2^{(j)} = \delta^{(ij)}$).

Then from equation (10.13) we have

$$\hat{\mathbf{Y}} = \frac{ia_1}{k - k_1}\mathbf{R}_1 + \sum_j \frac{ia_2^{(j)}}{k - k_2}\mathbf{R}_2^{(j)} + \frac{ia_3}{k - k_3}\mathbf{R}_3, \tag{10.22}$$

with

$$a_1 = \frac{{}^{\mathrm{T}}\mathbf{L}_1\mathscr{A}^1\mathbf{Y}(0)}{{}^{\mathrm{T}}\mathbf{L}_1\mathscr{A}^1\mathbf{R}_1},\ a_2^{(j)} = \frac{{}^{\mathrm{T}}\mathbf{L}_2^{(j)}\mathscr{A}^1\mathbf{Y}(0)}{{}^{\mathrm{T}}\mathbf{L}_2^{(j)}\mathscr{A}^1\mathbf{R}_2^{(j)}},\ a_3 = \frac{{}^{\mathrm{T}}\mathbf{L}_3\mathscr{A}^1\mathbf{Y}(0)}{{}^{\mathrm{T}}\mathbf{L}_3\mathscr{A}^1\mathbf{R}_3}.$$

It follows that $\hat{\mathbf{Y}}$ is the Laplace transform of the following function

$$\mathbf{Y}(x) = a_1\mathbf{R}_1\mathrm{e}^{-ik_1x} + \sum_j a_2^{(j)}\mathbf{R}_2^{(j)}\mathrm{e}^{-ik_2x} + a_3\mathbf{R}_3\mathrm{e}^{-ik_3x}. \tag{10.23}$$

Now, in Π_+, because of Proposition 10.1, assuming $\mathrm{Im}\,\omega > 0$ implies $|\mathbf{Y}(x)| \to \infty$ as $x \to \infty$, which violates the boundary condition (10.8). Therefore, a self-sustaining mode of instability requires $a_1 = a_2^{(i)} = a_3 = 0$ ahead of the shock.

In Π_- one has instead $\operatorname{Im} k_1 > 0$, $\operatorname{Im} k_2 > 0$, $\operatorname{Im} k_3 < 0$ and therefore, in order to satisfy the aforementioned boundary condition, the only requirement is $a_3 = 0$. A self-sustaining mode of instability requires only $a_3 = 0$ behind the shock and therefore

$$^{\mathrm{T}}\mathbf{L}_3 \mathscr{A}^1 \mathbf{Y}(0) = 0. \tag{10.24}$$

Equation (10.24) can be rewritten in the form

$$^{\mathrm{T}}\mathbf{L}_3 \mathscr{A}^1 \delta\mathbf{U}(0) = 0, \quad \text{in } \Pi_-, \tag{10.25}$$

where $\delta\mathbf{U}(0)$ is the field's perturbation behind the shock, calculated from equations (10.10). The vector $\mathscr{A}^1\delta\mathbf{U}(0)$ is given by

$$\mathscr{A}^1\delta\mathbf{U}(0) = \begin{bmatrix} u^1\delta\rho + \rho\delta u^1 \\ (e+p)u^1\delta u^1 + \Gamma^2\delta p \\ (e+p)u^1\delta u^2 \\ (e+p)u^1\delta u^3 \\ u^1(\delta p - a^2\delta\rho) \end{bmatrix},$$

where the quantities $\delta\rho$, δu^i, δp are evaluated at $x = 0$. A simple calculation shows that

$$\mathscr{A}^1\delta\mathbf{U}(0) = \begin{bmatrix} \delta M^1 \\ \Gamma^2(\delta T^{11} - v\delta W^1) \\ \delta T^{12} \\ \delta T^{13} \\ \dfrac{1}{\rho\varepsilon_p}(\Gamma\delta W^1 - u^1\delta T^{11} - f\delta M^1) \end{bmatrix},$$

with $f = 1 + \varepsilon + \dfrac{p}{\rho}$, hence from equations (10.10) one gets

$$\mathscr{A}^1\delta\mathbf{U}(0) = i\delta\varphi \begin{bmatrix} \omega[\tilde{\rho}] \\ -v\omega\Gamma^2(\tilde{f}[\tilde{\rho}] - [p]) \\ l[p] \\ m[p] \\ \dfrac{\Gamma\omega}{\rho\varepsilon_p}(v^2\tilde{f}[\tilde{\rho}] - [p]) \end{bmatrix},$$

with $\tilde{\rho} = \Gamma\rho$, $\tilde{f} = \Gamma f$.

The left eigenvector $^{\mathrm{T}}\mathbf{L}_3$ has components

$$^{\mathrm{T}}\mathbf{L}_3 = \left(a^2, -\frac{c_s^2 k_3 + \omega v}{\Gamma\ \ \Omega}, -\frac{c_s^2 l}{\Gamma\Omega}, -\frac{c_s^2 m}{\Gamma\Omega}, 1\right).$$

For a nontrivial perturbation, $\delta\varphi \neq 0$ and equation (10.25) is equivalent to

$$g(\zeta) = \tilde{A}\zeta^2 + \tilde{B}\zeta + \tilde{C} = 0, \tag{10.26}$$

where

$$\tilde{A} = 1 - \tilde{H} - \tilde{K}M^2, \quad \tilde{B} = (\tilde{K} - 1)v^2 + \tilde{H}, \quad \tilde{C} = (\tilde{K} - 1)(1 - v^2),$$
$$\tilde{H} = M^2(\tilde{K} - 1)p'_\varepsilon/\rho, \quad \tilde{K} = v_+/v, \quad M = |v|/c_s, \tag{10.27}$$

and

$$\zeta = \frac{\Omega}{kv},$$

with $p_\varepsilon = \dfrac{\partial p(\varepsilon, \rho)}{\partial \varepsilon} \neq 0$.

From $\operatorname{Im}\omega > 0$, $\operatorname{Im} k_3 < 0$, it follows that $\operatorname{Im}\Omega > 0$. Also, it immediately checks that $\operatorname{Re} k_3 < 0$.

In order to study the solutions of equations (10.26) we need the following lemma characterizing the allowed range of the variable ζ.

LEMMA 10.1. *The range of the variable ζ in the complex ζ-plane consists of:*

(i) *the segment $\mathcal{I}$ of the real line* $1 \leq \zeta < z_0 = \dfrac{c_s}{|v|\Gamma^2(1 - |v|c_s)}$;

(ii) *the open half disk $\mathcal{D}$ (lying in the lower part of the complex plane) of center and radius, respectively,*

$$z_1 = \tfrac{1}{2}(1 + 1/\mathcal{M}^2), \quad R_1 = \tfrac{1}{2}(-1 + 1/\mathcal{M}^2),$$

where $\mathcal{M}$ is the proper downstream Mach number

$$\mathcal{M} = \frac{\Gamma|v|}{\Gamma_s c_s}. \tag{10.28}$$

Proof. The acoustic branch of the characteristic equation (10.18) can be written as

$$\mathcal{G} \equiv (\mathcal{M}^2 + v^2)\Omega^2 - 2kv^3\Omega - k^2v^2(1 - v^2) = v^2(l^2 + m^2),$$

whence one obtains the following restrictions

$$\operatorname{Re}\mathcal{G} = (\mathcal{M}^2 + v^2)(\Omega_R^2 - \Omega_I^2) - 2v^3(k_R\Omega_R - k_I\Omega_I)$$
$$- v^2(1 - v^2)(k_R^2 - k_I^2) \geq 0, \tag{10.29}$$

$$\operatorname{Im}\mathcal{G} = (\mathcal{M}^2 + v^2)\Omega_R\Omega_I - v^3(k_R\Omega_I + k_I\Omega_R) - v^2(1 - v^2)k_Rk_I = 0, \tag{10.30}$$

where $\Omega = \Omega_R + i\Omega_I$, $k = k_R + ik_I$.

It is convenient to introduce the following variables

$$x = \frac{\Omega_R}{vk_R}, \quad y = \frac{\Omega_I}{vk_I}, \quad r = \frac{k_I}{k_R}.$$

Since it is not restrictive to assume that $\operatorname{Re}\omega \geq 0$, from the definitions one has the following limitations on the variables x, y, r:

$$x \geq 1, y > 1, \quad 0 < r \leq \infty.$$

Then equation (10.30) reads

$$(\mathscr{M}^2 + v^2)xy - v^2(x + y) - (1 - v^2) = 0, \tag{10.31}$$

which represents a hyperbola, which in parametric form can be described by

$$\begin{aligned} x &= b + \nu\hat{z}, \\ y &= b + \hat{z}/\nu, \quad M^* \leq \nu < 1/M^*, \end{aligned} \tag{10.32}$$

where

$$b = \frac{v^2}{\mathscr{M}^2 + v^2}, \quad \hat{z} = \frac{c_s(1 - v^2)}{|v|(1 - v^2c_s^2)}, \quad M^* = \frac{v(1 - c_s^2)}{c_s(1 - v^2)}.$$

The restriction (10.29) can be put in the form

$$[(x - b)^2 - \hat{z}^2]/r - r[(y - b)^2 - \hat{z}^2] \geq 0$$

and this is verified iff $x > b + \hat{z}$, which is equivalent to $\nu \geq 1$.
Finally, $\zeta = \zeta_R + i\zeta_I$ can be written in the parametric form

$$\zeta_R = b + \hat{z}\frac{\nu + r^2/\nu}{1 + r^2},$$

$$\zeta_I = -\hat{z}r(\nu - 1/\nu)r/(1 + r^2),$$

which, in terms of ν and the angle $\theta = 2\arctan r$, is rewritten as

$$\begin{aligned} \zeta_R &= b + \hat{z}[\tfrac{1}{2}(\nu + 1/\nu) + \tfrac{1}{2}(\nu - 1/\nu)\cos\theta], \\ \zeta_I &= -\hat{z}\tfrac{1}{2}(\nu - 1/\nu)\sin\theta, \\ 1 &\leq \nu < 1/M^*, \quad 0 < \theta \leq \pi \end{aligned} \tag{10.33}$$

and is easily seen to represent the set $\mathscr{I} \cup \mathscr{D}$. Q.E.D.

Now we can state the main stability result.

THEOREM 10.1. *Let*

$$F = 1 - v^2 + M(1 - vv_+) - \tilde{H}. \tag{10.34}$$

then:

(i) *if* $F<0$ *then the step-shock is unstable:*
(ii) *if* $F>0$ *then the step-shock is stable against linear perturbations.*

Proof.
(i) It is easy to check that

$$g(1)=\tilde{K}(1-M^2)>0,$$

$$g(z_0)=\frac{c_s(c_s-|v|)}{\Gamma^2 v^2(1-|v|c_s)}F<0,$$

hence there is a real solution $\zeta^*\in\mathscr{I}$ of equation (10.26) and this corresponds to a mode of instability.

(ii) In this case $g(1)>0$ and $g(z_0)>0$ and, therefore, if the roots of equation (10.26) are real then there are either two or no roots belonging to $\mathscr{I}$. Also, we note that if $\tilde{A}\tilde{C}<0$ then the roots (which are real in this case) have different signs, hence there is no root in $\mathscr{I}$. Also, if the coefficients $\tilde{A},\tilde{B},\tilde{C}$ have the same sign then the roots are either both negative or both with a negative real part and in neither case do they belong to the region $\mathscr{I}\cup\mathscr{D}$. Therefore, it remains to be investigated the cases when $\tilde{A}\tilde{C}>0$ and the coefficients do not have the same sign, which leaves the cases

(1) $\tilde{A}>0,\tilde{B}<0,\tilde{C}>0$;
(2) $\tilde{A}<0,\tilde{B}>0,\tilde{C}<0$.

(1) This case is possible only when $p'_\varepsilon<0,\tilde{K}-1>0$.

The coefficients $\tilde{A},\tilde{B},\tilde{C}$ satisfy the relationship

$$\tilde{A}+\tilde{B}+\tilde{C}\mathscr{M}^2=1-M^2. \tag{10.35}$$

Let

$$\alpha=-\tilde{A}/\tilde{B},\quad \gamma=-\tilde{C}/\tilde{B},\quad \alpha>0,\gamma>0.$$

Then equation (10.26) is written

$$h(\zeta)\equiv a\zeta^2-\zeta+\gamma=0. \tag{10.36}$$

We can distinguish two cases:
(1a) $\tilde{\Delta}=1-4\alpha\gamma>0$. Then the roots are real. A necessary and sufficient condition for them to lie in $\mathscr{I}$ is that the minimum of $h(\zeta)$ belongs to $\mathscr{I}$ and this condition is

$$1<1/2\alpha<z_0. \tag{10.37}$$

Also, from (10.35), the condition $\tilde{C} > 0$ is written

$$\alpha + \gamma \mathscr{M}^2 - 1 > 0 \tag{10.38}$$

and it is easily seen that the inequalities (10.37)–(10.38) and $\tilde{\Delta} > 0$ are not compatible.

(1b) $\tilde{\Delta} < 0$. Then the solution of equation (10.36) with negative imaginary part is

$$\zeta_R = 1/2\alpha, \quad \zeta_R = -\sqrt{-\tilde{\Delta}}/2\alpha$$

and by imposing that it belongs to the half disk $\mathscr{D}$ gives the condition

$$\alpha + \mathscr{M}^2\gamma - \tfrac{1}{2}(1 + \mathscr{M}^2) < 0,$$

which is incompatible with (10.38).

(2) This case is possible only if $p'_\varepsilon < 0, \tilde{K} - 1 < 0$.

We distinguish two cases:

(2a) $D = \tilde{B}^2 - 4\tilde{A}\tilde{C} > 0$. Then the roots are real. However, for $\zeta \in \mathscr{I}$, it is $g(\zeta) > 0$ because $g(\zeta)$ in this case is concave and takes positive values at the end points of $\mathscr{I}$.

(2b) $D = \tilde{B}^2 - 4\tilde{A}\tilde{C} < 0$. Then for a complex root ζ one has

$$\operatorname{Re} g(\zeta) = g(\zeta_R) - \tilde{A}\zeta_I^2 > g(\zeta_R) > 0$$

because $\zeta_R \in \mathscr{I}$. Q.E.D.

10.2. Corrugation stability of a relativistic shock

In inertial coordinates (t, x^i), let $\Sigma(t)$ be the moving surface representing a shock wavefront.

Taking the jump of the field equations (10.1) across $\Sigma(t)$ and using the first order compatibility relations (3.23)–(3.25) yields

$$\begin{aligned}\mathscr{A}^0_- \mathscr{D}[[\mathbf{U}]] + (\mathscr{A}^i_- \nu_i - v_\Sigma \mathscr{A}^0_-)\mathbf{Y}^1 + [[\mathscr{A}^0]](\partial_t \mathbf{U})_+ \\ + [[\mathscr{A}^i]](\partial_i \mathbf{U})_+ + \mathscr{A}^i_- \tilde{\partial}_i [[\mathbf{U}]] = 0,\end{aligned} \tag{10.39}$$

where $\mathscr{D}$ is the Thomas displacement derivative, ν_i is the normal to Σ, v_Σ is the normal speed of propagation of Σ, $\mathbf{Y}^1 = [[\nu^i \partial_i \mathbf{U}]]$, and $\tilde{\partial}_i = \pi_i^j \partial_j$, with $\pi_i^j = \delta_i^j - \nu_i \nu^j$ the projection tensor onto $\Sigma(t)$.

We shall assume that the Rankine-Hugoniot relations can be solved giving

$$[[\mathbf{U}]] = G(\mathbf{U}_+, \nu_i, v_\Sigma). \tag{10.40}$$

Substituting equation (10.40) into equation (10.39), multiplying by the left eigenvector $\mathbf{L}_{-}^{(k)}$ of the matrix $\mathscr{A}^i_{-} - \lambda\mathscr{A}^0_{-}$ corresponding to the eigenvalue $\lambda_{-}^{(k)}$ (taken to be simple), and assuming propagation into a constant state, yields

$$^{\mathrm{T}}\mathbf{L}_{-}^{(k)}\left\{\mathscr{A}^0_{-}\frac{\partial G}{\partial v_\Sigma}\mathscr{D}v_\Sigma + \mathscr{A}^0_{-}\frac{\partial G}{\partial v_i}\mathscr{D}v_i + \mathscr{A}^i_{-}\frac{\partial G}{\partial v_\Sigma}\tilde{\partial}_i v_\Sigma \right.$$

$$\left. + \mathscr{A}^i_{-}\frac{\partial G}{\partial v_i}\tilde{\partial}_i v_j\right\} + (\lambda_{-}^{(k)} - v_\Sigma)\pi^1 = 0, \tag{10.41}$$

where

$$\pi^1 = {}^{\mathrm{T}}\mathbf{L}_{-}^{(k)}\mathscr{A}^0\mathbf{Y}^1. \tag{10.42}$$

For a right propagating acoustic shock, in an inertial frame in which the fluid ahead of the wavefront is at rest, the jump conditions have been analyzed in Section 8.2, and can be written in the form of equations (8.35)–(8.37), that is,

$$\rho = \rho_0\frac{v_\Sigma}{\Gamma_b(v_\Sigma - v_b)}, \tag{10.43a}$$

$$p = \frac{p_0 + \rho_0 v_\Sigma v_b}{1 - v_\Sigma v_b}, \tag{10.43b}$$

$$\frac{1+\varepsilon}{\Gamma_b} = \frac{p_0\,\rho_b}{\rho_0\,v_\Sigma} + 1 + \varepsilon_0. \tag{10.43c}$$

Henceforth in this section, the quantities ahead of the front will be denoted with the subscript 0, and the quantities behind the front are without a subscript (except v_b, Γ_b).

By using equations (10.43) in order to evaluate $\dfrac{\partial G}{\partial v_\Sigma}$ and $\dfrac{\partial G}{\partial v_i}$ and substituting into equation (10.41), one gets

$$\left\{\rho a^2\left(v_b + \frac{1}{c_s}\right)\frac{\mathrm{d}(\Gamma_b v_b)}{\mathrm{d}v_\Sigma} + \Gamma_b(1 + c_s v_b)\frac{\mathrm{d}p}{\mathrm{d}v_\Sigma}\right\}\mathscr{D}v_\Sigma$$

$$+ \Gamma_b v_b a^2 \rho\tilde{\partial}_i v^i + (\lambda_3 - v_\Sigma)\pi^1 = 0, \tag{10.44}$$

where λ_3 is given by expression (10.15).

For the step-shock considered in the previous section [equation (10.3)], equation (10.44) is automatically satisfied because $\overset{0}{v}_\Sigma$ is constant, v^i is constant, and π^1 vanishes.

Let Σ^0 be the initial plane wavefront (at $t = 0$) of the step-shock.

A particular class of perturbations perturbs the initial wavefront $\Sigma^0 \to \Sigma(0)$ in such a way that:

(i) the normal speed of propagation remains the same,

$$(v_\Sigma)_{t=0} = \overset{0}{v}_\Sigma$$

all over the perturbed surface $\Sigma(0)$;

(ii) the field ahead of the perturbed shock front remains constant.

The perturbed field behind $\Sigma(0)$ can be found from equations (10.43). The quantity π^1 can be assigned arbitrarily on $\Sigma(0)$ and corresponds to the arbitrariness in the profile of the perturbed field behind $\Sigma(0)$.

The quantity π^1 has a well-defined intrinsic meaning. In fact, under the transformation of field variables

$$\mathbf{U} = \psi(\mathbf{U}'),$$

with ψ a diffeomorphism, the field equations (10.1) are written

$$\mathscr{A}^{0'}\partial_t \mathbf{U}' + \mathscr{A}^{i'}\partial_i \mathbf{U}' = 0,$$

where

$$\mathscr{A}^{0'} = \mathscr{A}^0 C, \quad \mathscr{A}^{i'} = \mathscr{A}^i C,$$

with $C = \nabla_{\mathbf{U}'}\psi$ a nonsingular matrix.

Then one easily finds that

$$\pi^{1'} = \pi^1 + {}^{\mathrm{T}}\mathbf{L}^{(k)}_{-}\mathscr{A}^0_{-} C_{-}[[C^{-1}]]v^i(\partial_i \mathbf{U}')_{+}$$

and, therefore, *for propagation into a constant state π^1 is invariant.*

Whitham's intuitive definition of stability can be made precise as follows.

DEFINITION 10.1. *The step-shock* (10.3) *is said to be corrugationally stable if for perturbations satisfying* (i) *and* (ii) *and for $|\pi^1|$ sufficiently small one has:*

$$(\mathscr{D}v_\Sigma)(\tilde{\partial}_i v^i) < 0 \tag{10.45}$$

in an interval $[0, \bar{t}]$.

The condition (10.45) is a precise formulation of the requirement that the shock acceleration decreases where the front is expanding and increases where the front is converging. In fact, $\tilde{\partial}_i v^i$ measures the expansion of a bundle of rays (normal curves to the wavefront) (Fig. 10.1). Notice also that the above definition is relativistically invariant.

A necessary condition for corrugation stability is that the requirement (10.45) *be satisfied at* $t = 0$ *and for perturbations* $\pi^1 = 0$ (this latter statement having an invariant meaning, independent of the choice of the field variables).

Then, from equation (10.44) it follows that

$$v_b \Theta > 0, \tag{10.46}$$

where

$$\Theta = \left(v_b + \frac{1}{c_s}\right)\frac{\mathrm{d}(\Gamma_b v_b)}{\mathrm{d}v_\Sigma} + \frac{\Gamma_b(1 + c_s v_b)}{\rho a^2}\frac{\mathrm{d}p}{\mathrm{d}v_\Sigma}. \tag{10.47}$$

A straightforward calculation shows that

$$\Theta = \frac{\Gamma_b^3(1 + c_s v_b)}{c_s}\left\{\frac{\mathrm{d}v_b}{\mathrm{d}v_\Sigma}\left(1 + \frac{v_\Sigma - v_b}{c_s(1 - v_b v_\Sigma)}\right) + \frac{v_b}{c_s v_\Sigma}\frac{v_\Sigma - v_b}{(1 - v_b v_\Sigma)}\right\},$$

which can also be written in the form

$$\Theta = \frac{\Gamma_b^3(1 + c_s v_b)(v_\Sigma - v_b)}{c_s^2(1 - v_b v_\Sigma)}\left\{\frac{v_b}{v_\Sigma} + \left(\frac{\mathrm{d}v_b}{\mathrm{d}v_\Sigma}\right)\left(1 + c_s\frac{(1 - v_b v_\Sigma)}{v_\Sigma - v_b}\right)\right\}. \tag{10.48}$$

The inequality (10.46) is equivalent to

$$\frac{v_b^2}{v_\Sigma}\left\{1 + \frac{v_\Sigma}{v_b}\left(\frac{\mathrm{d}v_b}{\mathrm{d}v_\Sigma}\right)\left(1 + c_s\frac{1 - v_b v_\Sigma}{v_\Sigma - v_b}\right)\right\} > 0, \tag{10.49}$$

Fig. 10.1. Sketch of the intuitive definition of corrugation stability.

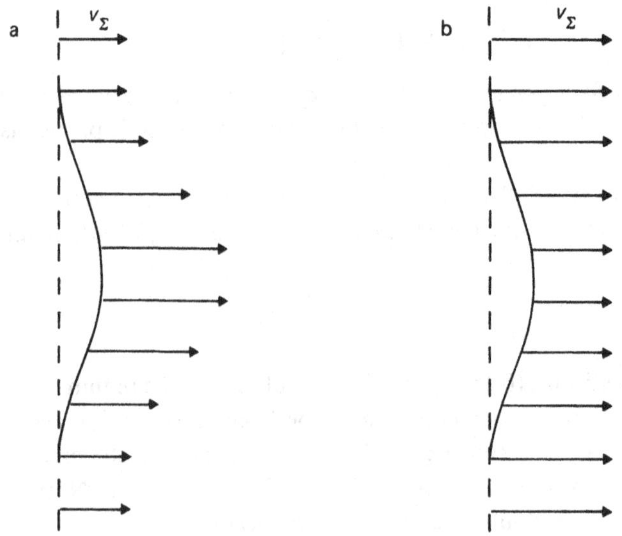

which reduces to

$$\Phi \equiv \frac{v_\Sigma}{v_b}\frac{dv_b}{dv_\Sigma}\left(1 + c_s\frac{1 - v_b v_\Sigma}{v_\Sigma - v_b}\right) + 1 > 0. \tag{10.50}$$

The relationship between v_b, v_Σ and speeds v, v_+ relative to the shock front used in the previous section is

$$v = \frac{v_b - v_\Sigma}{1 - v_b v_\Sigma}, \quad v_+ = -v_\Sigma. \tag{10.51}$$

In order to compare the corrugation stability criterion (10.50) with the linear stability one it is convenient to express the quantity F in the inertial frame in which the fluid ahead of the shock is at rest.

It is easily seen that

$$\frac{(1 - v_b v_\Sigma)^2}{(1 - v_\Sigma^2)} F \equiv F_b \equiv 1 - v_b^2 + \frac{v_\Sigma - v_b}{c_s} - \frac{1}{\rho\varepsilon_p}\frac{v_b(v_\Sigma - v_b)}{c_s}. \tag{10.52}$$

The equation for the Taub adiabat is [equation (8.7)]

$$f^2 - f_0^2 = (p - p_0)(\tau + \tau_0). \tag{10.53}$$

By differentiating equation (10.53), keeping constant the state ahead, one has

$$2f\,df = (\tau_0 + \tau)dp + (p - p_0)d\tau, \tag{10.54}$$

which gives

$$\left\{2(1 + \varepsilon) + \frac{p + p_0}{\rho}\right\}d\varepsilon + \left\{(1 + \varepsilon)(p + p_0) + \frac{2pp_0}{\rho}\right\}d\left(\frac{1}{\rho}\right)$$
$$= \left\{\tau_0 - \tau + \frac{p - p_0}{\rho^2}\right\}dp. \tag{10.55}$$

Now

$$d\varepsilon = \varepsilon_\rho d\rho + \varepsilon_p dp. \tag{10.56}$$

From the first law of thermodynamics

$$T\,dS = d\varepsilon - \frac{p}{\rho^2}d\rho = \left(\varepsilon_\rho - \frac{p}{\rho^2}\right)d\rho + \varepsilon_p dp$$

one obtains

$$\varepsilon_p\left(\frac{\partial p}{\partial \rho}\right)_S = \frac{p}{\rho^2} - \varepsilon_\rho,$$

whence

$$\varepsilon_\rho = \frac{p}{\rho^2} - \varepsilon_p f c_s^2. \tag{10.57}$$

Using (10.56), (10.57), and the jump conditions (10.43), equation (10.55) yields

$$f\left\{-1 + \frac{2 - v_b v_\Sigma - v_b^2}{v_b(v_\Sigma - v_b)} c_s^2 \rho \varepsilon_p\right\} \mathrm{d}\left(\frac{1}{\rho}\right)$$

$$= \frac{\Gamma_b^2}{\rho_0^2 v_\Sigma^2}\left\{(1 - v_b v_\Sigma)^2 - \frac{v_\Sigma - v_b}{v_b}(2 - v_b v_\Sigma - v_b^2)\rho\varepsilon_p\right\}\mathrm{d}p. \tag{10.58}$$

Differentiating (10.43a) along the Taub adiabat gives

$$\rho_0 \frac{\mathrm{d}(1/\rho)}{\mathrm{d}p} = \frac{v_b \Gamma_b}{v_\Sigma^2} \frac{\mathrm{d}v_\Sigma}{\mathrm{d}p} - \frac{\Gamma_b^3}{v_\Sigma}(1 - v_\Sigma v_b)\frac{\mathrm{d}v_b}{\mathrm{d}p}, \tag{10.59}$$

and, by differentiating (10.43b),

$$(1 - v_\Sigma v_b)\mathrm{d}p = (e_0 + p)(v_\Sigma \mathrm{d}v_b + v_b \mathrm{d}v_\Sigma). \tag{10.60}$$

Let

$$X = \frac{v_\Sigma \, \mathrm{d}v_b}{v_b \, \mathrm{d}v_\Sigma}. \tag{10.61}$$

Then, from (10.43) it follows that

$$\frac{\mathrm{d}v_\Sigma}{\mathrm{d}p} = \frac{(1 - v_\Sigma v_b)^2}{v_b(1 + X)(e_0 + p_0)}, \tag{10.62}$$

$$\frac{\mathrm{d}v_b}{\mathrm{d}p} = \frac{\mathrm{d}v_b}{\mathrm{d}v_\Sigma}\frac{\mathrm{d}v_\Sigma}{\mathrm{d}p} = X\frac{v_b}{v_\Sigma}\frac{\mathrm{d}v_\Sigma}{\mathrm{d}p} = \frac{X}{v_\Sigma v_\Sigma}\frac{(1 - v_\Sigma v_b)^2}{(1 + X)(e_0 + p_0)}. \tag{10.63}$$

Substituting equations (10.62)–(10.63) into equation (10.59) yields

$$\rho_0 \frac{\mathrm{d}(1/\rho)}{\mathrm{d}p} = \frac{v_b \Gamma_b}{v_\Sigma} \frac{1 - v_b v_\Sigma}{(p - p_0)(1 + X)}\left(1 - \frac{1 - v_b v_\Sigma}{1 - v_b^2} X\right). \tag{10.64}$$

By comparing this equation with (10.58) one gets

$$\frac{(1 - v_b v_\Sigma)^2 - \dfrac{v_\Sigma - v_b}{v_b}(2 - v_\Sigma v_b - v_b)^2 \rho\varepsilon_p}{-1 + \dfrac{2 - v_\Sigma v_b - v_b^2}{v_b(v_\Sigma - v_b)} c_s^2 \rho\varepsilon_p}$$

$$= \frac{1 - v_\Sigma v_b}{1 + X}(1 - v_b^2 - (1 - v_\Sigma v_b)X), \tag{10.65}$$

and solving with respect to X,

$$X = \frac{\dfrac{(1-v_b^2)(v_\Sigma - v_b)}{v_b(1-v_b v_\Sigma)}\dfrac{1}{M^2} + \dfrac{v_\Sigma - v_b}{v_b} - (1 - v_b v_\Sigma)\dfrac{1}{\rho \varepsilon_p}}{\dfrac{v_\Sigma - v_b}{v_b}\left(\dfrac{1}{M^2} - 1\right)}, \qquad (10.66)$$

where

$$M = \frac{v_\Sigma - v_b}{1 - v_\Sigma v_b}\frac{1}{c_s}.$$

In terms of M and X the corrugation stability criterion (10.50) is written

$$\Phi = \left(1 + \frac{1}{M}\right)X + 1 > 0. \qquad (10.67)$$

Substituting equation (10.66) into equation (10.67) gives

$$\Phi = 1 - v_b^2 + \frac{v_\Sigma - v_b}{c_s} - \frac{v_b(v_\Sigma - v_b)}{c_s^2}\frac{1}{\rho \varepsilon_p} = F_b. \qquad (10.68)$$

Therefore the two sufficient conditions for instability (linear and corrugation) coincide.

Another equivalent formulation of the stability criterion is the following. From the jump relations equations (8.5) and (8.16) one has

$$m^2 = \frac{p - p_0}{\tau - \tau_0} = (\rho_0 \Gamma_\Sigma v_\Sigma)^2. \qquad (10.69)$$

Differentiating the above equation along the Taub adiabat and making use of (10.62) and of the jump relations in the form (9.69)–(9.70) gives

$$\frac{1 + X}{2\Gamma_\Sigma^2} = \frac{1 - v v_\Sigma}{1 + m^2 \dfrac{d\tau}{dp}} \qquad (10.70)$$

and substituting into (10.50) yields

$$M\Phi = \frac{\Gamma_\Sigma^2[1 - 2v_\Sigma v + v_\Sigma^2 + 2(v_\Sigma - v)/c_s] - m^2 d\tau/dp}{1 + m^2 d\tau/dp}. \qquad (10.71)$$

Therefore the inequality $\Phi > 0$, since $M > 0$, is equivalent to

$$-1 < m^2 \frac{d\tau}{dp} < \frac{(1 - v_\Sigma v)(1 + 2M) + v_\Sigma(v_\Sigma - v)}{1 - v_\Sigma^2}. \qquad (10.72)$$

Remark 1. The nonrelativistic limit of the above criterion coincides with the usual one (Erpenbeck, 1962).

Remark 2. From equation (10.72) it is easy to see that *compressive weak shocks are always stable.* In fact, since, for weak shocks, the Taub and Poisson adiabat have a second order contact at (p_0, τ_0), we can substitute τ'_p for $d\tau/dp$ along the shock adiabat. Then $\tau'_p \leq 0$ (relativistic causality) implies that the second half of the inequality (10.72) is satisfied because the right-hand side is positive. The first half of the inequality reads

$$\frac{d\tau}{dp} - \frac{\tau - \tau_0}{p - p_0} = \tfrac{1}{2}(\tau''_p)_0[[p]] + O([[p]]^2)$$

along the Taub adiabat in the neighborhood of (τ_0, p_0). This inequality is satisfied as a consequence of the compressibility assumption (8.17a).

Remark 3. It is interesting to investigate the stability condition for barotropic fluids. We start from the jump relations (8.31b), (8.34),

$$\Gamma^2(e+p)(v_\Sigma - v_b) = (p + e_0)v_\Sigma,$$
$$\Gamma^2(e+p)(v_\Sigma - v_b)v_b = p - p_0$$

from which we obtain

$$v_\Sigma v_b = \frac{p - p_0}{p + e_0}$$

and

$$(e - e_0)v_\Sigma = (p_0 + e)v_b.$$

From these equations we obtain

$$v_\Sigma^2 = \frac{(p - p_0)(p_0 + e)}{(e - e_0)(p + e_0)}, \qquad (10.73a)$$

$$v_b^2 = \frac{(p - p_0)(e - e_0)}{(p_0 + e)(p + e_0)}. \qquad (10.73b)$$

Also,

$$v_\Sigma \frac{dv_b}{dv_\Sigma} + v_b = \frac{c_s^2(e_0 + p_0)}{(p + e_0)^2}\frac{de}{dv_\Sigma}$$

and

$$(v_\Sigma - v_b)\frac{de}{dv_\Sigma} = e_0 - e + (p_0 + e)\frac{dv_b}{dv_\Sigma},$$

whence, by using equation (8.26a),

$$\frac{dv_b}{dv_\Sigma} = \frac{e - e_0}{p_0 + e} \frac{c_s^2(p_0 + e)(e - e_0) + (p - p_0)(p + e_0)}{c_s^2(p_0 + e)(e - e_0) - (p - p_0)(p + e_0)}, \tag{10.74}$$

from which

$$X = \frac{c_s^2 + v_-^2}{c_s^2 - v_-^2}, \tag{10.75}$$

where

$$v_- = \frac{v_b - v_\Sigma}{1 - v_b v_\Sigma}$$

is the fluid speed behind the shock as measured in the shock frame.

The corrugation stability criterion (10.50) can be written as

$$\frac{v_\Sigma}{v_b} \frac{dv_b}{dv_\Sigma}\left(1 - \frac{c_s}{v_-}\right) + 1 > 0. \tag{10.76}$$

By inserting the expression (10.75) into (10.76) we find that the latter inequality is verified iff

$$c_s > |v_-|.$$

Therefore, for a barotropic fluid, the corrugation stability condition is equivalent to one of the Lax conditions (subsonic flow behind the flow). For a $\mathscr{C}^2$ differentiable state equation the above condition implies the Weyl condition (8.27), as can be seen from PROPOSITION 8.12. It is easy to prove that for cold matter fluids the Weyl condition (8.27) implies the above Lax condition (Russo and Anile, 1987). In this case the state equation can be put in the form

$$p = p(\rho), \quad e = e(\rho).$$

One can prove that

$$p'_\tau = \rho^2 \frac{p'_e}{p'_e - 1}$$

and hence the condition $c_s > |v_-|$ is equivalent to

$$\frac{p'_\tau(1 - M^2) + M^2\rho^2}{p'_\tau - \rho^2} > 0,$$

and by employing relativistic causality, $\tau'_p < 0$, and the relationship

$$\rho^2 \frac{M^2}{1 - M^2} = \frac{p - p_0}{\tau - \tau_0}$$

[obtained by employing equation (8.26a)], one finds

$$p'_\tau < \frac{p - p_0}{\tau - \tau_0},$$

which is the convexity condition and is equivalent to

$$p''_\tau > 0,$$

which in its turn is easily seen to be equivalent to the Weyl condition.

By using the results of Proposition 8.12 then we have that for a smooth ($\mathscr{C}^2$) state equation $p = p(e)$ the corrugation stability condition is implied by the Weyl condition

$$W = 2p'_e(1 - p'_e) + (e + p)p''_e > 0.$$

For barotropic fluids of the cold matter type, for which the state equation can be given parametrically as

$$p = p(\rho), \quad e = e(\rho),$$

the Weyl condition is also necessary for corrugation stability. In this case it is easy to see that the relativistic Weyl condition is more restrictive than the nonrelativistic one (which is also equivalent to the corrugation stability condition for a nonrelativistic shock). In fact, let

$$W_c = 2p'_\rho + \rho p''_\rho.$$

Then

$$\frac{W_c}{f} = W + 3(p'_e)^2$$

and therefore $W_c < 0$ implies $W < 0$ (classical instability entails relativistic instability).

The corrugation condition (10.50) or equivalently the linear stability one is not always easy to apply because it requires the explicit solution of the jump conditions across the shock. A weaker sufficient condition, which is much easier to apply, is the following one, which is the relativistic extension of the Bethe criterion (Bethe, 1942).

PROPOSITION 10.2. *If*

$$\frac{1}{\rho \varepsilon_p} < \frac{\rho a^2}{p} \tag{10.77}$$

then the linear stability condition (Theorem 10.1) holds.

Proof. We have

$$F = 1 - v^2 - \frac{v}{c_s}(1 - v_0 v) - \frac{1}{\rho \varepsilon_p c_s^2}(v_0 v - v^2).$$

Now, from equation (10.77)

$$F > 1 - v^2 - \frac{v}{c_s}(1 - v_0 v) - (v_0 v - v^2)\frac{\rho f}{p}.$$

Finally, from equations (8.26)

$$F > \frac{(e - e_0) - (p - p_0)}{(e - e_0)(e + p_0)} p_0 \left(1 + \frac{e}{p}\right) - \frac{v}{c_s}(1 - v_0 v) > 0. \qquad \text{Q.E.D.}$$

Remark 4. An interesting interpretation of the linear stability condition is that its violation would entail the possibility of shock breakup. In fact, it can be proved (Russo and Anile, 1987) that in this case there is another solution of the jump relations, consisting of a weaker shock followed by a contact discontinuity and a backward traveling rarefaction simple wave.

Remark 5. An interesting stability effect arises for ultrarelativistic shocks. From equation (10.44) we have

$$\mathscr{D} v_\Sigma = -\frac{\Gamma_b v_b}{\Theta} \tilde{\partial}_i v^i$$

for $\pi^1 = 0$, with Θ given by equation (10.47). In the case of a barotropic fluid with a constant speed of sound c_s one finds, after some manipulations,

$$\mathscr{D} v_\Sigma = -\frac{c_s^2 v_\Sigma (v_\Sigma - c_s)^2 (v_\Sigma + c_s)}{v_\Sigma (1 - c_s^2) - c_s \Gamma_\Sigma^{-2}} \Gamma_\Sigma^{-2} \tilde{\partial}_i v^i$$

from which it is apparent that, as $v_\Sigma \to 1$,

$$\mathscr{D} v_\Sigma = 0(\Gamma_\Sigma^{-2})$$

This phenomenon, which is similar to the effect described in Section 9.1, shows clearly that an ultrarelativistic shock tends to preserve its corrugations.

10.3. Further stability considerations

In Section 10.1 we investigated the existence of an instability mode which is exponentially growing with time and vanishing at spatial infinity. In this section we will investigate the existence of undamped bounded stationary modes. We shall limit ourselves to perturbations of the shock surface and of the field behind the shock. We consider perturbations of the kind (10.7), that is,

$$\delta \mathbf{U} = \mathbf{Y}(x)\mathrm{e}^{-i(\omega t + ly + mz)},$$

with ω, l, m real, $\mathbf{Y}(x)$ bounded, and, without loss of generality, $\omega \geq 0$. Then $\mathbf{Y}(x)$ is given by equation (10.23)

$$\mathbf{Y}(x) = a_1 \mathbf{R}_1 \mathrm{e}^{-ik_1 x} + \sum_j \overset{(j)}{a_2} \overset{(j)}{\mathbf{R}_2} \mathrm{e}^{-ik_2 x} + a_3 \mathbf{R}_3 \mathrm{e}^{-ik_3 x}.$$

The interpretation of the various modes is not straightforward. Naively one would interpret the k_3-mode as representing an acoustic wave incident upon the shock and the other modes as the reflected waves. However, a difficulty arises for the k_1-mode because from equation (10.21) one sees that for $l^2 + m^2$ sufficiently large one has $k_1 < 0$. An unambiguous physical interpretation can be obtained by observing that, in the shock rest frame, acoustic waves are dispersive and, therefore, in order to define the incident and reflected wave, one must resort to the concept of group velocity $\mathbf{v}_g$ given by

$$\mathbf{v}_g = -\left(\frac{\partial \omega}{\partial k}, \frac{\partial \omega}{\partial l}, \frac{\partial \omega}{\partial m}\right).$$

From the dispersion relation equation (10.18) one finds

$$\frac{\partial \omega}{\partial k} = -\frac{\mathcal{M}^2 \Omega v - k v^2}{\mathcal{M}^2 \Omega + v^2 \omega}, \tag{10.78a}$$

$$\frac{\partial \omega}{\partial l} = \frac{l v^2}{\mathcal{M}^2 \Omega + v^2 \omega}, \tag{10.78b}$$

$$\frac{\partial \omega}{\partial m} = \frac{m v^2}{\mathcal{M}^2 \Omega + v^2 \omega}. \tag{10.78c}$$

It is easy to see that for the acoustic modes $\Omega > 0$ and therefore $\partial \omega / \partial l$, $\partial \omega / \partial m$ are positive. Now we will prove that the group speed corresponding to k_3 is directed toward the shock front, whereas that corresponding to k_1 is in the opposite direction and hence the modes corresponding to k_3 and k_1 can be interpreted as incident and reflected acoustic waves.

From the dispersion relation (10.21) it is easy to see that k_3, k_1 vary in the ranges

$$-\frac{(1 + c_s v)\omega}{v + c_s} \leq k_3 \leq \frac{v(1 - c_s^2)\omega}{c_s^2 - v^2}, \tag{10.79}$$

$$\frac{v(1-c_s^2)\omega}{c_s^2-v^2} \le k_1 \le \frac{1-vc_s}{c_s-v}\omega. \tag{10.80}$$

Now

$$(v_g)_x = \frac{v^2\omega}{c_s^2(1-v^2)(\mathscr{M}^2\Omega+v^2\omega)}\left(v(1-c_s^2)-(c_s^2-v^2)\frac{k}{\omega}\right)$$

and therefore

$$(v_g)_x > 0$$

for the k_3-mode, which represents the incident acoustic wave and

$$(v_g)_x < 0$$

for the k_1-mode, which represents the reflected acoustic wave.

The instability we will consider is that for which the reflected waves exist without an incoming incident wave. This instability is sometimes referred to as a spontaneous emission of acoustic waves (Fowels, 1976). Therefore, in order for this instability to exist we must require

$$a_3 = 0,$$

which leads to equation (10.26).

It is convenient to introduce the variable

$$\xi = \frac{k_3 v}{\omega + k_3 v}. \tag{10.81}$$

In terms of ξ, equation (10.26) is rewritten as

$$\tilde{C}\xi^2 + \tilde{B}\xi + \tilde{A} = 0, \tag{10.82}$$

with $\tilde{A}$, $\tilde{B}$, $\tilde{C}$ given by equation (10.27) and ξ varing in the range

$$\mathscr{M}^2 \le \xi \le \Gamma^2(M - v^2). \tag{10.83}$$

It is easy to check that in this interval the function

$$f(\xi) = \tilde{C}\xi^2 + \tilde{B}\xi + \tilde{A}$$

is monotonic.

Now we can state the following stability condition.

PROPOSITION 10.3.

(i) *If*

$$1 - v^2 - (\tilde{K} - 1)M^2 < \tilde{H} < 1 - v^2 + M(1 - vv_+) \tag{10.84}$$

then there is an unstable mode.

(ii) *If*

$$\tilde{H} < 1 - v^2 - (\tilde{K} - 1)M^2 \tag{10.85}$$

then there is no unstable mode.

Proof.

(i) It is easy to check that

$$f(\mathcal{M}^2) = (1 - \mathcal{M}^2)(1 - v^2 - (\tilde{K} - 1)M^2 - \tilde{H}) < 0,$$
$$f(\Gamma^2(M - v^2)) = \Gamma^2(1 - M)(1 - v^2 + M(1 - vv_+) - \tilde{H}) > 0.$$

Therefore, there exists a solution of $f(\xi) = 0$ in the allowed interval.

(ii) In this case $f(\mathcal{M}^2) > 0$. Also, it is easy to check that

$$f(\Gamma^2(M - v^2)) > 0$$

irrespective of the sign of $\tilde{H}$. Therefore, since $f(\xi)$ is monotonic in the given interval, it cannot have a zero. Q.E.D.

Remark 6. The condition (10.85) can be rewritten in the form

$$\tilde{F} = 1 - v^2 - \frac{vv_+ - v^2}{c_s^2}\left(1 + \frac{1}{\rho\varepsilon_p}\right) > 0. \tag{10.86}$$

By using equations (8.26a)–(8.26c), equation (10.86) is rewritten as

$$\tilde{F} = \frac{[(e - e_0) - (p - p_0)](e + p)}{(e + p_0)(e - e_0)}\left(1 - \frac{p - p_0}{(e + p)c_s^2}\left(1 + \frac{1}{\rho\varepsilon_p}\right)\right) > 0,$$

which, since $p - p_0 < e - e_0$, is equivalent to

$$(e + p)c_s^2 - p\left(1 + \frac{1}{\rho\varepsilon_p}\right) + p_0\left(1 + \frac{1}{\rho\varepsilon_p}\right) > 0, \tag{10.87}$$

that is,

$$\frac{\rho}{p}\left(\frac{\partial p}{\partial \rho}\right)_\varepsilon - 1 + \frac{p_0}{p}\left(1 + \frac{1}{\rho\varepsilon_p}\right) > 0. \tag{10.88}$$

This form of the criterion is formally identical to the one in nonrelativistic fluid dynamics (Kontorovich, 1957). From equation (10.88) a sufficient condition for stability can be obtained in the form

$$\frac{\rho}{p}\left(\frac{\partial p}{\partial \rho}\right)_\varepsilon \geq 1, \tag{10.89}$$

$$1 + \frac{1}{\rho\varepsilon_p} > 0, \tag{10.90}$$

which again coincides with the classical form (Kontorovich, 1957).

Remark 7. If F is given by (10.34) then it is easy to see that

$$F = \tilde{F} + \frac{|v|}{c_s}(1 - vv_+) + \frac{vv_+ - v^2}{c_s^2}. \tag{10.91}$$

Now, for a compressive shock, $|v_+| > |v|$ and therefore $F > \tilde{F}$. It follows that the reflection stability criterion is more restrictive than the linear stability one.

It is interesting to investigate the form of the reflection stability condition in the case of a barotropic fluid. The form of the criterion (10.87) is not directly applicable because the quantity $1/\rho\varepsilon_p$ is not defined for a barotropic fluid. This difficulty can be circumvented by noting that in equation (10.66) the quantity X on the left-hand side is well defined for a barotropic fluid and has the value given by equation (10.75). This relationship then gives the numerical value of the quantity $1/\rho\varepsilon_p$ for a barotropic fluid and it turns out to be c_s^2. Therefore, the reflection stability criterion (10.87) for a barotropic fluid is equivalent to

$$p'_e > \frac{p - p_0}{e + p_0}. \tag{10.92}$$

A sufficient condition for reflection stability is then

$$p'_e > \frac{p}{e}. \tag{10.93}$$

10.4. Examples

We shall first consider the case of cold matter fluids. In compact objects such as neutron stars the thermal energy is much lower than the Fermi energy of nucleons and the matter can be treated as cold (although it corresponds to temperatures of order 10^9 K). In such a medium shock waves could propagate and if they are moderately weak they would not alter the cold matter assumptions.

Cold fluids can be described by a barotropic state equation of the kind

$$p = p(\rho), \quad e = e(\rho), \tag{10.94}$$

which corresponds to assuming that the flow is isentropic. This assumption is consistent with the presence of moderately weak shock waves because the entropy jump across a weak shock is of third order in the shock strength [equation (8.17)].

The stability condition for a shock in a barotropic fluid reduces to the compressibility assumption (Remark 3), that is,

$$\tau_p'' > 0, \tag{10.95}$$

which, consistently, is also the stability condition for weak shocks. We will now investigate this property for several cold matter state equations of astrophysical interest.

One has

$$\tau_p'' = \frac{1}{\rho^2 (p_e')^3} W,$$

where W is defined by equation (8.27) as

$$W = 2(1 - p_e')p_e' + (e + p)p_e''.$$

Therefore, equation (10.95) is equivalent to the relativistic Weyl condition

$$W > 0. \tag{10.96}$$

We will consider the following cases.

i. Completely degenerate ideal Fermi gas

This equation of state is the crudest approximation to that appropriate for neutron star matter (Shapiro and Teukolsky, 1983). It can be written in parametric form as

$$e = \rho + \frac{m}{\lambda_N^3} \chi(x), \tag{10.97}$$

$$p = \frac{m_N}{\lambda_N^3} \phi(x), \tag{10.98}$$

where m_N is the nucleon mass, λ_N is the nucleon Compton wavelength, $x = p_F/m_N$, p_F is the Fermi momentum, and

$$\chi(x) = \frac{1}{8\pi^2}\{x(1 + x^2)^{1/2}(1 + 2x^2) - \ln[x + (1 + x^2)^{1/2}]\}, \tag{10.99}$$

$$\phi(x) = \frac{1}{8\pi^2}\{x(1 + x^2)^{1/2}(\tfrac{2}{3}x^2 - 1) + \ln[x + (1 + x^2)^{1/2}]\}. \tag{10.100}$$

The density ρ is given as a function of x by

$$\rho = \frac{m_N}{3\pi^2 \lambda_N^3} x^3 = \bar{\rho} x^3. \tag{10.101}$$

An easy calculation gives

$$p'_e = \frac{1}{3}\frac{x^2}{1 + x^2 + (1 + x^2)^{1/2}}, \qquad (10.102)$$

whence

$$0 < p'_e < \tfrac{1}{3}.$$

Also, one has

$$p''_e > 0$$

and therefore the stability condition (10.96) is verified. Also, the reflection stability criterion (10.93) is easily checked.

ii. Cold ideal n-p-e gas

The fluid consists of a degenerate gas of electrons, protons, and neutrons in equilibrium with the beta inverse reactions

$$e^- + p \to n + \nu$$

and the neutrinos are assumed to escape from the system (zero chemical potential). The energy density and pressure are given by

$$e = \frac{m_e}{\lambda_e^3}\chi(x_e) + \frac{m_p}{\lambda_p^3}\chi(x_p) + \frac{m_n}{\lambda_n^3}\chi(x_n), \qquad (10.103)$$

$$p = \frac{m_e}{\lambda_e^3}\phi(x_e) + \frac{m_p}{\lambda_p^3}\phi(x_p) + \frac{m_n}{\lambda_n^3}\phi(x_n), \qquad (10.104)$$

with an obvious extension of the notation used for the completely degenerate Fermi gas (Shapiro and Teukolsky, 1983). The parameters x_e, x_p, x_n are related by the thermodynamical equilibrium condition

$$m_e(1 + x_e^2)^{1/2} + m_p(1 + x_p^2)^{1/2} = m_n(1 + x_n^2)^{1/2}, \qquad (10.105)$$

expressing the balance of chemical potentials, and

$$m_e x_e = m_p x_p, \qquad (10.106)$$

expressing charge neutrality.

The density threshold for the beta inverse reaction is obtained by setting $x_n = 0$ in equation (10.105) and corresponds to the critical density

$$\rho_c \approx 1.2 \times 10^7 \text{ g/cm}^3,$$

where ρ is obtained by $\rho = e/c^2$ (c being the speed of light in cgs units).

From equations (10.105)–(10.106) it is possible to express x_p and x_n as

Fig. 10.2. (a) Continuous line: equation of state $p = p(\rho)$ for a cold *n-p-e* gas. Dashed line: pressure threshold for reflection instability. (b) Plot of the adiabatic index as function of ρ for a cold *n-p-e* gas.

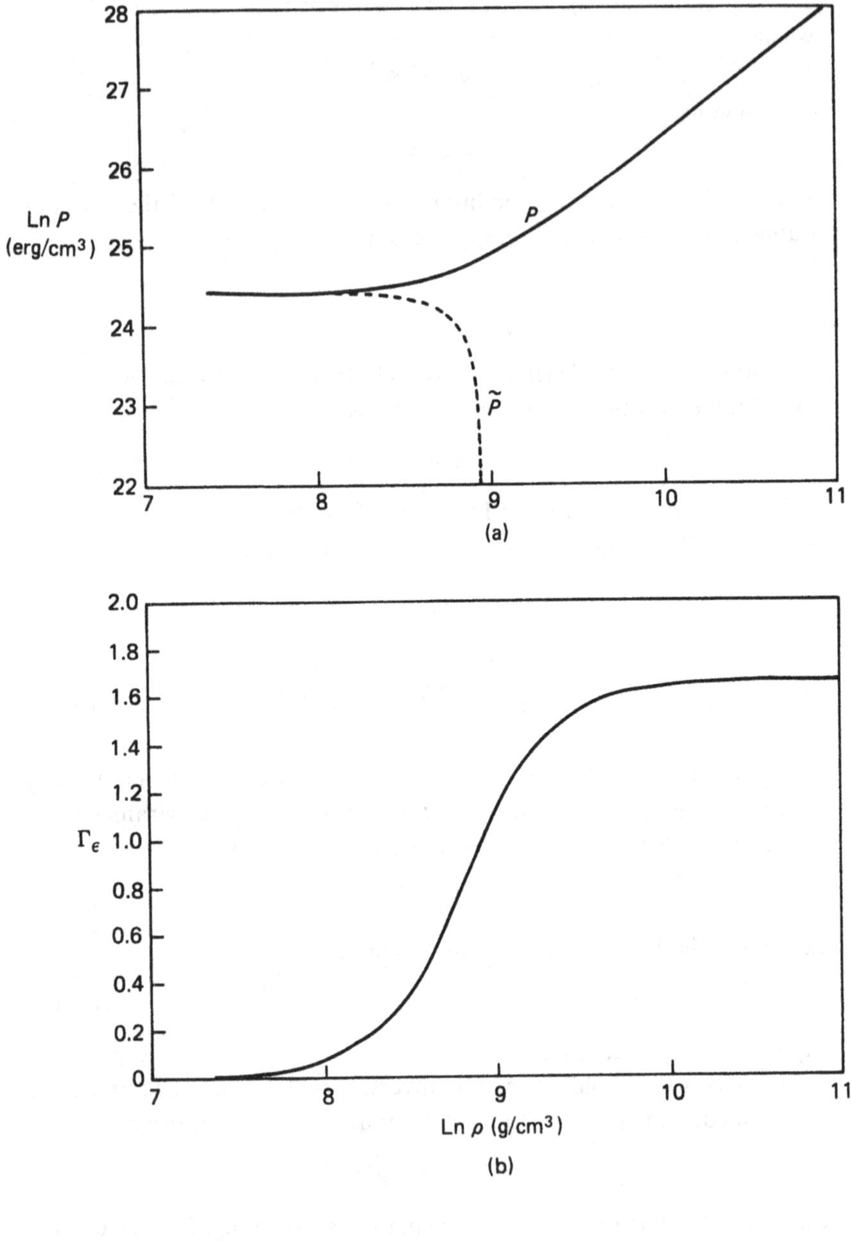

functions of x_e and obtain the state equations in parametric form. The stability condition (10.95) has been checked numerically for densities in the range $\rho_c \leq \rho \leq 10^{15}\,\text{g/cm}^3$ (Russo, 1988). For $\rho < \rho_c$ the matter state equation is given by a Thomas-Fermi model. At the point ρ_c the function $p(\rho)$ has a discontinuity in the derivative. Therefore, in this case the stability condition must be put in the original nonlocal form

$$c_s^2 > \frac{(p - p_0)(p + e_0)}{(e + p_0)(e - e_0)}. \tag{10.107}$$

One finds that if there is a shock across the density ρ_c the stability condition is violated (Russo, 1988). The density at which $W = 0$ is very close to the critical density, the difference being of order of $10^6\,\text{g/cm}^3$ (Fig. 10.2a).

Also, there is a range of densities for which the reflection stability condition (10.92) is violated (Russo, 1988). For $\rho < 8{,}5226 \times 10^8\,\text{g/cm}^3$ one has for the adiabatic index

$$\tilde{\Gamma} \equiv \frac{e}{p}\left(\frac{\partial p}{\partial e}\right),$$

$\tilde{\Gamma} < 1$ (Fig. 10.2b). Hence, for a shock with

$$p_0 < \frac{p - ep'_e}{1 + p'_e}$$

the inequality (10.92) is violated. In Fig. 10.2a we have also plotted the curve

$$\tilde{p} \equiv \frac{p - ep'_e}{1 + p'_e}.$$

iii. Harrison-Wheeler state equation

This equation of state, which has been used in several calculations of neutron star structure is an improvement upon the cold n-p-e gas in that it incorporates some effects of nuclear forces and of free neutrons at sufficiently high density (neutron drip at the density ρ_d) (Shapiro and Teukolsky, 1983). The corrugation stability criterion is verified (numerically) in the local form in the range where the state equation is smooth. At ρ_d the state equation curve has a discontinuity in its first derivative and it is possible to show that the stability condition in the form (10.107) is not verified across ρ_d (Fig. 10.3a) (Russo, 1988). In Fig. 10.3b we plot the adiabatic index $\tilde{\Gamma}$. It is seen that there are regions in which reflection instability may occur.

Fig. 10.3 (a) The Harrison-Wheeler state equation $p = p(\rho)$. The dotted line represents a region in which the Weyl condition $W > 0$ is violated. The lower line is the pressure threshold for reflection instability. (b) Adiabatic index for the Harrison-Wheeler state equation. The dotted line is as in part (a).

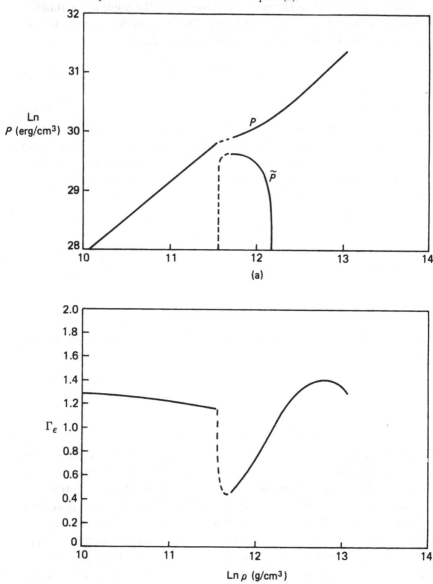

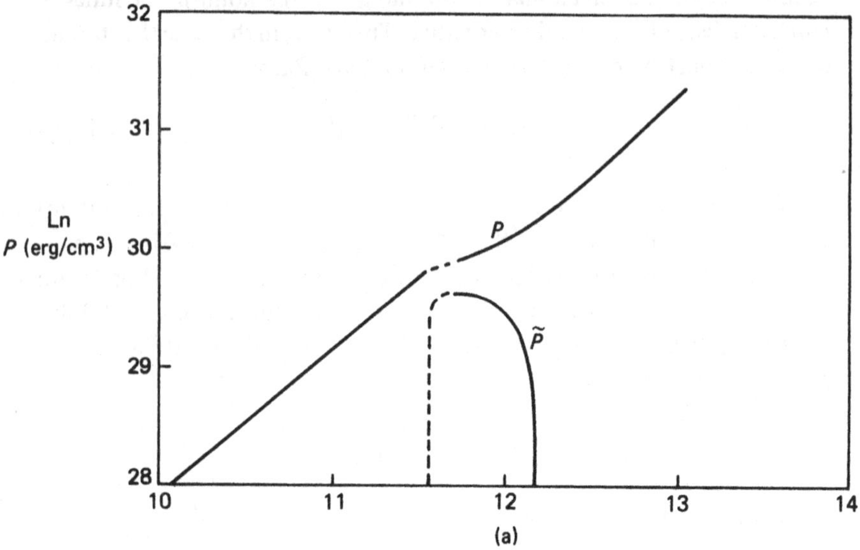

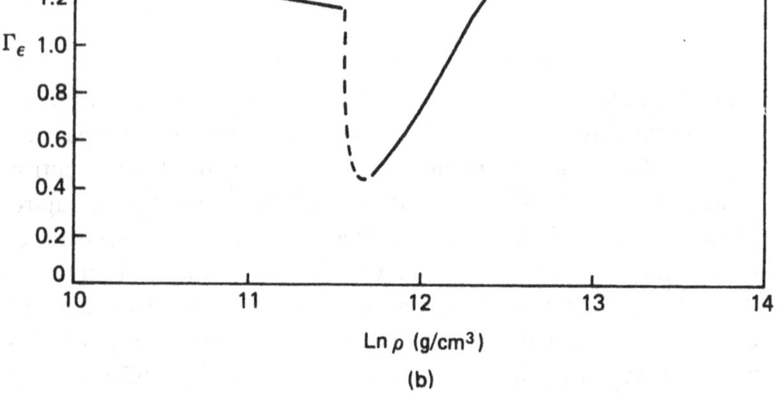

iv. Bethe, Baym, and Pethick state equation

This equation of state is a considerable improvement upon the Harrison-Wheeler one in that it includes more effects of nuclear interactions. In the neighborhood of the neutron drip point it provides an expression for the adiabatic index

$$\frac{\rho}{p}p'_\rho = \tfrac{4}{3}\{1 - K(\rho - \rho_d)^{1/2}\},$$

with K a positive constant. Hence, when $\rho \to \rho_d$ one has $p''_\rho \to -\infty$ and therefore there is a range where the corrugation stability condition does not hold. Likewise, there is a wider range in which $\tilde{\Gamma} < 1$ and the reflection stability condition is not verified.

Now we will consider some cases of hot matter fluids.

v. Polytropic gas

For this gas

$$p = (\gamma - 1)\rho\varepsilon$$

and the sufficient conditions (10.89)–(10.90) for reflection stability are verified.

vi. The Synge gas

For this gas

$$\varepsilon = G(z) - \frac{1}{z} - 1,$$

$$p = \frac{\rho}{z}.$$

From the properties of the function G (Synge, 1957) one has, as mentioned in Section 8.2, that both G and $G^2 - 2G/z$ are positive and monotonically decreasing. It follows that

$$G' + \frac{1}{z} < 0$$

and therefore the sufficient conditions (10.89)–(10.90) for reflection stability hold.

vii. State equations of nuclear fluid dynamics

Most of these state equations that have been used in the description of heavy ion collisions (Clare and Strottman, 1986) are of the form

$$p = \rho[\tfrac{2}{3}\varepsilon + g(\rho)]. \tag{10.108}$$

The first term arises from the nonrelativistic Fermi energy of the nucleus and $g(\rho)$ varies according to the particular model. Condition (10.90) is trivially satisfied and (10.89) gives

$$g'(\rho) \geq 0. \tag{10.109}$$

Three classes of models (soft, medium, hard) can be put in this form and it can be shown (Russo, 1988) that for values of the density above a threshold the inequality (10.109) is satisfied. Such a threshold corresponds to a very

Fig. 10.4. Partially ionized monoatomic gas in thermal equilibrium with radiation. Region in the density-temperature plane in which $\Gamma_\varepsilon < 1$.

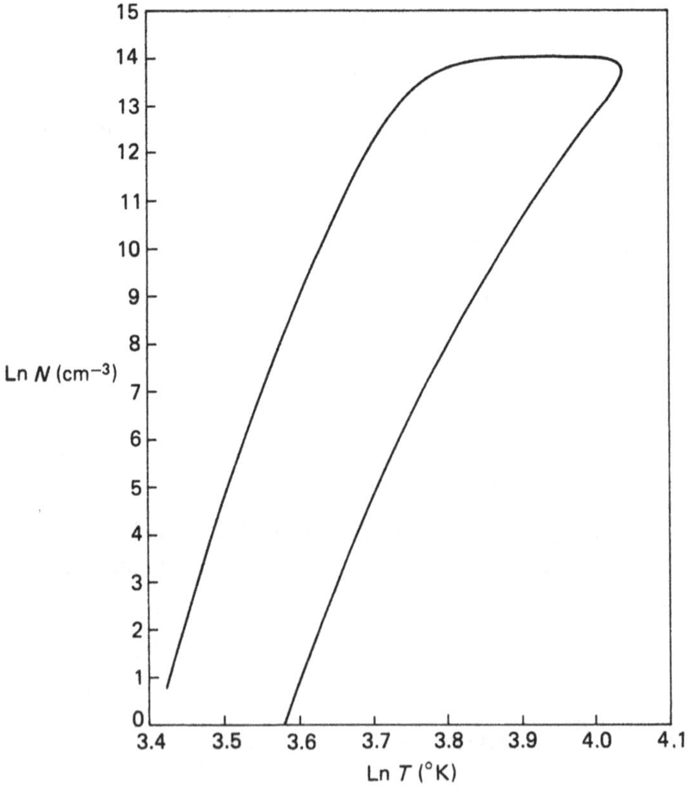

low density at which a hydrodynamical description is no longer adequate and therefore, for all applications, reflection stability is satisfied.

The Weyl condition $W > 0$ can also be checked for these state equations and is satisfied for densities in excess of a threshold density.

Russo (1988) has performed a detailed analysis of the various stability conditions for several state equations. In particular, he found that for a partially ionized monoatomic gas in thermal equilibrium with radiation there is a range of parameters (of interest for applications in astrophysics and plasma physics) in which

$$\Gamma_\varepsilon \equiv \frac{\rho}{p}\left(\frac{\partial p}{\partial \rho}\right)_\varepsilon < 1$$

and the reflection stability criterion is not satisfied (Fig. 10.4).

We have seen that the sufficient conditions for reflection stability (10.89)–(10.90) in many cases are also necessary. Such a criterion can be put in a suggestive form. The speed of sound c_s can be expressed by

$$c_s^2 = \frac{\rho}{e+p}\left(\frac{\partial p}{\partial \rho}\right)_S.$$

If we express the state equation in the form $p = p(\rho, \varepsilon)$ we obtain

$$c_s^2 = \frac{\rho}{e+p}\left(\frac{\partial p}{\partial \rho} + \frac{p}{\rho^2}\frac{\partial p}{\partial \varepsilon}\right)$$

and therefore the condition (10.90) is written

$$c_s^2 \geq \frac{p}{e+p}\left(1 + \frac{1}{\rho}\frac{\partial p}{\partial \varepsilon}\right).$$

Therefore it seems that the shock stability conditions operate so as to oppose relativistic causality ($c_s^2 \leq 1$). Stiffer state equations are more likely to satisfy the shock stability conditions but at the same time run the risk of violating relativistic causality (Russo, 1988).

References

Abramowicz, M. A., Calvani, M., and Nobili, L. (1980). Thick accretion disks with super-Eddington luminosities. *The Astrophysical Journal* **242**, 772–88.

Achterberg, A. (1983). Variational principle for relativistic magnetohydrodynamics. *Physical Review A* **28**, 2449–58.

Amendt, P. and Weitzner, H. (1985). Relativistically covariant warm charged fluid beam modeling. *The Physics of Fluids* **28**, 949–57.

Akhiezer, A. I., Akhiezer, I. A., Polovin, R. V., Sitenko, A. G., and Stepanov, K. N. (1975). *Plasma Electrodynamics*, vol II-2. Oxford: Pergamon Press.

Amsden, A. A., Harlow, F. H., and Nix, R. J. (1977). Relativistic nuclear fluid dynamics. *Physical Review C* **15**, 2059–71.

Anile, A. M. (1976). Geometrical optics in general relativity: a study of the higher order corrections. *Journal of Mathematical Physics* **17**, 576–84.

Anile, A. M. (1977). Nonlinear high-frequency waves in relativistic cosmology. *Rendiconti dell'Accademia Nazionale dei Lincei (Scienze fisiche)* **63**, 375–84.

Anile, A. M. (1982). A geometric characterization of the compatibility relations for regularly discontinuous tensor fields. *Le Matematiche* **37**, 105–19.

Anile, A. M. (1983). Gravitational focussing of shock waves in general relativity. In *Proceedings of the Journees Relativistes, Turin, 1983*, eds. Francaviglia, M., Benenti, S. and Ferraris, M. Bologna: Pitagora Editrice, 1–10.

Anile, A. M. (1984a). Nonlinear wave propagation in relativistic hydrodynamics and cosmology. In *General Relativity and Gravitation*, eds. Bertotti, B. et al. Amsterdam: Reidel, pp. 313–35.

Anile, A. M. (1984b). Evolution of relativistic shock waves in continuum mechanics. *Annales de l'Institut Henri Poincaré* **40**, 371.

Anile, A. M. and Breuer, R. (1974). Gravitational Stokes parameters. *The Astrophysical Journal* **189**, 39–49.

An'le, A. M. and Carbonaro, P. (1988). Weakly nonlinear waves in a cold relativistic plasma. A covariant approach. *Atti Seminario Matematico e Fisico dell'Universita di Modena*, in press.

Anile, A. M. and Greco, A. (1978). Asymptotic waves and critical time in general relativistic magnetohydrodynamics. *Annales de l'Institut Henri Poincaré* **29**, 257–72.

Anile, A. M., Miller, J. C., and Motta, S. (1980). Damping of relativistic shocks in an expanding universe. *Lettere al Nuovo Cimento* **29**, 268–72.

Anile, A. M., Miller, J. C., and Motta, S. (1983). Formation and damping of relativistic strong shocks. *Physics of Fluids* **26**, 1450–60.

Anile, A. M. and Moschetti, G. (1979). Conservation laws for geometrical optics in general relativistic refractive media. *Il Nuovo Cimento* **50B**, 194–8.

Anile, A. M., Moschetti, G., and Bogoyavlenski, O. I. (1987). Some solutions of Einstein's equations with shock waves. *Journal of Mathematical Physics* **28**, 2942–8.

Anile, A. M. and Muscato, O. (1983). Simple waves in relativistic magneto-fluid dynamics. *Lettere al Nuovo Cimento* **38**, 581–7.

Anile, A. M. and Muscato, O. (1988). Magneto fluid dynamic simple waves in special relativity. *Annales de l'Institut Henri Poincaré* **48**, 1–16.
Anile, A. M. and Pantano, P. (1977). Geometrical optics in dispersive media. *Physics Letters* **61A**, 215–18.
Anile, A. M. and Pantano, P. (1979). Foundation of geometrical optics in general relativistic dispersive media. *Journal of Mathematical Physics* **20**, 177–83.
Anile, A. M. and Pennisi, S. (1987). On the mathematical structure of test relativistic magneto-fluid dynamics. *Annales del l'Institut Henri Poincaré* **46**, 27–44.
Anile, A. M. and Pennisi, S. (1989). An improved relativistic warm plasma model. In *Relativistic Fluid Dynamics*, eds. Anile, A. M. and Choquet-Bruhat, Y. Berlin: Springer.
Anile, A. M. and Russo, G. (1986a). Generalized wavefront expansion I: higher order corrections for the propagation of shock waves. *Wave Motion* **8**, 243–58.
Anile, A. M. and Russo, G. (1986b). Corrugation stability for plane relativistic shock waves. *Physics of Fluids* **29**, 2847–52.
Anile, A. M. and Russo, G. (1987). Linear stability for plane relativistic shock waves. *Physics of Fluids* **30**, 1045–51.
Anile, A. M. and Russo, G. (1988). Generalized wavefront expansion II: the propagation of step-shocks. *Wave Motion*, 3–18.
Appl, S. and Camezind, M. (1988). Shock conditions for relativistic MHD jets. *Astronomy and Astrophysics* **206**, 258–68.
Asano, N. (1974). Wave propagation in nonuniform media. *Supplement of the Progress of Theoretical Physics* **55**, 52–79.
Barrow, J. D. (1983). Cosmology, elementary particles, and the regularity of the universe. *Fundamentals of Cosmic Physics* **8**, 83–199.
Barz, H. W., Csernai, L. P., Kampfer, B., and Lukacs, B. (1985). Stability of detonation fronts leading to quark-gluon plasma. *Physical Review D* **32**, 115–22.
Begelmann, M. C., Blandford, R. D., and Rees, M. (1984). Theory of extragalactic radio sources. *Review of Modern Physics* **15**, 255–350.
Bekenstein, J. D. and Oron, E. (1979). Interior magnetohydrodynamic structure of a rotating relativistic star. *Physical Review D* **189**, 2827–37.
Berger, M. and Berger, M. (1968). *Perspectives in Nonlinearity*. New York: Benjamin.
Bethe, H. A. (1942). The theory of shock waves for an arbitrary equation of state. Office of Scientific Development Report No. 545.
Bicak, J. and Hadrava, P. (1975). General relativistic radiative transfer theory in refractive and dispersive media. *Astronomy and Astrophysics* **44**, 389–99.
Blandford, R. D. and Königl, A. (1979). A model for the knots in the M87 Jet. *Astrophysical Letters* **20** 15–21.
Blandford, R. D. and McKee, C. F. (1976). Fluid dynamics of relativistic blast waves. *The Physics of Fluids* **19**, 1130–8.
Bogoyavlenski, O. I. (1978). General relativistic self-similar solutions with a spherical shock wave. *Soviet Physics, JETP* **46**, 633–40.
Boillat, G. (1965). *La Propagation des Ondes*. Paris: Gauthier-Villars.
Boillat, G. (1969). Ray velocity and exceptional waves: a covariant formulation. *Journal of Mathematical Physics* **10**, 452–4.
Boillat, G. (1970). Simple waves in N-dimensional propagation. *Journal of Mathematical Physics* **11**, 1482–3.
Boillat, G. (1973). Covariant disturbances and exceptional waves. *Journal of Mathematical Physics* **14**, 973–6.
Boillat, G. (1974). Sur l'existence et la recherche d'équations de conservation supplémentaires pour les systemes hyperboliques. *Comptes Rendues de l'Academie des Sciences, Paris* **278A**, 909–12.

Boillat, G. (1975). Chocs caractéristiques et ondes simples exceptionelles pour les systèmes conservatifs à integrale d'énergie: forme explicite de la solution. *Comptes Rendues de l'Academie des Sciences, Paris* **280A**, 1325–8.

Boillat, G. (1976). Sur une fonction croissante comme l'entropie et génératrice des chocs dans les systèmes hyperboliques. *Comptes Rendues de l'Academie des Sciences Paris* **283A**, 409–12.

Boillat, G. (1981). Caractéristiques complexes et instabilité des champs quasi-linéaires. *Rendiconti del Circolo Matematico di Palermo* **30**, 416–20.

Boillat, G. (1982a). Urti. In *Wave propagation*, ed. Ferrarese, G. Naples: Liguori, pp. 169–92.

Boillat, G. (1982b). Chocs avec contraintes et densité d'énergie convexe. *Comptes Rendues de l'Academie des Sciences. Paris* **295**, 747–50.

Breuer, R. A. and Ehlers, J. (1980). Propagation of high-frequency electromagnetic waves through a magnetized plasma in curved space-time. I. *Proceedings of the Royal Society, London* **A370**, 389–406.

Breuer, R. A. and Ehlers, J. (1981). Propagation of high-frequency electromagnetic waves through a magnetized plasma in curved space-time. II. Application of the asymptotic approximation. *Proceedings of the Royal Society, London* **A374**, 65–86.

Brill, D. R. and Hartle, J. B. (1964). Method of the self-consistent field in general relativity and its application to the gravitational geon. *Physical Review* **135**, B271–8.

Brio, M. (1987). Propagation of weakly nonlinear magnetoacoustic waves. *Wave Motion* **9**, 455–8.

Cabannes, H. (1970). *Theoretical Magneto-Fluid Dynamics*. New York: Academic Press.

Carioli, S. M. and Motta, S. (1984). The evolution of nonlinear primordial fluctuations. In *Proceedings of the VI Italian Meeting on General Relativity and Gravitation*, eds. Fabbri, R. and Modugno, M. Bologna: Pitagora Editrice, pp. 49–60.

Cattaneo, C. (1970). *Introduction à la Theorie Macroscopique des Fluides Relativistes*. Paris: College de France.

Cattaneo, C. (1978). Funzioni regolarmente discontinue attraverso una ipersuperficie: formula generale di compatibilita' geometrica. *Istituto Lombardo* (*Rend. Sc.*) **A112**, 139–49.

Cattaneo, C. (1981). *Elementi di Teoria della Propagazione Ondosa*. Bologna: Pitagora Editrice.

Chandrasekhar, S. (1939). *An Introduction to the Study of Stellar Structure*. Chicago: University of Chicago Press.

Chen, P. J. (1976). *Selected Topics in Wave Propagation*. Leyden: Noordhoff.

Chevalier, R. A. (1981). Hydrodynamic models of supernova explosions. *Fundamentals of Cosmic Physics* **7**, 1–58.

Chin, R. C. Y., Garrison, J. C., Levermore, C. D., and Wong, J. (1986). Weakly nonlinear acoustic instabilities. *Wave Motion* **8**, 537–59.

Chiu, C. K. and Gross, R. A. (1969). Shock waves in plasma physics. *Advances in Plasma Physics* **2**, 140–99.

Choquet-Bruhat, Y. (1968). *Géométrie Différentielle et Systèmes Extérieurs*. Paris: Dunod.

Choquet-Bruhat, Y. (1969a). Ondes asymptotiques et approchées pour des systèmes d'équations aux dérivées partielles nonlinéaires. *Journal de Mathematiques Pures et Appliques* **48**, 117–58.

Choquet-Bruhat, Y. (1969b). Construction de solutions radiative approachées des equations d'Einstein. *Communications on Mathematical Physics* **12**, 16–35.

Choquet-Bruhat, Y. (1973). Approximate radiative solutions of Einstein-Maxwell equations. In *Ondes et Radiations Gravitationelles*, ed. Lichnerowicz, A. Paris: Editions du C.N.R.S., pp. 81–6.

Choquet-Bruhat, Y. and Greco, A. (1983). High-frequency asymptotic solutions of Yang-Mills and associated fields. *Journal of Mathematical Physics* **24**, 377–9.

Choquet-Bruhat, Y. and Taub, A. H. (1977). High-frequency self-gravitating charged scalar fields. *General Relativity and Gravitation* **8**, 561–71.

Clare, R. B. and Strottman, D. (1986). Relativistic hydrodynamics and heavy ion reactions. *Physics reports* **141**, 177–280.

Cleymans, J., Gavai, R. V., and Suhonen, E. (1986). Quarks and gluons at high temperatures and densities. *Physics Reports* **130**, 217–92.

Coll, B. (1976). Fronts de combustion en magnetohydrodynamique relativiste. *Annales de l'Institut Henri Poincaré* **25**, 363–91.

Courant, R. and Friedrichs, K. O. (1976). *Supersonic Flow and Shock Waves*. New York: Springer.

Courant, R. and Hilbert, D. (1953). *Methods of Mathematical Physics*. New York: Interscience.

Davidson, R. C. (1974). *Theory of Nonneutral Plasmas*. Reading, Mass: Benjamin.

de Arajuro, M. E. (1986). On the assumptions made in treating the gravitational wave problem by the high-frequency approximation. *General Relativity and Gravitation* **18**, 219–3.

Deb Ray, G. and Chakraborty, T. K. (1978). Propagation of plane relativistic shock waves in the presence of a magnetic field. *Astrophysics and Space Science* **56**, 119–28.

De Felice, F. and Clarke, C. J. F. (1989). *Relativity on curved manifolds*. Cambridge: Cambridge University Press.

Demianski, M. (1985). *Relativistic Astrophysics*. Oxford: Pergamon Press.

De Groot, S. R., Van der Leeuw, W. A., and Van Weert, C. G. (1980). *Relativistic Kinetic Theory*. Amsterdam: North Holland.

Dewar, R. L. (1977). Energy-momentum tensors for dispersive electromagnetic waves. *Australian Journal of Physics* **30**, 533–75.

Dixon, G. (1978). *Special Relativity, the Foundation of Macroscopic Physics*. Cambridge: Cambridge University Press.

Dougherty, J. P. (1970). Lagrangian methods in plasma dynamics. Part I. General theory of the method of the averaged Lagrangian. *Journal of Plasma Physics* **4**, 761–85.

Dougherty, J. P. (1974). Lagrangian methods in plasma dynamics. Part 2. Construction of Lagrangians for plasmas. *Journal of Plasma Physics* **11**, 331–46.

D'yakov, S. P. (1954). *Zhur. Eksptl. i Teoret. Fiz* **27**, 288 (in Russian). (Translation: Atomic Energy Research Establishment, AERE Lib. Trans. 648, 1956.)

Ehlers, J. (1967). Zum übergang von der wellenoptik zur geometrischen optik in der allgemeinen Relativitätstheorie. *Zeitschrift fur Naturforschung* **22a**, 1328–32.

Ehlers, J. (1971). General relativity and kinetic theory. In *General Relativity and Cosmology*, ed. Sachs, R. K. New York: Academic Press, pp. 1–70.

Ellis, G. F. R. (1971). Theoretical cosmology. In *General Relativity and Cosmology*, ed. Sachs, R. K. New York: Academic Press, pp. 104–82.

Eltgroth, P. G. (1971). Similarity analysis for relativistic flow in one dimension. *The Physics of Fluids* **14**, 2631–5.

Eltgroth, P. G. (1972). Nonplanar relativistic flow. *The Physics of Fluids* **15**, 2140–4.

Erpenbeck, J. J. (1962). Stability of step-shocks. *Physics of Fluids* **5**, 1181–7.

Ferrarese, G. (1982). Introduzione alla meccanica relativistica dei continui con struttura scalare. *Rendiconti del Seminario Matematico dell'Universita' di Padova* **68**, 31–47.

Ferrari, A. and Tsinganos, K. (1986). Wind type flows and astrophysical jets. *Canadian Journal of Physics* **64**, 456–62.

Fisher, A. and Marsden, D. P. (1972). The Einstein evolution equations as a first order quasi-linear symmetric hyperbolic system. I. *Communications on Mathematical Physics* **28**, 1–38.

Fowels, G. R. (1981). Stimulated and spontaneous emission of acoustic waves from shock fronts. *Physics of Fluids* **24**, 220–7.

Friedlander, G. (1975). *The Wave Equation on a Curved Space-time.* Cambridge: Cambridge University Press.

Friedrichs, K. O. (1954). Symmetric hyperbolic linear differential equations. *Communications on Pure and Applied Mathematics* **7**, 345–92.

Friedrichs, K. O. (1974). On the laws of relativistic electromagneto-fluid dynamics. *Communications on Pure and Applied Mathematics* **27**, 749–808.

Friedrichs, K. O. (1978). Conservation equations and the laws of motion in classical physics. *Communications on Pure and Applied Mathematics* **31**, 123–31.

Friedrichs, K. O. and Lax, P. D. (1971). Systems of conservation equations with a convex extension *Proceedings of the National Academy of Sciences, U.S.A.* **68**, 1686–8.

Fujimara, F. S. and Kennel, C. F. (1979). Numerical solutions of the trans-relativistic shock relations *Astronomy and Astrophysics* **79**, 299–305.

Gardner, C. S. (1963). Comment on Stability of Step Shock, *Physics of Fluids* **6**, 1366–7.

Gardner, C. S. and Kruskal, M. D. (1964). Stability of plane magnetohydrodynamic shocks. *Physics of Fluids* **7**, 700–6.

Giambo', S. (1982). On some aspects of the evolution law for weak discontinuities in classical and relativistic cases. In *Onde e Stabilita' nei mezzi continui, Quaderno del G.N.F.M. (C.N.R.)*, eds. Anile, A. M., Motta, S., and Pluchino, S. Catania: Tipografia Universitaria, pp. 170–86.

Goldreich, P. and Weber, S. V. (1980). Homologously collapsing stellar cores. *The Astrophysical Journal* **238**, 991–7.

Granik, A. (1983). Shock damping in an ultrarelativistic gas. *The Physics of Fluids* **26**, 1763–8.

Greco, A. (1972). On the exceptional waves in relativistic magnetohydrodynamics (MHDR). *Rendiconti dell' Accademia Nazionale dei Lincei (Scienze Fisiche)* **52**, 507–12.

Gross, R.A. (1971). The physics of strong shock waves in gases. In *Physics of High-Energy Density*, eds. Caldirola, P. and Knoepfel, H. New York: Academic Press, pp. 245–277.

Gross, R. A., Chen, Y. G., Halmoy, E., and Moriette, P. (1970). Strong shock waves. *Physical Review Letters* **25**, 575–7.

Gyulassy, M., Kajante, K., Kurki-Suonio, H., and McLerran, L. (1984). Deflagrations and detonations as a mechanism of hadron bubble growth in supercooled quark-gluon plasmas. *Nuclear Physics* **B237**, 477–501.

Hadamard, J. (1903). *Leçons sur la Propagation des Ondes.* Paris: Herman.

Hartle, J. B. (1978). Bounds on the mass and moment of inertia of non rotating neutron stars. *Physics Reports* **46**, 201–47.

Hawking, S. W. and Ellis, G. F. R. (1973). *The Large Scale Structure of Space-time.* Cambridge: Cambridge University Press.

Hawley, J. F., Smarr, L. L., and Wilson, J. R. (1984a). A numerical study of nonspherical black hole accretion. I. Equations and test problems. *The Astrophysical Journal* **277**, 296–311.

Hawley, J. F., Smarr, L. L., and Wilson, J. R. (1984b). A numerical study of nonspherical black hole accretion. II. Finite differencing and code calibration. *The Astrophysical Journal Supplement Series* **55**, 211–46.

Hunter, J. K. and Keller, J. (1983). Weakly nonlinear high frequency waves. *Communications on Pure and Applied Mathematics* **36**, 547–69.

Ikeuchi, S., Tomisaka, K., and Ostriker, J. (1983). The structure and expansion law of a shock wave in an expanding universe. *The Astrophysical Journal* **265**, 583–96.

Isaacson, R. A. (1968). Gravitational radiation in the limit of high frequency, II: nonlinear terms and the effective stress tensor. *Physical Review* **166**, 1272–80.

Ishizuka, T. (1980). Propagation of a shock wave in general relativity. *Progress of Theoretical Physics* **63**, 1541–50.
Ishikuza, T. and Sakashita, S. (1980). Propagation of the general relativistic blast wave. *Progress of Theoretical Physics* **63**, 1945–9.
Israel, W. (1960). Relativistic theory of shock waves. *Proceedings of the Royal Society, London* **A259**, 129–43.
Israel, W. (1966). Singular hypersurfaces and thin shells in general relativity. *Il Nuovo Cimento* **44**, 1–14.
Israel, W. (1989). Relativistic thermodynamics and kinetic theory. In *Relativistic Fluid Dynamics*, eds. Anile, A. M. and Choquet-Bruhat, Y. Berlin: Springer.
Jeffrey, A. (1976). *Quasi-linear Hyperbolic Systems and Waves*. London: Pitman Books.
Jeffrey, A. and Kawahara, T. (1982). *Asymptotic Methods in Nonlinear Wave Theory*. London: Pitman Advanced Publishing Program.
Johnson, M. H. and McKee, C. F. (1971). Relativistic hydrodynamics in one dimension. *Physical Review D* **3**, 858–63.
Königl, A. (1980). Relativistic gas dynamics in two dimensions, *Physics of Fluids* **23**, 1083–90.
Kontorovich, V. M. (1957). Concerning the stability of shock waves. *Journal of Experimental and Theoretical Physics, U.S.S.R.* **33**, 1525–6.
Kosinski, W. (1986). *Field Singularities and Wave Analysis in Continuum Mechanics*. Chichester, England: Ellis Horwood Publishers.
Landau, L.D. and Lifshitz, E. M. (1959a). *Fluid Mechanics*. Elmsford, New York: Pergamon Press.
Landau, L. D. and Lifshitz, E. M. (1959). *Statistical Physics*. London: Pergamon Press.
Lanza, A., Miller, J. C., and Motta, S. (1982). Relativistic shocks in a Synge gas. *Lettere al Nuovo Cimento* **35**, 309–14.
Lanza, A., Miller, J. C., and Motta, S. (1985). Formation and damping of relativistic strong shocks in a Synge gas. *Physics of Fluids* **28**, 97–103.
Lanza, A., Miller, J. C., and Motta, S. (1987). Relativistic shocks in a gas of interacting particles and radiation. *Il Nuovo Cimento* **98**, 119–30.
Lessen, M. and Deshpande, N. V. (1967). Stability of magnetohydrodynamic shock waves. *Journal of Plasma Physics* **1**, 463–72.
Liang, E. P. T. (1977a). Relativistic simple waves: shock damping and entropy production. *The Astrophysical Journal* **211**, 361–76.
Liang, E. P. T. (1977b). Galaxies and entropy from nonlinear fluctuations: a simple wave analysis. *The Astrophysical Journal* **216**, 206–11.
Liang, E. P. T. and Baker, K. (1977). Damping of relativistic shocks. *Physical Review Letter* **39**, 191–3.
Liberman, M. A. and Velikovich, A. (1985). *Physics of Shock Waves in Gases and Plasmas*. Berlin: Springer.
Lichnerowicz, A. (1955). *Théories Relativistes de la Gravitation et de l'Electromagnétisme*. Paris: Masson and Cie.
Lichnerowicz, A. (1960). Ondes et radiations électromagnétiques et gravitationelles en relativité générale. *Annali di Matematica Pura ed Applicata* **50**, 1–95.
Lichnerowicz, A. (1967). *Relativistic Hydrodynamics and Magnetohydrodynamics*. New York; Benjamin.
Lichnerowicz, A. (1971). Ondes de chocs, ondes infinitesimales et rayons en hydrodynamique et magnetohydrodynamique relativistes. In *Relativistic Fluid Dynamics*, ed. C. Cattaneo. Rome: Cremonese, pp. 144–97.
Lichnerowicz, A. (1976). Shock waves in relativistic magnetohydrodynamics under general assumptions. *Journal lo Mathematical Physics* **17**, 2135–42.

Lovelace, R. V., Mehanian, C., and Mobarry, C. M. (1986). Theory of axisymmetric magnetohydrodynamic flows: disks. *The Astrophysical Journal Supplement Series* **62**, 1–37.

MacCallum, M. A. H. and Taub, A. H. (1973). The averaged Lagrangian and high-frequency gravitational waves. *Communications on Mathematical Physics* **30**, 153–69.

MacLellan, A. G. (1980). *The classical thermodynamics of deformable materials.* Cambridge: Cambridge University Press.

Madore, J. (1974). Faraday transport in a curved space-time. *Communications on Mathematical Physics* **38**, 103–10.

Maeda, K. (1986). Bubble dynamics in the expanding universe. *General Relativity and Gravitation* **18**, 931–51.

Maeda, K. and Oohara, K. (1982). General relativistic collapse of a rotating star with magnetic field. *Progress of Theoretical Physics* **68**, 567–79.

Majda, A. (1984). *Compressible Fluid Flow and Systems of Conservation Laws in Several Space Variables*, Applied Mathematical Sciences. New York: Springer.

Majorana, A. (1987). Analytical solutions of the Rankine-Hugoniot relations for a relativistic simple gas. *Il Nuovo Cimento* **98**, 111–8.

Majorana, A. and Anile, A. M. (1987). Magnetoacoustic shock waves in a relativistic gas. *The Physics of Fluids*, 3045–9.

Maugin, G. (1976). Conditions de compatibilité pour une hypersurface singulière en mécanique relativiste des milieux continus. *Annales de l'Institut Henri Poincaré* **24**, 213–41.

May, M. M. and White, R. H. (1967). Stellar dynamics and gravitational collapse. *Methods of Computational Physics* **7**, 219–58.

Miller, J. C. and Pantano, O. (1987). Relativistic hydrodynamics of the cosmological quark-hadron phase transition. In *General Relativity and Gravitational Physics* (Proceedings of the 7th Italian Conference), eds. Bruzzo U., Cianci R., and Massa, E. Singapore: World Scientific, pp. 357–71.

Miller, R. B. (1985). *Intense Charged Particle Beams.* New York: Plenum Press.

Misner, C. W., Thorne, K. S., and Wheeler, J. A. (1973). *Gravitation.* San Francisco: Freeman.

Mobarry, C. M. and Lovelace, R. V. E. (1986). Magnetohydrodynamic flows in Schwarzschild geometry. *The Astrophysical Journal* **309**, 455–66.

Moschetti, G. (1987). Una legge di conservazione dell'ottica geometrica in relativita' generale. *Le Matematiche* **39**, 81–5.

Muscato, O. (1987). Formalismo Lagrangiano in magneto-fluido dinamica relativistica. *Le Matematiche* **39**, 171–80.

Muscato, O. (1988a). Breaking of relativistic simple waves. *Journal of Fluid Mechanics* **196**, 223–39.

Muscato, O. (1988b). Relativistic magnetoacoustic simple wave solutions for a Synge gas. *Il Nuovo Cimento* **101B**, 39–51.

Newcomb, W. A. (1982). Warm relativistic electron fluid. *The Physics of Fluids* **25**, 846–51.

Osnes, E. and Strottman, D. (1986). Causal constraints on the nuclear equation of state, *Physics Letters* **166B**, 5–9.

Peebles, P. J. E. (1971). *Physical Cosmology.* Princeton, N.J.: Princeton University Press.

Peebles, P. J. E. (1980). *The Large-Scale Structure of the Universe.* Princeton, N.J.: Princeton University Press.

Pham Mau Quam (1969). *Introduction a la Géométrie des Variétés Différentiables.* Paris: Dunod.

Pichon, G. (1965). Étude relativiste de fluides visqueux et chargés, *Annales de l'Institut Henri Poincaré, Section A*, **2**, 21–85.

Pirani, F. A. E. (1957). Invariant formulation of gravitational radiation theory. *Physical Review* **105**, 1089–99.

Plohr, B., Glimm, J., and McBryan, O. (1983). Applications of front tracking to two dimensional gas dynamics calculations. In *Lectures Notes in Engineering*, vol. 3, eds. Chandra, J. and Flaherty, J. New York: Springer, pp. 180–91.

Rees, M. J. (1978). The M87 jet: internal shocks in a plasma beam?. *Monthly Notices of the Royal Astronomical Society*. 184, short communication, 61p–65p.

Ruggeri, T. and Strumia, A. (1981a). Main field and convex covariant density for quasi-linear hyperbolic systems. *Annales de l'Institut Henri Poincaré*, Physique théorique **34**, 65–84.

Ruggeri, T. and Strumia, A. (1981b). Convex covariant entropy density, symmetric conservative form, and shock waves in relativistic magnetohydrodynamics. *Journal of Mathematical Physics* **22**, 1824–33.

Russo, G. (1988). Stability properties of relativistic shock waves: applications. *The Astrophysical Journal* **334**, 707–21.

Russo, G. and Anile, A. M. (1987). Stability properties of relativistic shock waves: basic results. *Physics of Fluids* **38**, 2406–13.

Saini, G. L. (1976). Some basic results on relativistic fluid mechanics and shock waves. *Journal of Mathematical Analysis and Applications* **56**, 711–17.

Shapiro, L. S. and Teukolsky, S. A. (1983). *Black Holes. White Dwarfs, and Neutron Stars.* New York: Wiley.

Shapiro, P. R. (1979). Relativistic blast waves in two dimensions: I. the adiabatic case. *The Astrophysical Journal* **233**, 831–50.

Shapiro, P. R. (1980). Relativistic blast waves that accelerate. *The Astrophysical Journal* **236**, 958–69.

Shukla, P. K., Rao, N. N., Yu, M. Y., and Tsintsadze, N. L. (1986). Relativistic nonlinear effects in plasmas. *Physics Reports* **138**, 1–149.

Siambis, J. G. (1979). Adiabatic equations of state for intense relativistic particle beams. *The Physics of Fluids* **22**, 1372–4.

Singh, H. N. (1984). Propagation of a shock wave in relativistic magnetofluids. *Journal of Physics A* **17**, 1547–54.

Sloan, J. H. and Smarr, L. L. (1986). General relativistic magnetohydrodynamics. In *Numerical Astrophysics, A Festschrift in honor of J. R. Wilson*, ed. Centrella, J. Chicago: The University of Chicago Press, pp. 52–68.

Sobel, M. I., Siemens, P. J., Bondorf, J. P., and Bethe, H. A. (1975). Shock waves in colliding nuclei. *Nuclear Physics* **A251**, 502–29.

Sod, G. A. (1978). A survey of several finite difference methods for systems of nonlinear hyperbolic conservation laws. *Journal of Computational Physics* **27**, 1–31.

Steinhardt, P. J. (1982). Relativistic detonation waves and bubble growth in false vacuum decay. *Physical Review D* **25**, 2074–85.

Stix, T. H. (1962). *The Theory of Plasma Waves*. New York: McGraw-Hill.

Strottman, D. (1989). Relativistic hydrodynamics and heavy ion reactions. In *Relativistic Fluid Dynamics*, eds. Anile, A. M. and Choquet-Bruhat, Y. Berlin: Springer.

Synge, J. L. (1956). *Relativity: The Special Theory*. Amsterdam: North Holland.

Synge, J. L. (1957). *The Relativistic Gas*. Amsterdam: North Holland.

Synge, J. L. (1960). *Relativity: The General Theory*. Amsterdam: North Holland.

Taub, A. H. (1948). Relativistic Rankine–Hugoniot equations. *Physical Review* **74**, 328–34.

Taub, A. H. (1978). Relativistic fluid mechanics. *Annual Review of Fluid Mechanics* **10**, 301–32.

Taussig, R. (1973). Shock wave production of relativistic plasmas. In *Dynamics of Ionized Gases*, eds. Lighthill, M. J., Imai, I., and Sato, H. Tokyo: Tokyo University Press, pp. 113–30.

Thomas, T. Y. (1957). Extended compatibility conditions for the study of surfaces of discontinuity in continuum mechanics. *Journal of Mathematics and Mechanics* **6**, 311–22.

Thompson, K. W. (1986). The special relativistic shock tube. *Journal of Fluid Mechanics* **171**, 365–75.

Thorne, K. S. (1973). Relativistic shocks: the Taub adiabat. *The Astrophysical Journal* **179**, 897–907.

Thorne, K. S. (1981). Relativistic radiative transfer: moment formalism. *Monthly Notices of the Royal Astronomical Society* **194**, 439–73.

Toepfer, A. J. (1971). Finite temperature relativistic electron beams. *Physical Review A* **3**, 1444–52.

Truesdell, C. and Toupin, R. A. (1960). *The Classical Field Theories. In Handbuch der Physik*, Bd. III/1 Berlin: Springer 266–793.

Van Riper, K. A. (1979). General relativistic hydrodynamics and the adiabatic collapse of stellar cores. *The Astrophysical Journal* **232**, 558–71.

Vishniac, E. T. (1982). Nonlinear effects on cosmological perturbations. I. The evolution of adiabatic perturbations. *The Astrophysical Journal* **253**, 446–56.

Weinberg, S. (1972). *Gravitation and Cosmology*. New York: Wiley

Whitham, G. B. (1974). *Linear and Nonlinear Waves*. New York: Wiley.

Yodzsis, P. (1971). Some general relations in relativistic magnetohydrodynamics. *Physical Review D* **3**, 294–5.

Zel'dovich, Ya. B. and Novikov, I. D. (1971). *Relativistic Astrophysics. vol 1. Stars and Relativity*. Chicago: The University of Chicago Press.

Index